A Primer of Genome Science

SECOND EDITION

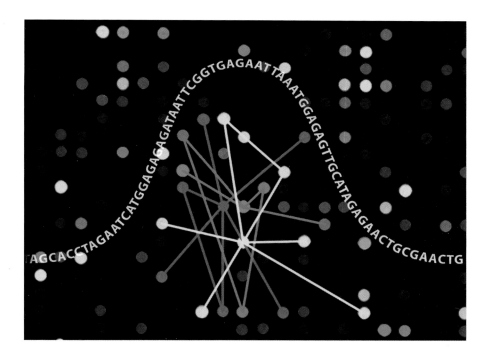

GREG GIBSON
SPENCER V. MUSE

North Carolina State University

SINAUER ASSOCIATES, INC. PUBLISHERS
Sunderland, Massachusetts, 01375

The Cover

The cover represents the five major areas of genome research covered in the book. DNA sequencing, analysis of polymorphism, and variation are represented by the normal distribution of the letters A, C, G and T, which spell "A Primer of Genome Science" in codons. The background image of an Agilent Technologies long oligonucleotide microarray represents gene expression profiling, while the network of linkages between spots of similar hue symbolizes research in proteomics and computational biology, all of which lead to a holistic picture of the interactions among biological components. Microarray image on cover reprinted with permission of Agilent Technologies.

A Primer of Genome Science, Second Edition

Copyright © 2004 by Sinauer Associates, Inc. All rights reserved. This book may not be reproduced in whole or in part without permission of the publisher. For information address:
Sinauer Associates, Inc.
23 Plumtree Road
Sunderland, MA 01375
U.S.A.

FAX 413-549-1118
publish@sinauer.com, orders@sinauer.com
World Wide Web site at http://www.sinauer.com/genomics

Notice of Trademarks

All trademarked names appearing in this book are used in an editorial fashion, to the benefit of the trademark owners, and with no intent to infringe upon these trademarks.

Library of Congress Cataloging-in-Publication Data

Gibson, Greg.
  A primer of genome science / Greg Gibson and Spencer V. Muse.— 2nd ed.
      p. ; cm.
  Includes bibliographical references and index.
   ISBN 0-87893-232-1 (pbk. : alk. paper)
 1. Genomes. 2. Bioinformatics. 3. Genomes—Research—Methodology. 4. Genetics.
   [DNLM: 1. Genome. 2. Gene Expression. 3. Genomics. QH 447 G448p 2004] I. Muse, Spencer V., 1966- II. Title.
   QH447.G534 2004
   572.8'6—dc22
                          2004024285

6 5 4 3 2

*To Ann and Alan Gibson and to John Goebel,
for opening our minds to science.*

Contents

2   Genome Sequencing and Annotation

3   SNPs and Variation

4 Gene Expression and the Transcriptome

5 Proteomics and Functional Genomics

6 Integrative Genomics

Preface

This book is an introduction to genome science intended for senior undergraduates or for graduate students who are new to the field. Four years ago, we were inspired to write this text as a result of our experiences teaching introductory graduate classes in functional genomics (GG) and bioinformatics (SM) at North Carolina State University. The general features of the field of genome science seem to have settled down around genome-scale sequencing, analysis of variation, gene expression profiling, proteomics, and metabolomics. Increasingly, computational biology and statistical analysis play a central role in genome research, and we endeavor to capture this trend toward a merger of molecular and theoretical biology. Another major theme is that genome research is not restricted to human biology or just a handful of biomedical model organisms: agricultural, ecological, behavioral and evolutionary genomics are emerging as dynamic sub-disciplines. By combining the basics of bioinformatics and experimental methods in one text, we hope to encourage students to develop a familiarity with genomics that will be useful whether they pursue careers in science, law, journalism, marketing, or wherever their interests may lead.

For the second edition, we have revised most of the text, updating it with some key new approaches where appropriate. Each of the first five chapters now include four worked exercises, such as conducting a web-based search or performing a simple statistical analysis. There are also four new boxes that explain the basic bioinformatic principles behind gene ontology, coalescent theory, statistics for microarray analysis, and interaction networks. Organizationally, we have moved the chapter on Variation (now Chapter 3) to immediately follow the chapter on Genome Sequencing (Chapter 2), reflecting the increased prominence of the study of polymorphism in modern genetics. The chapter on Gene Expression Profiling (Chapter 4) has been substantially overhauled, reflecting a more settled view of how experimental design is central to this area of genome science. The only chapter that has been left pretty much as it was in the first edition is the last one, on Integrative Genomics—not because metabolomics and *in silico* genomics haven't changed, but rather because they are evolving so rapidly that we felt that it would be better to wait a few more years before giving this area a necessary overhaul.

We assume that the reader is familiar with the content of typical "300 level" undergraduate courses in genetics, and so have eschewed an opening chapter that introduces the basic genetic terminology and concepts such as enhancers, exons and introns, cDNA, the central dogma, and the chromosomal basis of inheritance. Most readers will probably benefit from having a copy of a standard genetics text available for reference, and some will wish to graduate quickly to any of the many new detailed treatments of each of the main topics covered in single chapters in this book.

We recognize that students will come to the discipline with vastly different educational backgrounds. Some will know how to program neural networks but be unaware of the distinction between SNPs and indels; some will have used microarrays but never have heard of a mapping function, let alone a Markov chain. Our strategy is to describe the core experimental methods side-by-side with the analytical approaches. We emphasize throughout the importance of statistics and analysis in the design and interpretation of genomic data. Some readers will undoubtedly be frustrated by our decision not to concentrate on the biological impact of genomic data, but that decision was made for two reasons: the first is that there is so much in flux that this would be premature; the second is that this task is better left to individual teachers, who can put their own emphasis on things in the context of class discussions.

In addition to the new exercises in the text, some much more complicated exercises involving large datasets are provided online. Use of the World Wide Web is an essential part of instruction in genomics, which also depends on computers for laboratory information management, database establishment, data sharing, and statistical analysis. Consequently, the Web site for this book provides, among other things, a series of related DNA sequence, microarray and SNP datasets that can be downloaded and analyzed at will. The descriptive titles for these exercises can be found at the conclusion of Chapters 2 through 5.

In the final analysis, there is no substitute for reading the primary literature. It is unlikely that an undergraduate course will cover all of the topics, or that most graduate instructors will want to. This *Primer* is not meant to be used as a reference so much as an aid, in conjunction with classroom instruction, for the building of a foundation from which forays can be made into specific topics from psychogenomics to motif detection, from comparison of genome sequences to modeling of metabolism. Undoubtedly there are topics that we have missed, and discussions that may be out of date before long, but given an understanding of the basic approaches in genome science, there is no reason why these cannot be explored independently.

It has been our opinion that genomics is too broad a topic to be accommodated simply by adding a chapter to standard textbooks. Furthermore, to do so misses the central reality that the view of biology is different from the perspective of the whole genome, as opposed to individual genes. We imagine that some students drawn to the genomic way of doing biology will seek to incorporate mathematics and computer science into their research from the beginning of their training, while others will seek to integrate classical molecular genetics and cell biology into their work. If this book helps make both of these tasks a little easier, it will have done its job.

Acknowledgments

We would both like to acknowledge the contributions of the students in our classes in helping to bring the structure of the book together, as well as the input from instructors who have adopted the *Primer* in their classes over the past few years. A number of colleagues reviewed and otherwise helped shape specific chapters, including Lisa Goering, Trisha Wittkopp, Erika Zimmerman, David Pollock, Sergey Nuzhdin, Steven Carr, David Bird, Dahlia Nielsen, Matt Rockman and Greg Wray, Andy Clark, Patricia Hurban, and several anonymous reviewers (none of whom, of course, bears any responsibility for errors of omission or commission). Greg Gibson would particularly like to thank Walter Gehring, David Hogness, Cathy Laurie, Trudy Mackay, and Julian Adams, who helped build scaffolds and bridge gaps. Spencer Muse adds his special appreciation to Bruce Weir for a decade of guidance and advice.

It was an absolute pleasure working with S. Mark Williams at Pyramis Studio in Durham, who rendered all of the figures with care and creativity, and with Andy Sinauer, Sydney Carroll, Chris Small, Janice Holabird, Jefferson Johnson, Susan McGlew, and everyone at Sinauer Associates, Inc., whose efficiency and enthusiasm so quickly turned a possibility into a reality.

Finally, we thank our wives, Diana and Cindy, from the bottoms of our hearts, for understanding and so much more. Sydney, Paddington, Marcy, Quinn, Cooper, Incense, Calvin, Bernie, and Oliver played their supporting roles superbly as well.

Greg Gibson and Spencer Muse
Raleigh, North Carolina
November, 2004

1 *Genome Projects: Organization and Objectives*

Genome science is the study of the structure, content, and evolution of genomes. Initially driven by high-volume nucleotide sequencing, technological advances have enabled the field to grow rapidly, and genome science, or "genomics," now also encompasses analysis of the expression and function of both genes and proteins. The related disciplines of bioinformatics and computational biology are increasingly integrated with empirical analyses. Genome projects are now in place for well over 100 different organisms. This chapter provides an overview of these projects as well as an introduction to some of the central concepts in genome science.

The Core Aims of Genome Science

All genome projects share a common set of aims, subsets of which are emphasized according to available resources, economic and scientific goals, and the biological attributes of the organism under study. The major aims of individual genome projects are:

1. **To establish an integrated Web-based database and research interface.** Access to the enormous volume of data generated by genome researchers is facilitated by ever-evolving Web interfaces. Many Web sites initially grew out of the efforts of a small group of leaders at a single research institution, but such sites demand millions of dollars for upkeep and innovation, leading to the creation of mirror sites and organizations dedicated to collecting and linking data, to quality control, and to presenting the data in a useful manner. Rather than serving as a storage bin for sequence and other data, most sites are now built on state-of-the-art relational databases. Many of these use a common resource of innovative software for data searches and online analysis that have been developed under the umbrella of the Generic Model Organism Database (GMOD) project.

2. **To assemble physical and genetic maps of the genome.** The location of genes in a genome can be specified both according to physical distance and relative position defined by recombination frequencies. This information is crucial for comparing the genomes of related species and for putting together phenotypic and genetic data (particularly when mapping disease loci). Genetic maps are used in animal and plant breeding, and in numerous areas of basic biological research. The art of map making has been honed by geneticists for the better part of a century, but genomicists are extending it to more and more species, with greater and greater accuracy.

3. **To generate and order genomic and expressed gene sequences.** High-volume sequencing is one of the defining features of genome science. The basic procedure used is the same one developed by Fred Sanger two decades ago, except that it has been heavily automated both at the levels of running reactions and reading sequences. Sequencing whole genomes is now generally performed in a "shotgun" fashion: millions of short pieces of DNA sequence are computationally assembled into **contigs,** scaffolds, and eventually whole chromosomes. The term "contig" refers to a set of sequence fragments that have been ordered into a contiguous, linear stretch on the basis of sequence overlaps at the fragment ends. A set of contigs constitutes the "scaffold" of a whole genome sequence. Because much of a genome consists of noncoding DNA, genomicists are also interested in sequencing large numbers of **cDNA clones.** Such sequences are cloned from mRNA transcripts (see Chapter 2) and thus identify expressed genes and exon boundaries. Only one end of a cDNA need be sequenced to identify a clone, and these fragments are called **expressed sequence tags,** or **ESTs.** Because of alternative splicing and errors in the construction of cDNA libraries, there is not a one-to-one correspondence between ESTs and genes, but EST collections are thought to give a good first approximation of the diversity of genes expressed in a tissue.

4. **To identify and annotate the complete set of genes encoded within a genome.** Having the complete sequence of a genome is only the first step toward characterizing its gene content. The genes encoded within the sequence must then be identified using a combination of experimental and bioinformatic strategies. These include alignment of cDNA and genomic sequences, looking for sequences that are similar to those already identified in other genomes (both of these procedures rely on DNA alignment and comparison algorithms such as BLAST), and applying gene-finding software that recognizes DNA features that are associated with genes, such as open reading frames (ORFs), transcription start and termination sites, and exon/intron boundaries. Once a gene has been identified, it must be **annotated,** which entails linking its sequence to genetic data about the function, expression, and mutant phenotypes of the protein associated with the locus, as well as to comparative data from homologous proteins in other species.

5. **To characterize DNA sequence diversity.** It has been known for some time that all genomes are full of polymorphisms: sequence sites at which two or more variants are found in natural populations. **Single-nucleotide polymorphisms** are called **SNPs,** and it is generally assumed that most quantitative genetic variation—that is, the heritable component of variation in characters such as size, shape, yield, and disease susceptibility—should be traceable to SNPs or to insertion/deletion polymorphisms. Characterization of the distribution of SNPs is a crucial first step in efforts to find associations between SNP variation and phenotypic variation. Another fundamentally important quantity that remains to be characterized is the level of **haplotype** structure due to linkage disequilibrium (LD). A haplotype is a set of adjacent polymorphisms found on a single chromosome, while LD refers to nonrandom associations between sites. The farther apart two sites are, the more they tend to assort independently (i.e., randomly), but there is great variation in the distances involved, from tens of bases to tens of kilobases. Disease locus mapping now generally utilizes detailed knowledge of LD. Further, SNP variation, along with variation in repetitive sequences such as microsatellites, can be an invaluable tool for inferring relationships between individuals, in forensics, in evolutionary studies of the history of a species, and in studies of population structure.

6. **To compile atlases of gene expression.** Important clues to gene function can be gleaned from analyzing profiles of transcription and protein synthesis. Traditional methods for characterizing gene expression include Northern blots and in situ hybridization (and, where an antibody exists, Western blots and immunohistochemistry). Genomic methods rely more on detecting a tag corresponding to each gene in a library of hundreds of thousands of sequenced fragments. EST sequencing, as well as methods such as SAGE and differential display, can also be used for gene discovery. Because many genes are only expressed in a few tissues, libraries are prepared from each tissue of interest, or from different stages of development or infection, or in the presence of different toxins and other environmental agents.

Once a collection of unique ESTs (a "unigene set") has been assembled, nanomolar amounts of each gene fragment can be spotted onto a microarray or synthesized as oligonucleotides on a silicon chip, and hybridized to labeled cDNA prepared from different tissues. These microarray, or gene chip, approaches to expression allow scanning of the relative levels of expression of large fractions of the genome under hundreds of different conditions. Analyzing patterns of covariation in gene expression provides information about the regulation of gene expression, and can yield clues to unknown gene function as a result of "guilt by association." Gene expression is also a powerful tool in the identification of candidate genes for any given process, since a gene must (usually!) be expressed in a tissue in order to be doing something to it.

7. **To accumulate functional data, including biochemical and phenotypic properties of genes. Functional genomics** refers to a panoply of approaches under development to ascertain the biochemical, cellular, and/or physiological properties of each and every gene product. These include near-saturation mutagenesis (that is, screening hundreds of thousands of mutants to identify genes that affect traits as diverse as embryogenesis, immunology, and behavior), high-throughput reverse genetics (methods to systematically and specifically inactivate individual genes), and elaboration of genetic tools. **Proteomics,** a core element in functional genomics, includes methods for detecting protein expression and for detecting protein-protein interactions. Another subfield, **structural genomics,** seeks to elucidate the tertiary structure of each class of protein found in cells. **Pharmacogenomicists** are particularly interested in studying the interactions between small molecules (i.e., potential drugs) and proteins, both in vitro and in the context of living organisms. Research on model organisms such as mice, fruitflies, nematodes, various plants, and yeast is a crucial component of functional genomics.

8. **To provide the resources for comparison with other genomes.** Just as nothing in biology makes sense except in the light of evolution, nothing in genomics makes sense except in the light of comparative data. Comparative maps allow genetic data from one species to be used in the analysis of another, because local gene order along a chromosome tends to be conserved over millions of years—a phenomenon known as **synteny.** Even without synteny, the conservation of gene function is now known to be extensive enough that studies of the genetics of neuronal signaling or heart development in fruitflies can tell us much about the same processes in a primate. The generation of online resources is being prioritized in order to enhance an individual researcher's ability to use data being generated anywhere in the world, from any organism.

This chapter will describe the core elements of each of these objectives for the major genome projects in animals, plants, and microbes of medical, agricultural, and basic biological interest. Before doing so, it is essential to define the concepts of physical and genetic maps, as these two resources are central to all of genome science.

Mapping Genomes

Genetic Maps

A **genetic map** is a description of the relative order of genetic markers in linkage groups in which the distance between markers is expressed as units of recombination. The genetic markers are most often physical attributes of the DNA (such as sequence tags, simple repeats, or restriction enzyme polymorphisms), but may include phenotypes associated with Mendelian loci. In diploid organisms, genetic maps are typically assembled from data on the co-segregation of genetic markers either in pedigrees or in the progeny of controlled crosses. The standard unit of genetic distance is the **centiMorgan (cM),** an expression of the percentage of progeny in which a recombination event has occurred between two markers. Named for Thomas Hunt Morgan, 1 cM is an arbitrary unit equivalent to a recombination frequency of 0.01. In human autosomal euchromatin, 1 cM is approximately 1,000 kb, which is twice as long as the equivalent parameter for *Drosophila*.

While there is a high variance, individual chromosomes in many animal species tend to be on the order of 100 cM in length, indicating that one crossover occurs per chromosome per generation. Markers on different chromosomes have a 50-50 chance of co-segregating, and thus are 50 cM apart—which is the threshold for assigning markers to the same linkage group. Two markers that recombine 50% of the time may nevertheless be on the same linkage group, so long as they are joined by a third marker that shows less than 50% recombination with each of the markers (Figure 1.1). With sufficient markers, the number of linkage groups and the number of chromosomes should be the same for any given organism.

Figure 1.1 Assembling a genetic map. (A) Starting with a pair of different parental genomes, represented as green and blue chromosomes, a series of recombinant chromosomes are generated in controlled crosses. (B) The genotypes of multiple molecular or phenotypic markers in the recombinant individuals are determined, from which a table showing the frequency of recombinants between each marker is drawn up. (C) Software is then used to calculate the most likely genetic map from the entire data. In this hypothetical example, two linkage groups are inferred, one of which (top) is longer than 50 cM. The best estimate of the genetic distance in cM between each pair of markers is indicated on the right. Genetic maps can be assembled from pedigree data using similar principles.

Software for the assembly of genetic maps is freely available from a number of sources (see http://linkage.rockefeller.edu/soft/list.html for a comprehensive listing of genetic analysis software), perhaps the most popular of which is the Mapmaker/QTL program (Lander et al. 1987). Linkage data are converted to map distances by virtue of a **mapping function** that adjusts for the observation that the probability of crossovers leading to recombination does not increase linearly with physical distance. The two most common mapping functions bear the names of their developers, Haldane and Kosambi. The Kosambi mapping function also adjusts for the phenomenon of **interference,** whereby the presence of one crossover reduces the likelihood of another crossover in the vicinity.

Multiple factors lead to high variation in the correspondence between physical and genetic distances, including the variability of recombination rate along a chromosome (in most species, centromeres and telomeres are less recombinogenic than general euchromatin); genome-wide recombination rate variation within and between species; gain and loss of repetitive DNA; and the low resolution of genetic maps due to small sample sizes of progeny used to establish linkage.

Several methods have emerged for generating precise genetic maps in humans and most model organisms.

- **Recombinant inbred lines (RIL)** are collections of highly inbred lines generated by sib-mating or selfing of individual lines derived from the cross of a pair of genetically divergent inbred parents. RIL have proven particularly useful in mapping plant genomes. Seed for each line can be stored and disseminated to large numbers of workers, who are able to add markers to the existing maps at will.

- Construction of human, mouse, and many other vertebrate genetic maps has been greatly facilitated by the development of **radiation hybrid (RH) mapping,** in which fragments of chromosomes from the organism of study are incorporated into a panel of hamster fibroblast cell cultures (Deloukas et al. 1998). Species-specific PCR amplification is then used to ascertain which loci are present in each line, and the frequency at which markers co-segregate is an indicator of the physical distance between the markers. There is no need for the markers to be polymorphic in the study species. Mapping functions unique to RIL and RH mapping have been developed that allow precise ordering of loci, as well as comparison with physical data. Advanced RH panels can also be generated starting with irradiation of a hamster cell line that contains a whole single chromosome of the species being mapped, which allows more fine-scale mapping.

- **Single sperm typing** is a PCR-based method for actually counting the number of crossovers between two markers, with the potential to accurately define recombination fractions as small as 0.001 over just tens of kilobases. Evidence is emerging that there is variation in the distribution of crossovers along a gene, so that recombination may be focused on hotspots that could have a major effect on the structure of polymorphism (Jeffreys et al. 2004).

- Recombination rates can also be estimated indirectly from population genetic data, using advanced statistical algorithms to assess the significance of variation in the level of linkage disequilibrium along a locus in a population of sequences.

EXERCISE 1.1 *Constructing a genetic map*

Suppose that an orange-tree breeder has decided to begin to assemble a genetic map based on four recessive loci—*thickskin, reddish, sour,* and *petite*—named after the phenotypes of homozygotes. After identifying two true-breeding trees that are either completely wild-type or mutant for all four loci, the breeder crosses them, and then plants an orchard of F_2 trees. Based on the following frequencies of mutant classes, determine which loci are likely to be on the same chromosome and which are the most closely linked.

EXERCISE 1.1 TABLE 1

Normal	402	Thick	18
Petite	127	Red, sour, and petite	12
Sour	115	Red	11
Thick and red	108	Thick and petite	10
Thick, red, and petite	42	Thick, red, sour, and petite	8
Thick, red, and sour	41	Thick and sour	7
Sour and petite	38	Red and petite	3
Red and sour	24	Thick, sour, and petite	2

ANSWER: First confirm that each mutant segregates with approximately Mendelian ratios by adding up the total number of each phenotype. One-quarter of the total number of 968 trees, around 242, should show each recessive phenotype, and this seems to be the case since there are 242 *petite*, 249 *reddish*, 247 *sour*, and 236 *thickskin* orange-producing trees. Given these ratios, we would expect that unlinked loci would segregate independently, such that just 60 or so trees ($\frac{1}{4} \times \frac{1}{4} \times 968$) produce each double-mutant class, but would expect that completely linked loci would have close to 240 double mutants. Adding the number of double mutants gives the following results:

Thick and red	199	Thick and petite	62	Thick and sour	58
Red and petite	65	Red and sour	85	Sour and petite	60

These numbers indicate that the *thickskin* and *reddish* loci are quite closely linked, and that the *reddish* and *sour* loci are also probably linked, though less closely than *thickskin* and *reddish*. Since *thickskin* and *sour* appear to segregate independently, the *reddish* locus must lie between them. *Petite* maps to a different linkage group.

Physical Maps

A physical map is an assembly of contiguous stretches of chromosomal DNA—contigs—in which the distance between landmark sequences of DNA is expressed in kilobases. The ultimate physical map is the complete sequence, which allows physical distances to be defined in nucleotides, notwithstanding variance among individuals due to insertion/deletion polymorphism. Physical maps provide a scaffold upon which anonymous polymorphic markers can be placed, thereby facilitating finer scale linkage mapping; they confirm linkages inferred from recombination frequencies, and resolve ambiguities about the order of closely linked loci; they enable detailed comparisons of regions of synteny between genomes; and they can be a first step in the assembly of a complete genome sequence.

Two general strategies can be used to assemble contigs:

1. *Alignment of randomly isolated clones based on shared restriction fragment length profiles.* Clones range in size from 1,000s of kilobases (yeast artificial chromosomes, YACs) to 100s of kilobases (bacterial artificial chromosomes [BACs], or P1 clones) to kilobases (plasmid or bacteriophage clones). Restriction profiling (see Figure 2.10) has been automated using a combination of robotics, which reduces variance and human error while increasing throughput, and continually improving software for profiling. DNA fragments that have been separated by gel electrophoresis and visualized using standard dyes are imaged digitally, then placed in size bins that facilitate probabilistic comparison of profiles.

2. *Hybridization-based approaches.* In these techniques, labeled probes are used to identify which of an arrayed set of clones are likely to be contiguous. In **chromosome walking,** a set of adjacent clones are isolated using the terminal sequence of the first clone as the first probe to pull out a set of overlapping clones that share the probe sequence. One of these is then chosen to identify a new terminal probe for the second step (Figure 1.2) and so forth. The extent of overlap of sets of clones identified by hybridization is also determined by comparing the restriction enzyme profiles of each clone. Contigs can be extended by end-sequencing particular clones, leading to the identification of **sequence-tagged sites (STSs),** which in turn can be used as hybridization probes to either extend the chromosome walk or fill in gaps in the contig. The assembly of contigs and even whole genome physical maps is discussed in more detail in Chapter 2.

Cytological Maps

One historically prevalent aid in the alignment of physical and genetic maps is the use of cytological maps, as shown in Figure 1.3. **Cytological maps** are the banding patterns observed through a microscope on stained chromosome spreads. Traditional cytological preparations have included the salivary gland polytene chromosomes of insects and Giemsa-banded mammalian metaphase karyotypes. The demonstration that certain mutant phenotypes or medical conditions correlate with the deletion or rearrangement of chromosome sections provided some of the first evidence that chromosomes are the genetic material, and human karyotype mapping remains a vital tool in diagnosing a range of human disorders.

In situ hybridization of cloned DNA fragments to the chromosomes allows alignment with the physical map as well. Human chromosomes are divided into bands, with numbering on the small (petite, p) and long (non-petite, q) arms from the centromere to the telomere, giving rise, for example, to the nomenclature that yields 7p15 as the signifier for the location of the *Hox* complex in the middle of the short arm of chromosome 7. Search engines give researchers a wide range of options for the visual display of cytological, physical, and genetic map data.

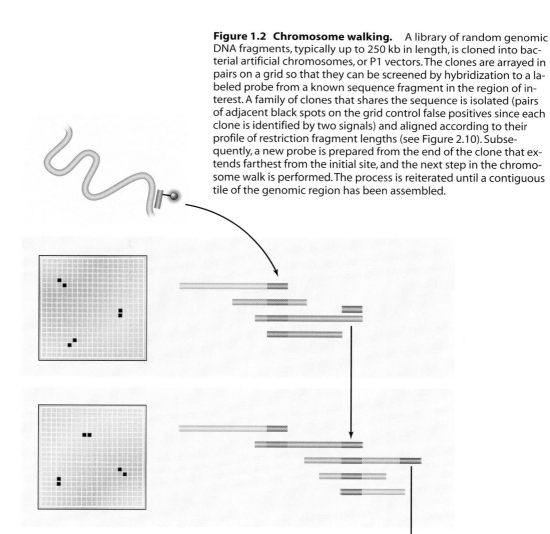

Figure 1.2 Chromosome walking.   A library of random genomic DNA fragments, typically up to 250 kb in length, is cloned into bacterial artificial chromosomes, or P1 vectors. The clones are arrayed in pairs on a grid so that they can be screened by hybridization to a labeled probe from a known sequence fragment in the region of interest. A family of clones that shares the sequence is isolated (pairs of adjacent black spots on the grid control false positives since each clone is identified by two signals) and aligned according to their profile of restriction fragment lengths (see Figure 2.10). Subsequently, a new probe is prepared from the end of the clone that extends farthest from the initial site, and the next step in the chromosome walk is performed. The process is reiterated until a contiguous tile of the genomic region has been assembled.

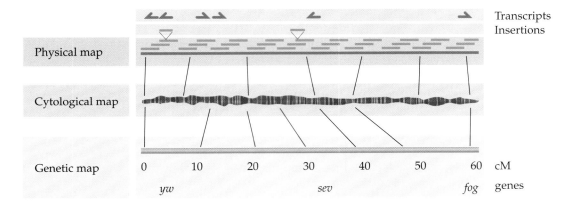

Figure 1.3 Alignment of cytological, physical, and genetic maps.   A cytological map is a representation of a chromosome based on the pattern of staining of bands, in this case a schematic of the polytene X chromosome of the fruitfly *Drosophila melanogaster*. Genetic (bottom), cytological (middle), and physical (top) maps can be aligned. The physical map includes the location of transcripts and sites of insertions and deletions as determined by standard molecular biology. Since recombination rates vary along a chromosome, typically being reduced near the telomere and centromere, distances between genetic, physical, and cytological markers are not uniform.

Comparative Genomics

In genomic parlance, the term "synteny" has come to refer to the conservation of gene order between chromosome segments of two or more organisms. Physical maps provide the most direct means for characterization of the extent of synteny, because highly conserved loci that can confidently be regarded as **homologs** (that is, derived from a common ancestral locus) provide anchoring landmarks. However, appropriate phylogenetic methods (discussed in Chapter 2) must be used to distinguish true homologs— properly known as **orthologs**—from **paralogs,** which are similar genes that arose as a result of duplication in one or both lineages subsequent to an evolutionary split. In many cases, entire complexes of genes have undergone multiple rounds of duplication, which complicates efforts to distinguish orthology from paralogy. This is not just of academic interest, but can be crucial when comparative physical maps are used to suggest the location of a locus with medical or other phenotypic effects based on its linkage to a cluster of syntenic loci.

The extent of conservation of gene order is an inverse function of the time since divergence from the ancestral locus, because chromosomal rearrangement is required to break up linkages. Rates of divergence vary considerably at all taxonomic levels. A notable example is the Japanese pufferfish, *Fugu rubripes*. Chosen for study largely on the basis of its unusually small genome size for a vertebrate (7.5 times smaller than the human genome), *Fugu* appears to have achieved this state by virtue of extensive DNA loss accompanied by rearrangements that probably shortened the average length

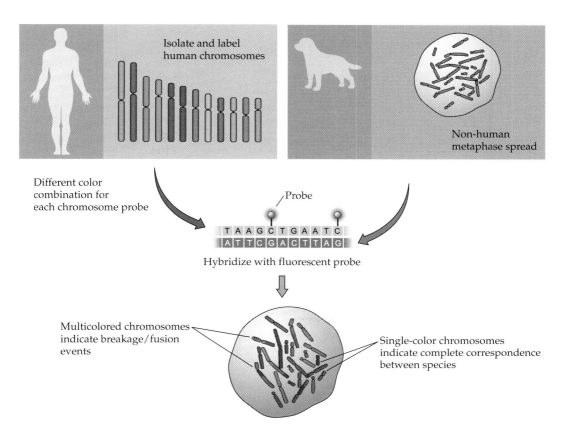

Figure 1.4 Chromosome painting. Chromosome painting uses the fluorescent in situ hybridization (FISH) technique to detect DNA sequences in metaphase spreads of animal cells. The fluorescently labeled hybrid karyotype (bottom) shows some chromosomes that are a single color, indicating complete correspondence with the human chromosomes, while multicolored chromosomes indicate that a chromosome breakage or fusion event has occurred that has contributed to chromosomal evolution between the species.

of syntenic regions. *Fugu* and other fish species show extensive gene order similarity with humans; one survey of over 500 zebrafish genes found that between 50% and 80% of these occur in conserved homology segments consisting of two or more genes in the same order as is found in the human genome (Barbazuk et al. 2000). Early indications are that monocotyledenous grasses and dicotyledenous species also retain segments of conserved gene order, covering hundreds of kilobases that are likely to be common to all plants (Mayer et al. 2001; also see Figure 1.17).

Chromosome fusions and splits, reciprocal translocations, and inversions all contribute to rearrangement of gene order. Within the great apes, each chromosome is thought to have been afflicted by an average of two or three rearrangements, many of which can be observed directly in karyotypes inferred from the banding patterns of mitotic chromosome spreads. This method has been enhanced dramatically by the development of the "chromosome painting" technique (Figure 1.4), in which each chromosome of one

Figure 1.5 Synteny between cat and human genomes. Ideograms for each of the 24 human chromosomes (22 autosomes plus the two sex chromosomes), shown on the right in each pair, are aligned against color-coded representations of corresponding cat chromosomes, as determined by a combination of FISH and RH mapping. Cat chromosomes are assigned to six groups (A–F) of 2–4 chromosomes each. The top row shows 12 autosomes that are essentially syntenic along their entire length, except for some small rearrangements that are not shown. The bottom row shows 10 autosomes that have at least one major rearrangement. The two sex chromosomes are also essentially syntenic between cat and human. (After Murphy et al. 2000.)

species is separately labeled with a set of fluorescent dyes that produces a unique hue, and hybridized to chromosome spreads of the other genome. Chromosome painting has been used to define regions of synteny covering segments on the order of one-tenth of a chromosome arm between dog or cat and human (Figure 1.5; Murphy et al. 2000). In addition to anchoring sequence scaffolds to cytological maps, this method is an important component of efforts to perform comparative mapping of species for which well-resolved physical maps are unavailable.

High-resolution comparative physical mapping is beginning to show that even within megabase-long stretches of synteny between human and mouse that may contain hundreds of genes, local inversions and insertions/deletions involving one or a few genes may not be uncommon. Families of genes organized in tandem clusters are particularly likely to diverge in number, organization, and coding content between species (Dehal et al. 2001). Considerable size variation in intergenic "junk" DNA is also apparent. While complete genome sequence alignment will provide the ultimate comparative mapping method, techniques based on chromosome painting and the alignment of landmark genes are likely to remain essential for drawing parallels between model genetic organisms and species of agricultural and medical importance (Pennacchio 2003).

The Human Genome Project

Genome science originated with a set of objectives that was conceived primarily to facilitate molecular biological research. With technological advances and increased involvement of private industry, the pursuit of these objectives has emerged as a discipline in its own right. Historians and philosophers of science will undoubtedly debate the relationship between the acquisition of information versus knowledge, and critics will continue to lament the absence of hypothesis-driven research in genomics, but the field is unique in having a clearly defined set of criteria by which its impact can be judged. By the objective criteria of goal attainment, the Human Genome Project in particular has been one of the most successful publicly funded scientific endeavors in history. In the remainder of this chapter we explore its objectives and achievements, as well as those of mouse, invertebrate, plant, and microbial genome projects.

Objectives

The objectives of the publicly funded Human Genome Project have been established in a series of 5-year plans authored by leading figures of the U.S. National Institutes of Health and Genome Research Institute (NHGRI)* (Collins and Galas 1993; Collins et al. 1998; Collins et al. 2003). These objectives are summarized in Tables 1.1 and 1.2. Most of the goals of the first and second plans were achieved ahead of time and under budget, including completion of the first-draft human genome sequence. Achievements beyond those originally included in the plans, such as the emergence of structural genomics and linkage disequilibrium mapping of disease loci, will help shape the discipline over the next decade.

The initial objectives of the HGP can be summarized briefly as follows. *First,* the generation of high-resolution genetic and physical maps that will help in the localization of disease-associated genes. *Second,* the attainment of sequence benchmarks, leading to generation of a complete genome sequence by the year 2005. (A draft version was achieved in 2000 by both the privately and publicly financed initiatives, but completion of the "finished" sequence—namely, an error rate of less than 1 in 10,000 nucleotides—of each chromosome has taken several more years.) *Third,* identification of each and every gene in the genome by a combination of bioinformatic identification of open reading frames (ORFs), generation of voluminous

---

*That NIH is instrumental in designing the Human Genome Project should surprise no one. In the early days of the HGP, the DOE was also heavily involved, which might be more puzzling. After the United States used the atomic bomb in World War II, the U.S. Congress charged DOE's predecessor agencies, the Atomic Energy Commission and the Energy Research and Development Administration, with "studying and analyzing genome structure, replication, damage, and repair and the consequences of genetic mutations, especially those caused by radiation and chemical by-products of energy production." The DOE's initial planning for a human genome reference sequence started in 1986, the NIH's in 1987. The two groups merged their planning beginning in 1990.

TABLE 1.1 *Initial Goals of the Human Genome Project*

From the First 5-Year Plan: 1993–1998

1. THE GENETIC MAP
   Complete 2 to 5 cM map by 1995
   Develop new technology for rapid and efficient genotyping

2. THE PHYSICAL MAP
   Complete STS map to 100 kb resolution

3. DNA SEQUENCING
   Develop approaches to sequence highly interesting regions on Mb scale
   Develop technology for automated high throughput sequencing
   Attain sequencing capacity of 50 Mb per year; sequence 80 Mb by 1998

4. GENE IDENTIFICATION
   Develop efficient methods for gene identification and placement on maps

5. TECHNOLOGY DEVELOPMENT
   Substantially expand support for innovative genome technology research

6. MODEL ORGANISMS
   Finish STS map of mouse genome to 300 kb resolution
   Obtain complete sequence of biologically interesting regions of mouse
     genome
   Finish sequences of *E. coli* and *S. cerevisiae* genomes
   Substantial progress on complete sequencing of *C. elegans* and
     D. melanogaster

7. INFORMATICS
   Continue to create, develop, and operate databases and database tools
   Consolidate, distribute, and develop software for genome projects
   Continue to develop tools for comparison and interpretation of genome
     information

8. ETHICAL, LEGAL, AND SOCIAL IMPLICATIONS (ELSI)
   Continue to identify and define issues and develop policy options
   Develop and disseminate policy regarding genetic testing
   Foster greater acceptance of human genetic variation
   Enhance public and professional education programs on sensitive issues

9. OTHER
   Training of interdisciplinary genome researchers
   Technology transfer into and out of genome centers
   Outreach

expressed sequence tag (EST) databases, and collation of functional data including comparative data from other animal genome projects. And *fourth,* compilation of exhaustive polymorphism databases, in particular of single-nucleotide polymorphisms (SNPs), to facilitate integration of genomic and clinical data, as well as studies of human diversity and evolution.

TABLE 1.2 *A Blueprint for the Future of the Human Genome Project*

Grand Challenges in the Third 5-Year Plan: 2003–2008

I. GENOMICS TO BIOLOGY

1. Comprehensively identify the structural and functional components encoded in the human genome
2. Elucidate the organization of genetic networks and protein pathways and establish how they contribute to cellular and organismal phenotypes
3. Develop a detailed understanding of the heritable variation in the human genome
4. Understand evolutionary variation across species and the mechanisms underlying it
5. Develop policy options that facilitate the widespread use of genome information in both research and clinical settings

II. GENOMICS TO HEALTH

1. Develop robust strategies for identifying the genetic contributions to disease and drug response
2. Develop strategies to identify gene variants that contribute to good health and resistance to disease
3. Develop genome-based approaches to prediction of disease susceptibility and drug response, early detection of illness, and molecular taxonomy of disease states
4. Use new understanding of genes and pathways to develop powerful new therapeutic approaches to disease
5. Investigate how genetic risk information is conveyed in clinical settings, how that information influences health strategies and behaviors, and how these affect health outcomes and costs
6. Develop genome-based tools that improve the health of all

III. GENOMICS TO SOCIETY

1. Develop policy options for the uses of genomics in medical and non-medical settings
2. Understand the relationships between genomics, race, and ethnicity, and the consequences of uncovering these relationships
3. Understand the consequences of uncovering the genomic contributions to human traits and behaviors
4. Assess how to define the ethical boundaries for uses of genomics

A list of ancillary, but no less important objectives, includes: support for and fostering of technological innovation in the domains of sequencing methodology and gene expression analysis; bioinformatic research and application; database establishment and standardization; and support for model systems genome science. A significant portion of the HGP budget

Figure 1.6   The architecture of the Human Genome Project in the twenty-first century.   This visual metaphor shows how the three major themes for future genome research are founded on six pillars of genome resources. (From Collins et al. 2003; courtesy of Darryl Leja, National Human Genome Research Institute, National Institutes of Health, Department of Health and Human Services.)

in the United States has also been set aside for research on the ethical, legal, and social implications of genetic research (the ELSI project; see Box 1.1).

The current blueprint for the future of the HGP focuses more on conversion of genomic data into biological knowledge than achievement of specific objectives. The visual metaphor in Figure 1.6 pictures genomics informing general biology, public health, and social issues, utilizing a foundation of six research pillars: genomic resources, technology development, computational biology, training, ELSI, and education. Fifteen grand challenges listed in Table 1.2 range from elucidation of the structure of genetic networks through prediction of susceptibility to disease, to development of policy options for ensuring equal access to the benefits of genome research.

The Content of the Human Genome

Completion of the first draft of the human genome sequence was announced at a press conference in May of 2000, but publication of this milestone was delayed until February of 2001 (IHGSC 2001; Venter et al. 2001). The intervening months were used to initiate a phase of refinement of the sequence assembly, including gap closure and verification of ambiguous aspects, along with gene annotation and prediction. The published draft genome fell well short of a definitive statement of the total gene content of the human genome, primarily because identifying coding sequences within random higher eukaryotic DNA sequences is not a trivial process. Estimates of the number of genes in the human genome have ranged from a low of 20,000 to a high in excess of 120,000 genes. The current consensus conters around 25,000 genes, as described in the publication of the finished euchromatic human genome sequence (IHGSC 2004).

This figure is only two times greater than the gene contents of two invertebrate multicellular eukaryotes—the fruitfly *Drosophila melanogaster* and the nematode worm *Caenorhabditis elegans*—and is a mere five times greater than that of a unicellular eukaryote, the yeast *Saccharomyces cerevisiae* (Table 1.3). Much of the increase in size can be attributed to two rounds of whole genome duplication, although independent expansion of certain gene families is also apparent. The complement of metabolic enzymes is essentially no different from that of other eukaryotes, while the numbers of genes involved in regulatory functions such as transcription, signal transduction, and intercellular signaling has expanded in proportion to the whole genome increase. First analyses confirmed that there are no dramatic differences in gene content between humans and other mammals such as the mouse; in fact the human genome is notable for a *reduction* in some gene classes, such as olfactory receptors.

Several landmarks in the generation of the human genome sequence need to be mentioned. The first high-resolution genetic map of the complete genome was published in September of 1994, and established the existence of 23 linkage groups (one per chromosome) with 1,200 markers at an average of 1 cM intervals (Gyapay et al. 1994). This map was quickly followed by a physical map generated from 52,000 sequence tag sites (STS) at approximately 60 kb intervals that formed the scaffold for the top-down public sequencing effort (Hudson et al. 1995). A database of 30,000 expressed se-

TABLE 1.3 *Comparison of Gene Content in Some Representative Genomes*

	E. coli	S. cerevisiae	Drosophila[a]	A. thaliana[a]	M. musculus	H. sapiens
Genome size[a] (Mb)	4.6	12.0	120+	115+	2,500+	3,000+
Number of genes[b]	4,300	6,250	13,600	25,500	30,000	50,000
Average gene density (kb)	1.1	1.9	8.8	4.5	80	60
Number of gene families[c]	2,500	4,500	8,000	11,000	10,000	10,000

[a]*Drosophila*, human, and *Arabidopsis* genome sizes indicate sequenced euchromatin, excluding heterochromatin.
[b]Approximate number of genes estimated from original annotation. The human estimate has recently halved to 25,000.
[c]Number of gene families rounded to 500.

BOX 1.1 The Ethical, Legal, and Social Implications of the Human Genome Project

The ELSI Program was established in January of 1990 after a combined working group of the National Institutes of Health and the Department of Energy recommended that research and education relating to the ethical, legal, and social implications of human genome research be incorporated as an essential component of the entire project from its inception. The National Human Genome Research Institute (NHGRI) now commits 5% of its annual budget to ELSI, funding three types of activity: regular R01 research grants, R25 education grants, and intramural programs at the NIH campus in Bethesda, Maryland. Web sites describing the program can be found at http://www.genome.gov/10001618 and http://www.ornl.gov/sci/techresources/Human_Genome/research/elsi.html, the latter documenting activities funded by the DOE.

The original mission statement of the project had four major objectives:

- Anticipation of the implications for individuals and society of sequencing the human genome
- Examination of the ethical, legal, and social implications of obtaining the sequence
- Stimulation of public discussion of the issues
- Development of policy options that would ensure beneficial use of HGP information

More practically, research is centered around four main subject areas:

- Privacy and fairness in the use and interpretation of genetic information (by insurers, employers, courts, schools, adoption agencies, and the military, among others)
- Clinical integration of new genetic technologies
- Issues relating to design and implementation of clinical research, including consent, participation, and reporting
- Public and professional education

One pressing social and individual issue that has attracted wide attention and generated great concern is the privacy and confidentiality of genetic information. This topic is particularly prominent in countries such as Iceland and Estonia, where government-sponsored databases of medical records (including both genetic histories dating back centuries and newly obtained genotype data) have been supplied to medical research companies. Another area of concern is the psychological impact and potential for stigmatization inherent in the generation of genetic data, particularly in the context of racial mistrust and socioeconomic differences in gathering of and access to genetic information. Reproductive issues can arise relating to informed consent and the rights of parents to know versus their fear of knowing, as well as potential moral (and possibly even legal) obligations once data has been obtained.

The program works toward implementing clear and uniform standards for informed consent and the conduct of clinical research, against a backdrop of federal reluctance to regulate and public uncertainty about the new technology. Another goal is the education of self-help groups, educators, and the media about distinguishing between very complex statistical associations and oversimplified assertions about the nature-nurture debate, biological determinism, and heritability. Philosophical discussions cover issues ranging from the basis of human responsibility, to the human right to "play God" with genetic material, to the meaning of free will in relation to genetically influenced behaviors.

In the more "practical" arena are studies of economic, safety, and environmental issues in relation to the release of genetically modified organisms (GMOs) at a time when public mistrust of science is increasing, particularly in Europe; the commercial and legal issues associated with the patenting of genetic material, procedures, and data, including international treaties and obligations as well as the right to free access to published data; and the forensic implications of DNA profiles in legal issues from paternity testing to the presumption of innocence.

Web pages and documents providing more detail on each of these issues can be accessed through Web links at the ORNL site. In addition, at least two major university institutes have established useful Web sites relating to their involvement in the ELSI project. The University of Kansas Medical Center site (http://www.kumc.edu/gec/prof/geneelsi.html) provides a series of policy papers put out by the American Society of Human Genetics and the American College of Medical Genetics on issues as diverse as fetal screening, cancer testing, testing for late-onset psychological disease, genetics and adoption, population screening of at-risk populations, and eugenics. It also provides an extensive set of links to legislative acts dealing, for example, with health insurance portability, Americans with disabilities, and birth defects prevention, as well as links to other internet resources established by public policy institutes and journals. Medical genetics courses and education opportunities, including the 30 or so genetic counseling programs in the United States, are also listed.

The Lawrence Berkeley Laboratory's ELSI Project site (http://www.lbl.gov/Education/ELSI/ELSI.html) is more concerned with the development of educational materials and is developing a series of teaching modules on topics such as breast cancer screening, genetic patents and intellectual property, and personal privacy and medical databases.

In 1998, the ELSI Research Planning and Evaluation Group issued a new set of goals as part of the NHGRI reevaluation of the future of the Human Genome Project. The five new major aims are:

1. Examine issues surrounding the completion of the human genome sequence and the study of human genetic variation: How will SNP mapping affect our understanding of race and ethnic diversity? How can we balance individual rights with ongoing research needs? Are there new concerns relating to the commercialization of human genetics? and, How can we best educate professionals and the general public alike about the implications of human genetic variation?

2. Examine issues raised by the integration of genetic technologies and knowledge into health care and public health activities: Will genetic testing promote risky behavior or intolerance? What are the social implications of pharmacogenomics, the tailoring of treatments for complex conditions to genotype? Will genetic factors be overemphasized merely because they can be more objectively defined than environmental ones? and, What will be the impact of genomics on health care provision and insurance issues, morbidity and mortality, and reproductive behavior?

3. Examine issues raised by the integration of knowledge about gene-environment interactions in nonclinical settings: Are there conditions under which genetic testing should not be considered? What issues are raised by the storage of blood and other tissue samples by the military and police, among other groups? What legal issues arise in relation to adoption and child custody, and are they affected by prior exposure to cultural and environmental variables and pathogens? and, Can we identify potential abuses of genetic information in the workplace, in classrooms, and by the media?

4. Explore interactions between new genetic knowledge and philosophy, theology, and sociology: Will our appreciation of the place of humans in relation to other living creatures change? What are the implications of behavioral genetics for traditional notions of self, responsibility, and spirituality? and, Is lengthening the human life span likely and/or desirable?

5. Explore how socioeconomic factors and concepts of race and ethnicity influence the use and interpretation of genetic information, the utilization of genetic services, and development of policy: How are individual views about the impact of genetics influenced by ethnic and social factors? Will particular communities be more vulnerable to abuse or more likely to benefit from genomics? and, What are the most effective strategies to ensure that genetic counseling is provided in a culturally sensitive and relevant manner?

quence tags, initially thought to represent the first collection of unique genes, was reported by Adams et al. in 1995, and several commercial consortia were also assembling orders-of-magnitude-greater EST databases by that time. A collection of more than 3,000 SNPs was described by Wang et al. in 1998, but this figure had exceeded 1.8 million mapped SNPs by the middle of 2004 (ISMWG 2001; http://snp.cshl.org)—in essence providing polymorphic markers at 2 kb intervals and placing 85% of all exons within 5 kb of a SNP. The first draft-chromosome sequence was also published in 2000 for the smallest human chromosome, chromosome 21 (Hattori et al. 2000), and finished quality whole chromosomes have been published at regular intervals since 2003. New builds of the genome sequence are updated on the public Web sites almost monthly, ensuring constant integration of the latest sequence information.

Two questions are often asked in relation to the content of the human genome: Whose genome was sequenced? and, When can we regard it as finished? The answer to the first question is that the sequence is derived from a collection of several libraries obtained from a set of anonymous donors. In the mid 1990s, ethical concerns were raised over the ramifications of one individual contributing the complete sequence. The issue was resolved to general satisfaction by reconstructing the core libraries. Both the International Human Genome Sequencing Consortium (IHGSC) and the private firm Celera Genomics reported that they assembled their sequence from multiple libraries of ethnically diverse individuals, although one particular individual's DNA contributed three-quarters and two-thirds of the raw sequence, respectively, as shown in Figure 1.7. Both groups adhered to strict privacy and consent guidelines when enrolling donors, whose identities remain unknown to researchers and unavailable to the general public.*

The IHGSC made initial libraries from individuals of both sexes, but as reported in their paper, by chance only males were used in the assembly of the first draft. The ethnic identity of the eight individuals whose samples were included is unknown, since the final samples were chosen at random. By contrast, the Celera sample included at least one individual from each of four ethnic groups, as well as both males and females. Race is not regarded as a concern, both because ongoing genome diversity surveys will be far more informative with regard to identification of racial differences, and because it is well known that the overwhelming majority of human sequence variation is shared across the human races.

As to when the human sequence will be "finished," the question of what constitutes an acceptable error rate is defined pragmatically as 1 in 10,000 bases, as this level is a minimum required for consistent open reading frame identification. In fact, human polymorphism is an order of magnitude greater than this, meaning that there are expected to be at least 10 SNPs for

---

*However, J. Craig Venter admitted to CBS' *60 Minutes* in May of 2002 that in fact his own DNA contributed substantially to the Celera sequence. He and his wife, Claire Fraser, continued the family tradition by sequencing their own poodle, Shadow, as the first-draft canine genome sequence.

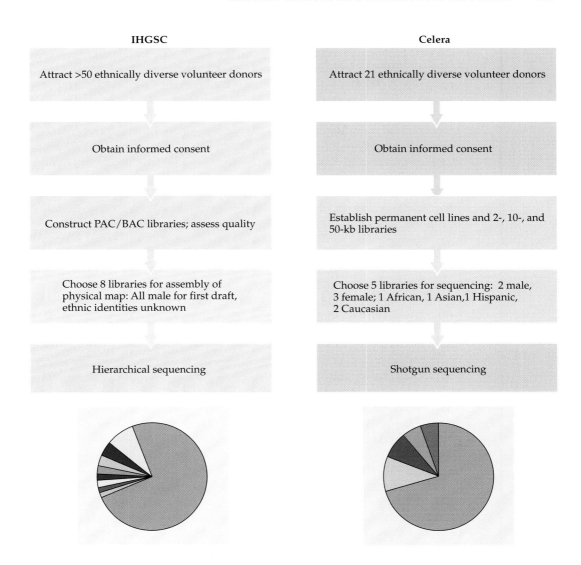

Figure 1.7 Whose genome was sequenced? The contributions of individual donors to the draft human genome sequences, as reported in February of 2001 by IHGSC and Celera Genomics. The size of each shaded sector in the pie charts is in proportion to the amount of sequence contributed by a single individual. Note that 8% of the IHGSC sequence was derived from clones of unpublished origin. Identification of SNP diversity, discussed in Chapter 3, uses a much wider sample of human variation.

each sequencing error. The complete sequence of 99% of human euchromatin has now been published to an estimated error rate of approximately 1 event in 100,000 bases (IHGSC 2004). It is, however, recognized that the complete sequence contains numerous deleterious mutations, since all genomes harbor several "lethal equivalents" that segregate in heterozygous form; however, these too will eventually be identified in human diversity surveys. Extensive tracts of heterochromatin, mostly associated with centromeres that may account for as much as 20% of the total genome, will probably never be sequenced. Since the completion of the first draft, much of the sequencing focus of the HGP has been on characterizing human diversity. The International HapMap project described in Chapter 3 is intended to map all of the major haplotypes in the human genome and characterize their distribution among populations, as a step toward identification of human disease susceptibility factors. Agencies in Canada, China, Japan, Nigeria, the United Kingdom, and the United States have developed a public resource (http://www.hapmap.org) that allows researchers around the world to access the latest information concerning the structure of variation in the human genome.

Internet Resources

Central internet resources for all genome projects are coordinated by the **National Center for Biotechnology Information (NCBI)** under the auspices of the United States National Institutes of Health and National Library of Medicine (http://www.ncbi.nlm.nih.gov), and by project Ensembl (http://www.ensembl.org), a collaboration between the European Bioinformatics Institute (EMBL-EBI) and the Sanger Center in the UK. These resources were established in the late 1980s as general repositories for molecular biology information, and have expanded to fill roles in database management, computational software provision, and dissemination of biomedical information (Figure 1.8; Wheeler et al. 2000; Birney et al. 2004). Both sites have online tutorials and comprehensive descriptions of all of the services and resources that they provide, and Exercise 1.2 below provides an opportunity for you to compare their genome browsers.

The NCBI site's guide to **Human Genome Resources** and the **EBI's Ensembl Genome Browser** provide entry points for ongoing human genomic research. Clickable chromosome karyotypes provide immediate access to high-resolution physical maps of any segment of the genome, allowing several hundred kilobases of finished sequence covering an interval of interest to be downloaded in a few seconds. Alternatively, loci can be searched by name or accession number, and search tools are being developed that identify all genes of a particular family or predicted structure. Several alternate genome viewers are available, each of which has particular advantages and features (Box 1.2), perhaps the most notable of which is the University of California at Santa Cruz (UCSC) Genome Browser (http://genome.cse.ucsc.edu). Also note that commercial sites and data mining packages have been developed by companies such as Incyte Genomics, Celera, Rosetta Inpharmatics, Informax, and LION Biosciences. These are available only to individuals with access to institutional or government subscriptions, but in some aspects they provide more detailed annotation and support than is offered by the public-access viewers.

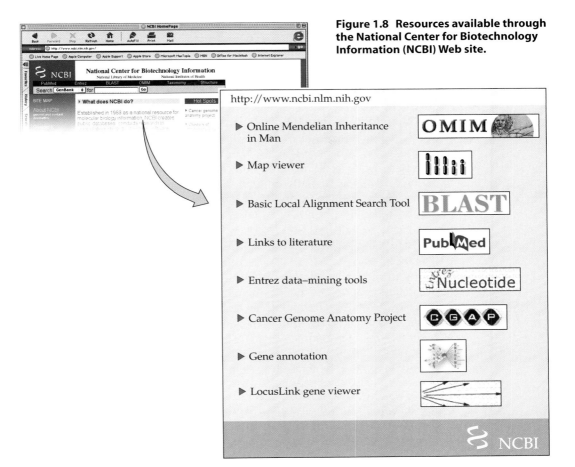

Figure 1.8 Resources available through the National Center for Biotechnology Information (NCBI) Web site.

http://www.ncbi.nlm.nih.gov

▶ Online Mendelian Inheritance in Man

▶ Map viewer

▶ Basic Local Alignment Search Tool

▶ Links to literature

▶ Entrez data–mining tools

▶ Cancer Genome Anatomy Project

▶ Gene annotation

▶ LocusLink gene viewer

EXERCISE 1.2 *Use the NCBI and Ensembl genome browsers to examine a human disease gene*

Choose a human disease of interest to you, and then use the OMIM site to identify a gene that is implicated in the etiology of the disease. Then use both the NCBI and Ensembl genome browsers to answer the following questions about the gene:

a. What are the various identifiers (aliases) for your gene?

b. Where is the gene located on the chromosome (cytologically and physically)?

c. What is the Reference Sequence (RefSeq) for the gene?

d. How many exons are there in the major transcript, and how long is it?

e. What is known about the function of the gene?

f. Do the two annotations agree? Which browser do you prefer, and why?

(continued on next page)

Exercise 1.2 (continued)

SAMPLE ANSWER: Go to http://www.ncbi.nlm.nih.gov, click on OMIM, and type "asthma" into the search field. Scroll through the genes until you find one of interest—for example, Interleukin 13 (IL13), and click on the blue number ("*147683"). This page gives a lot of textual information, but you can also link out to various sites, including Human Gene Mutation Database (HGMD) or Entrez Gene, which tells you that the GeneID for IL13 is 3596 and the cytological location is 5q31 (chromosome 5, long arm, position 132.1 Mb). The gene is also called ALRH, P600, and IL-13. The RefSeq is NM_002188, an mRNA sequence. There appear to be 4 exons in the major transcript, which is 1,282 bp long and encodes a 146 amino acid protein (34 of which are for the signal peptide). The protein is an immunoregulatory cytokine that regulates B-cell maturation and differentiation and the inflammatory response in the presence of allergins. All of this information can also be obtained from http://www.ensembl.org simply by typing "IL-13" into the search engine after selecting the Human Genome Browser. The Ensembl gene identifier is ENSG00000169194, and the gene view indicates that the transcript spans almost 3 kb. A number of Gene Ontology functions are listed as well.

Medical and clinical data are integrated with genome data through the **Online Mendelian Inheritance in Man (OMIM)** Web site (Figure 1.9), which grew out of an exhaustive catalog of human genetic disorders (McKusick 1998; Hamosh et al. 2000). OMIM is a searchable database that provides text summarizing recent genetic research in response to a query about a particular disease, as well as links to MedLine (a database of research abstracts), GenBank sequence entries (see Box 1.2), and other information. These data are primarily intended for physicians and human geneticists, but a less technical "Genes and Disease" site also exists within the NCBI resource that compiles information according to disease types such as muscle, metabolism, cardiovascular, and psychological disorders. OMIM lists in excess of 15,000 known disease-causing Mendelian disorders, and provides links to genes that have been implicated in the etiology of complex multifactorial diseases.

A major objective of the NCBI is to allow researchers in remote locations to be able both to submit their own data for inclusion in the growing databases, or simply to query and search data submitted by others. Increasingly, software for performing complex bioinformatic analyses is incorporated into Web sites so that virtual experiments can be conducted directly over the internet without the need to download voluminous databases. For example, the GEO BLAST tool allows a researcher to search for all genes in the gene expression database that have similar sequences, and then compare levels of expression of the genes across species and experimental conditions. Clearly, the utility of these resources is heavily dependent on the quality of the data they contain.

Figure 1.9 Resources available through the Online Mendelian Inheritance in Man (OMIM) Web site.

Animal Genome Projects

The International Sequencing Consortium (ISC; http://www.intlgenome.org) maintains a database of animal and plant genome sequencing projects supported mainly by funding agencies in the United States, England, France, Japan, and Canada. In addition to the projects described below, sequencing is well advanced for mosquitoes, honeybees, water fleas, a frog, a fruit bat, sea squirts and urchins, cows, a possum and a wallaby, and several primates; some of these organisms are shown in Figure 1.10. At the National Human Genome Research Institute (NHGRI), the decision to commit the tens of millions of dollars required for any new genome is made by a council of senior genome scientists who consider the arguments made in a 10-page "white paper" submitted by proponents of the organism. They weigh the expected impact of the sequence on enabling biomedical research and the annotation of sequence function (hence the wide phylogenetic representation of organisms), based on the idea that analysis of patterns of sequence conservation and evolution will be an indispensable tool in making sense of the human genome. A half-dozen large sequencing centers then bid for the contract to generate the sequence, with the expectation that a draft genome can be produced for most animals within 3–6 months. The NHGRI has a policy of rapid public dissemination of the data so that each community can immediately begin the task of genome annotation.

BOX 1.2 GenBank Files

There are as many ways to present the structure and annotation of a gene or sequence as there are genomics-related Web sites. A basic problem is that there is often no single correct structure for a given gene, due to alternative splicing and transcription start sites, the small errors that inevitably occur during cDNA cloning, and the fact that the algorithms underlying gene structure prediction software are imperfect. Furthermore, all genomes are full of polymorphism.

Thus the same gene may be represented by multiple different sequences or annotations in the genome databases. The National Center for Biotechnology Information (NCBI) has adopted a policy of accepting all sequences that are submitted to GenBank, but has developed a standard form of annotation that carries the implication "user beware." Hand curation by experts is required before any one sequence is elevated to the status of "RefSeq," or reference sequence; all other sequences should be regarded as supporting evidence.

All Entrez nucleotide files in GenBank share a number of features, which we illustrate using the human *HoxA1* file that you can access simply by typing the identifier 11421562 into the search field (select "Nucleotide" in the pulldown menu) at http://www.ncbi.nlm.nih.gov, and then click on the accession number (for example, XM_004915).

1. The top few lines define the **locus,** the Gene Index number (GI:11421562); the length of the sequence; whether it is derived from mRNA or a genomic clone; the date of submission; the **definition,** including the species from which the gene was obtained; the common name of the gene; the complete systematic position of the organism; and, where possible, the source tissue.

2. The locus is followed by the **reference,** including names of the **authors,** the **journal** and the **title** of the article in which the sequence was or will be published, and any historical notes on updates that have been directly submitted to the NCBI.

3. The meat of the file is the **features** section, which has subheadings describing the known extent of the gene; the coding sequence (CDS), including the predicted protein sequence; and miscellaneous features (**misc_feature**) such as intron-exon boundaries, identified protein domains, variations, mutations (if the sequence is derived from a mutant strain), and alternate transcripts. All of these are associated with links—for example, to OMIM, LocusLink, and the PFAM protein domain database file. Longer sequence scaffold files that span hundreds of kilobases (for example, *HoxA1* is contained within PAC file **AC_004079**) annotate each predicted gene, and also include pointers to sequence-tagged sites, repeats, and other miscellaneous features.

4. Next comes the sequence itself in blocks of 10 bases, 60 bases per row, with a running tally of the site number at the beginning of each row. Users have the option of displaying the sequence in FASTA format (which is simply an uninterrupted sequence of letters following a header) or downloading it as a text file by clicking on the "Display" and "Send to" boxes at the top of the page. XML and ASN.1 files can be used to export the file to certain bioinformatics applications.

There are numerous ways to view the structure of the gene graphically. Starting from the GenBank nucleotide file, link out to the EntrezGene file (you may have to click on the LocusID: 3198, then the blue GENE button). This will display a very low resolution image of the transcripts associated with the gene (in the case of Human *HOXA1*, two transcripts are seen). Alternatively, from the LocusLink page click on the brownish MAP button to bring up a MapView, and then the sv link next to the gene name, which will bring up a sequence view image similar to the one shown on the next page. The nucleotide and amino acid sequences are presented along with mRNA evidence in blue and the extent of protein isoforms in pink. A toggle tool allows you to move along the chromosome and zoom in or out to get a better picture of the chromosomal region.

Other browsers can be accessed directly, or through the NCBI pages. For example, from the LocusLink file, click the black UCSC or *e!* buttons to find the UC Santa Cruz and Ensembl genome browser representations, respectively. Ensembl shows four levels of graphical view: the whole chromosome, an overview of the genomic interval, a detailed view of the gene region, and a basepair/amino acid view. The image below shows how certain features (for example, SNPs and Ensembl transcripts) appear after they are selected from a pull-down Features menu in the detailed view window. The UCSC browser is particularly useful for comparing the alignments of a gene across multiple mammalian species. Most new browsers allow the user to choose a variety of tracks that display different types of evidence for transcripts, gene models, and other features of interest.

Chimpanzee
Pan troglodytes

Domestic dog
Canis familiaris

Tammar wallaby
Macropus eugenii

Rhesus macaque
Macaca mulata

Pufferfish
Tetraodon nigroviridis

Sea urchin
Stronglylocentrotus purpuratus

Sea squirt
Ciona savignyi

Red four beetle
Tribolium castanum

Mosquito
Anopheles gambiae

Honeybee
Apis mellifera

Figure 1.10 A gallery of animal genome sequencing projects.

Rodent Genome Projects

Almost a century of genetic research on mice and rats has ensured that these small mammals will occupy a central place in genome research. Three major advantages of rodent research are:

1. The existence of a large number of mutant strains that, combined with the potential for whole genome mutagenesis, will eventually lead to genetic analysis of every identified locus in the genome.

2. The existence of a panel of approximately 100 commonly used laboratory mouse strains with well-characterized genealogy—a formidable resource for the study of genetic variation and complex quantitative traits.

3. The evolutionary position of rodents, which are sufficiently divergent at the DNA sequence level from humans that the existence of conserved sequence blocks is generally an indicator of functional constraint, yet are sufficiently close to humans that many aspects of development, physiology, and the genetics of disease are shared.

The first public draft of the mouse genome sequence was published in late 2002 (MGSC 2002), and that of the rat followed within 18 months (RGSPC 2004). Like the human, these genome sequences are constantly updated on their respective NCBI and Ensembl browsers.

Although mutant strains of mice have been a productive source of material for genetic research for the better part of a century, functional genomic analysis of this model organism has been stimulated by three major advances achieved in the 1990s:

- First, the technology for targeted mutagenesis by homologous recombination of the wild-type locus with a disrupted copy has become routine. Time and expense are the only major obstacles to reverse genetics (moving from gene to phenotype), a strategy that is only likely to increase in popularity as gene expression profiling and comparative mapping define candidate genes for numerous traits.

- Second, several groups around the world have embarked on **saturation random mutagenesis programs**—that is, screens conducted on such a large scale that a point is reached where most new mutations occur in loci already defined by an existing mutation. Dominant mutations are much easier to identify than recessive ones, but even recessive mutations can be recovered in F_3 **designs,** in which the researcher looks for 1/4 of the grand-progeny to show an aberrant phenotype in a sibship, as will be discussed in Chapter 5.

- Third, the expense associated with colony maintenance has led to emergence of "phenomic" analysis, in which mutagenized lines are subject to batteries of biochemical, physiological, immunological, morphological, and behavioral tests in parallel by large research consortia. This approach offers the best prospect for large-scale identification of genes required for non-lethal phenotypes, thereby supplementing traditional analysis of embryogenesis, skeletogenesis, and skin and coat defects.

Figure 1.11 Resources available through the Mouse Genome Informatics (MGI) Web site.

Reflecting the centrality of laboratory-bred strains in mouse genetics, much of the genomic resources for mice are organized through a Web site at the Jackson Laboratory in Bar Harbor, Maine: http://www.informatics.jax.org. This Mouse Genome Informatics (MGI) site (Figure 1.11) includes a central role for physical and genetic maps, as well as search engines that allow searches by key word, accession numbers, expression patterns, and genomic location. Tumor biology is also prominently displayed, reflecting the high incidence of tumor classes specific to particular strains of mice. Online comparison of the mouse and human genome sequences facilitates identifica-

Figure 1.12 Mouse-human synteny and sequence conservation.   Conservation of ▶ gene order and DNA sequence between the human and mouse genomes is observed at three levels. (A) Blocks of synteny between mouse chromosome 11 and parts of five different human chromosomes are indicated (see the mouse-human Homology View map at **www.ncbi.nlm.nih.gov/Homology** for more detail). (B) Enlarged view of a small region corresponding to the human 5q31 interval. In this approximately 1-Mb region, there is almost perfect correspondence in the order, orientation, and spacing of 23 putative genes, including four interleukins. Within the region, 245 conserved sequences of more than 100 bp with 70% identity were detected, many of which fall in noncoding regions (red arrows). (C) Enlargement of the alignment of 50 kb that includes the genes *KIF-3A*, *IL-4*, and *IL-13*. Blue dots show the distribution of conserved sequences (sequences more than 100 bp long with from 50%–100% identity) between mouse and human. Two of the conserved blocks (red bars, with the indicated levels of identity) fall between genes, whereas most of the others (blue bars) are in the introns and exons of the genes. Such alignments are readily prepared online using PipMaker (**http://nog.cse.psu.edu/pipmaker**; Schwartz et al. 2000). (B and C after Loots et al. 2000, Figures 1 and 2.)

tion of likely regulatory sites and supports gene annotation, as demonstrated by Loots et al.'s (2000) characterization of a coordinate regulator of interleukin expression (Figure 1.12).

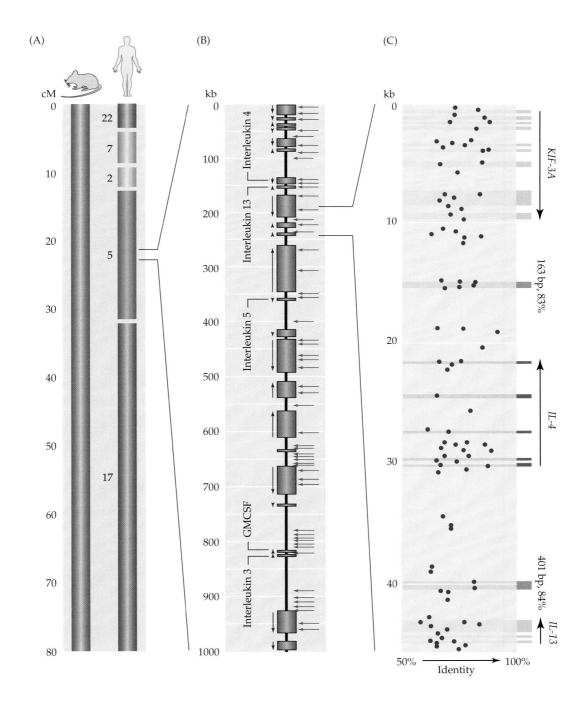

Two features unique to the mouse site are a comprehensive listing of molecular probes used in recombination mapping, and a database of strains, many of which can be obtained for nominal fees by appropriately accredited researchers. The molecular probes listing includes clones that can be used to prepare labeled probes for RFLP mapping, and oligonucleotide primer sequences that can be used to prime PCR reactions to amplify microsatellites or sequences that include an ever increasing set of SNP markers. The standard laboratory strains, as well as a collection of recombinant inbred lines derived from several of them, have already been typed for a high density of molecular markers, facilitating quantitative trait mapping as well as mapping of new Mendelian mutations as they are generated. Listings of probes and markers can be downloaded as tables in various formats, or searched online.

EXERCISE 1.3 *Compare the structure of a gene in a mouse and a human*

Using either the NCBI or ENSMBL Browser, explore the structure of the gene you used in Exercise 1.2 in a mouse and a human (and, if possible, in other vertebrates).

SAMPLE ANSWER: From the Ensembl Gene Report for Human IL13 (ENSG00000169194), scroll down to the "Orthologue Prediction," and click on the "View all genes in MultiContigView." At the top of the screen, the whole chromosome view shows you that IL13 is on mouse chromosome 11 and rat chromosome 10. The default "Detailed view" is of 200 kb, but you have the option of zooming in or out, or moving along the locus. The image below compares a human, a mouse, and a rat at 10 kb resolution, with light and dark shading representing stretches of relatively low or high sequence conservation.

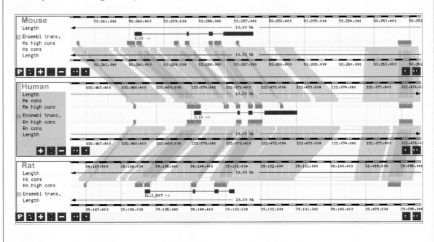

Other Vertebrate Biomedical Models

Two additional vertebrate genomes were fully sequenced in 2004, the chicken (*Gallus gallus*) and the dog (*Canis familiaris*). The major motivations for sequencing both species are biomedical: chickens are an important model for oncogenesis and virology (which also has enormous agronomic implications), while dogs are emerging as an essential model for a wide range of complex diseases such as asthma, parasite infection, cancer, arthritis, diabetes, and behavioral disorders. From a biological standpoint, the genomes are also expected to boost research into avian evolution, and of course the impact of artificial selection on breed diversity. More Mendelian disorders have been identified in dogs than in any other vertebrate, and many of these are breed-specific. Interestingly, much effort was put into identifying an extremely inbred dog before sequencing began so as to minimize the impact of polymorphism on assembly of the genome sequence (the lucky canine turned out to be a boxer named Tasha). It is possible to distinguish breeds on the basis of polymorphic microsatellite sequences (Parker et al. 2004), and the degree of sharing and differentiation of the genetic material should facilitate dissection of the obvious morphological, physiological, and behavioral differences between chihuahuas and greyhounds, or beagles and sheepdogs (Ostrander and Kruglyak 2000).

For the purposes of genetic analysis of vertebrate development, the zebrafish *Danio rerio* is an excellent complement to the mouse. Rapid and transparent embryogenesis, ease of culture, the existence of a dense genetic map, and a preponderance of available cell biological tools have established it as a model for the study of embryogenesis, neurogenesis, and organogenesis in particular. Saturation mutagenesis screens have uncovered thousands of genes that are required for the proper development of organs such as the heart and eye; of the musculature and appendages; for axon guidance; and for body symmetry and simple behaviors. In addition, new screens for maternal-effect loci are expected to help dissect fundamental aspects of pattern formation in the embryo. The zebrafish genome project can be accessed through the Zebrafish Information Network (http://zfin.org), which provides tools for functional genomic analysis, including extensive anatomical and gene expression atlases, high-density polymorphism databases for fine-structure recombination mapping, and online strain resources. Extensive sequence analysis of two evolutionarily divergent pufferfish with unusually small genomes, *Tetraodon nigroviridis* and *Fugu rubripes,* is also well advanced (Elgar et al. 1999; Roest Crollius et al. 2000), while genetic maps are under development for a variety of ecologically and commercially important fish species, including sticklebacks, cichlids, and salmonids.

Plans are also in place to sequence several nonhuman primates, including the rhesus macaque (*Macaca mulatta*) and the chimpanzee (*Pan troglodytes*). Practical motivation for this effort stems largely from a desire for better understanding of the origins of diversity in the immune system as well as mechanisms of pathogen resistance, including human susceptibility to the HIV virus. To students of evolution, however, the genomes of our closest relatives will be goldmines. Comparison of human and chimp sequences has already identified many of the genes that seem to have been positively selected during the evolution of *Homo sapiens* (Clark et al. 2003) and has demonstrated that we are differentiated from chimps by small deletions up to 10 kb in length, which occur on average every 500 kb along chromosome 21 (Frazer et al. 2003). Our appreciation of the significance of human genetic diversity will undoubtedly benefit from comparison to other primate outgroups, in ways that will only become clear over the next two decades but will be as profound as the knowledge of the human genome sequence alone.

Animal Breeding Projects

Genome projects to enhance animal and plant breeding efforts are being pursued, building on decades of classical genetic analysis. The Online Mendelian Inheritance in Animals (OMIA) database (**www.angis.org.au/ Databases/BIRX/omia**), maintained by a group at the University of Sydney, brings together linkage data and genetic maps for over a dozen species of agricultural importance. The site provides search engines that allow researchers to access data on inheritance patterns, molecular linkage, and, increasingly, molecular biology. Information can be accessed by disease or trait, as well as by species.

Although whole genome sequencing of the cow and pig is well advanced, sequence analysis of farm animals has been concentrated on the development of high-density genetic maps and EST databases that aim to identify tags homologous to genes that have been identified in model organisms. Characterization of polymorphism within and among breeds will just as importantly support advancement of quantitative genetic analysis, a discipline that has its origins in animal and plant breeding, and is of ever increasing economic importance. Although Mendelian loci that lead to disease and mortality are a great burden on agriculture, the future benefits of breeding programs lie in improvements in yield, infectious disease resistance, adaptation to climatic conditions, and improved food quality, not to mention maximizing the benefits of transgenic technology. These goals will be met both through enhanced genetic map development and association studies using SNP technology, as discussed in Chapter 3.

Each of the major farm animal genome projects has its own genome site or sites. In the United States these are supported by the Agricultural Research Service of the U.S. Department of Agriculture, and by the initiative of individual research groups, many associated with veterinary schools. The Roslin Institute in Edinburgh has overseen the establishment of a series of ArkDBs (**http://www.thearkdb.org**) (Hu et al. 2001), now providing genome resources for 10 species: cat, chicken, cow, deer, horse, pig, salmon, sheep, tilapia, and turkey. Most of these have nodes, or alternate sites, organized in Europe, Japan, and Australia. Each site provides continuously updated chromosome maps, data from radiation hybrid mapping panels, as well as marker databases that enable researchers anywhere in the world to carry out linkage studies with the most current marker densities. Organism-specific resources include information on common breeds, meeting and workshop announcements, and links that facilitate comparative genome analyses.

Invertebrate Model Organisms

The Generic Model Organism Database (GMOD), illustrated in Figure 1.13, is a coordinated effort of the mammalian, invertebrate, and plant genome communities to standardize Web tool construction and implementation and to provide open source software for database management. The Apollo and

Figure 1.13 The GMOD project.   The Generic Model Organisms Database is a collaboration between numerous organizations (indicated in the lower panel) with the aim of providing open source software for genome database construction. Some of the initial applications available for download are indicated on the right.

http://www.flybase.org

Figure 1.14 ***Drosophila* gene annotation.**   A typical GBrowse view shows 11 anno-tated genes in a 40-kb region of cytological band 43E of *Drosophila melanogaster*, cen-tered on the *saxophone* gene. Each gene either has a number beginning with CG, or is identified by its standard name (e.g., *sax*). The predicted structure of each set of tran-scripts is shown, along with the orientation on the chromosome. Further annotation tracks indicating the sites of transposable element insertions (one is shown in pink), re-peat elements, tRNAs, and predicted genes in *Drosophila simulans*, can be called up at will through a list of "tracks" below the image on the webpage (not shown) **http://www.flybase.org/cgi-bin/gbrowse-fb**.

GBrowse visualization tools (Figure 1.14) provide a familiar environment for examining gene structure, while other modules facilitate exploration of gene ontology and gene expression. Some of the computational issues involved in relational and object-oriented database schemas are discussed in Box 1.3.

The first genomes of multicellular eukaryotes to be sequenced completely were those of the nematode worm *Caenorhabditis elegans* (C. elegans Sequenc-ing Consortium 1998) and the fruitfly *Drosophila melanogaster* (Adams et al. 2000). These remarkable technological achievements were undertaken in part as proof-of-principle for considerably larger vertebrate genomes, in part for the intellectual excitement of learning what makes a complex organism tick, but primarily as support for traditional molecular genetic research. The subsequent sequencing of a dozen different *Drosophila* species was motivated largely as a model for how to use comparative genome data to functionally annotate genomes, while sequencing of several nematode species has more direct biomedical and agronomic importance from the perspective of parasite biology. Most of the tens of thousands of investigators studying these organisms spend up to several hours each day mining the genome data through the FlyBase (http://www.flybase.org) and WormBase (http://www.wormbase.org) Web sites.

Decades of genetic analysis have led to the molecular characterization of up to 20% of the complement of genes in these two organisms, providing an internal control for the accuracy of bioinformatic annotation procedures. Over 90% of the true genes in both species seem to have been identified, though generally their structures are not defined with high accuracy. A majority of the previously unidentified genes have been assigned a tentative function based on sequence similarity with members of the 100 or so known protein domains. Remarkably, between one-third and one-quarter of the predicted genes remain "orphans," with no known sequence similarity to genes in any other organism, and in many cases without functional data of any sort. Ongoing comprehensive EST sequencing and gene structure and mutational analysis, used in conjunction with improved bioinformatic algorithms, will be required to establish the definitive gene complement of these and other model genomes (Gopal et al. 2001; Reboul et al. 2001).

Comparison of the content of the two invertebrate genomes revealed a few surprises but also confirmed some long-held suspicions. Unexpectedly, there may be 50% more genes in the nematode genome (19,000) than there are in the fly genome (13,500), despite the fact that the fly is much more complex at several levels, including number of cells, number of cell types, and organization of the nervous system. Until recently, it was thought that the nematode represented an evolutionarily conserved, ancient mode of development, but phylogenetic revision suggests that it is most likely to be a highly derived molting protostome, or ecdysozoan. It is possible that an increase in gene number accompanied the evolution of its largely invariant mode of development. Some of the variance in gene number can be attributed to expansion and contraction of particular gene families, such as a surprising surplus of steroid-hormone receptors in the nematode, and the expansion of the olfactory receptor family in fruitflies. In any case, it is clear that there is no simple relationship between gene number and tissue complexity, or for that matter, between gene number and DNA content. Possibly the most profound revelation is the high degree of conservation of all of the major regulatory and biochemical pathways, most if not all of which are identifiable not only in both nematodes and flies but also in the unicellular eukaryote *Saccharomyces cerevisiae* and in vertebrate genomes (Chervitz et al. 1998).

From the point of view of functional genomics, a major impact of the invertebrate genome projects is the prospect of obtaining mutations in every single gene of the genomes. In flies, this is being achieved by a combination of saturation mutagenesis and construction of a library of overlapping defi-

BOX 1.3 Managing and Distributing Genome Data

As with many areas of science and technology, genome science has benefited greatly from advances in computing capabilities. Improved computational speed has been important, but a strong argument can be made that the growth of the internet has been even more crucial for genome scientists. In conjunction with the maturation of modern database technology, the World Wide Web has become the natural medium for managing and distributing genomic data.

The emergence of the internet allowed the creation of centralized data warehouses. Just as important, it led to the creation of shared public resources for searching and analyzing the contents of genomic databases. Full-featured Web sites such as those at NCBI (http://www.ncbi.nlm.nih.gov) and EBI (http://www.ensembl.org) provide immediate access to enormous amounts of data and analysis tools, free of charge, from anywhere on the globe. This is a dramatic change from the situation just over a decade ago, when the GenBank database was distributed by paid subscription in a small notebook full of 5.25" floppy disks!

Networking advances have also been important for within-laboratory data management. Automated capture devices allow data to be added to a lab's database immediately, and with little or no human intervention. Centralized **laboratory information management systems,** or **LIMS,** then allow users at multiple workstations (or even multiple geographic locations) to browse, edit, analyze, and annotate the data. These integrated systems, while they often do not offer cutting-edge analysis tools, provide a relatively seamless work environment and keep the user from reformatting data to satisfy the needs of individual analysis programs. New tools for internet-based data management are available through GMOD (http://www.gmod.org).

Although LIMS systems are invaluable time-savers in the lab, data format remains one of the annoyances of daily work in genome analysis. Sequence editors, databases, and statistical tools often require their own data format. Efforts to create standards exist, but have largely been unsuccessful because of the effort needed to retrofit existing software. Programs such as READSEQ, designed to convert among numerous popular formats, are essential items in the toolkit of genome scientists.

The core item for management and distribution of genomic data is a database system. Most databases can be classified as either **relational databases (RDB)** or **object-oriented databases (OODB).** These systems should be contrasted with familiar spreadsheet and flat file databases, such as text files and Microsoft Excel™, which store data and facilitate searches but are not designed for communication among files. Descriptions of the basic features of the different types of databases can be accessed at numerous Web sites, such as those provided by Microsoft and Oracle, or by the Object Management Group (http://www.omg.org).

Experts debate the relative merits of the two database philosophies. For our purposes, it is well established that RDBs are very effective for storing, searching, and distributing data that fits nicely into tabular for-

ciencies that remove every segment of each chromosome. Some elegant genetic trickery has been employed to enable targeted mutagenesis as well. In the nematode, saturation mutagenesis has been supplemented with RNAi technology, in which double-stranded RNA can literally be fed to the worms in their diet of *E. coli*, with the result that function of the corresponding gene is more often than not reproducibly reduced if not eliminated. The genetic

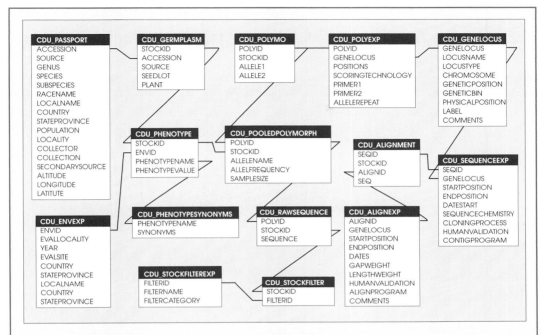

An example of a relational database (RDB) layout from the PanZea maize diversity project, indicating how items in one file are related to fields stored in different files.

mat. OODBs are particularly good at handling complex data structures, and are especially useful for performing analyses on sequence "objects." Descriptions of objects include information about the stored data, along with functions for operating on the data objects—a very efficient programming approach.

In contrast to simple techniques for storing data, such as spreadsheets, both RDBs and OODBs allow large amounts of data to be quickly retabulated, sorted, displayed, and queried. Query languages such as SQL (Structured Query Language) have been developed for fast and general searches of databases (explained in a tutorial available at **http://www.geocities.com/SiliconValley/Vista/2207/sg17.html**). Once the results of a database search are saved in files, scripting languages such as PERL eliminate the tedium of extracting and processing the relevant information returned by the search. PERL scripts can also be written to create data analysis pipelines for repetitive sequence analysis and management tasks. Sample scripts for biologists can be found at **http://bio.perl.org**.

resources are maintained in stock centers at the University of Minnesota (nematodes), and Indiana University and Sweden's Umeå University (flies), that are the lifeblood of invertebrate genetic research. Molecular resources such as SNP databases, probe and genomic clone collections, monoclonal antibody collections, and resources for microarray construction are supported by the respective genome projects.

Figure 1.15   Human disease genes in model organisms. Histograms showing the fraction of genes implicated in several classes of human diseases with orthologous genes in the fly, worm, and yeast genomes. Match significance is indicated by depth of color, with the darkest shade indicating the highest significance. Numbers in parentheses indicate the total number of genes in each disease class. (After Rubin et al. 2000.)

The publication of genome sequences has bolstered the proposition that invertebrates are not just models for the study of development and physiology, but can shed direct light on human disease (Rubin et al. 2000). More than 60% of a representative sample of 289 human genes that are mutated, amplified, or deleted in human diseases have an orthologous gene in the genome of *D. melanogaster*. The corresponding number is only slightly lower for *C. elegans*, and despite the fact that the yeast *S. cerevisiae* is unicellular, some 20% of human disease genes have orthologs even in that organism. Figure 1.15 shows the fraction of human disease genes in each of six categories that have orthologs in the fly, nematode, and yeast genomes, as detected by sequence similarity at three levels of significance within protein domains. Extrapolating from observations of the conservation of genetic interactions across the animal kingdom, we can confidently expect the genetic analysis of flies and worms to help uncover genes that interact with known disease-promoting loci. Several pharmaceutical companies are showing interest in invertebrate genomics for its potential to identify drugs that affect neural function. For example, screens for fluoxetine (Prozac™) resistance in nematodes (Choy and Thomas 1999) and alcohol tolerance in flies (Scholz et al. 2000) hint that these model organisms have a role to play in dissecting psychological disease. It is important to appreciate that there is no presumption that the invertebrate trait is the same as the mammalian trait; rather, the fact that molecular interactions between gene products can be conserved even when they affect distinct processes allows the functional comparison of genes across species.

Plant Genome Projects

Arabidopsis Thaliana

The first plant genome to be sequenced in its entirety was that of the model organism *Arabidopsis thaliana* (Arabidopsis Genome Initiative 2000), an achievement that was the result of a truly international consortium using a top-down approach of tiling across a physical map of cosmid clones. The *Arabidopsis* sequence was published in five installments, by chromosome, in *Nature* in 1999 and 2000.

The sequenced portion of 115 Mb is approximately the same size as that of the euchromatic portion of *Drosophila melanogaster*, but contains almost twice as many predicted genes (25,500). This gene complement seems to have evolved via two rounds of whole genome duplication followed by extensive shuffling of chromosomal regions and considerable gene loss. However, over 1,500 tandem arrays (generally two or three copies) of repeated genes have also been identified, with the result that the majority of genes are duplicated at some level and can be assigned to just 11,000 families. Some geneticists now regard this number as representative of the minimal complexity required to support multicellularity, and they believe it likely that all plant and animal genomes represent modifications of a basic "toolkit" of gene families that evolved more than 1 billion years ago.

In addition to the high level of segmental duplication (Figure 1.16), the complete sequence of the *Arabidopsis* genome reveals several features that distinguish plant from animal genomes. Over 800 nuclear genes may be of plastid descent, indicating that transfer of organellar DNA to the nucleus is probably an ongoing process; however, many of these genes have lost the protein plastid-targeting signal and so may now function in the cytoplasm. Transposable elements, including novel miniature elements such as MITEs and MULEs, account for at least 10% of the *Arabidopsis* genome; as in animals, these elements are concentrated in repetitive centromeric heterochromatin. Centromeric DNA is enriched for at least 40 families of repeats, and the genomic sequence opens up possibilities for functional characterization of centromeric as well as telomeric structural features.

Plant genomes contain several classes of genes that are either absent from, or under-represented in, animal genomes. The products of these "plant-specific" genes include:

- Enzymes required for cell wall biosynthesis.
- Transport proteins that move organic nutrients, inorganic ions, toxic compounds, metabolites, and even proteins and nucleic acids between cells.

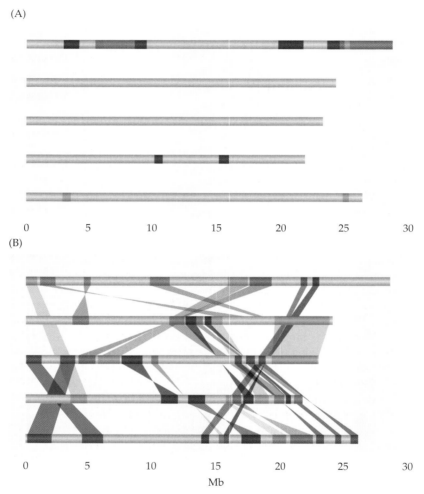

Figure 1.16 Chromosome duplications in the *Arabidopsis thaliana* genome.
Whole genome alignments of the model plant *Arabidopsis thaliana* indicate that there have been at least 30 segmental duplications within its genome. (A) Seven intrachromosomal duplications are shown as duplicated blocks of color within three of the five chromosomes; five duplications occur in the first chromosome, and the fourth and fifth chromosomes display one duplication apiece. (B) A schematic depiction of another two dozen interchromosomal segmental duplications. A twist in the band indicates that inversion accompanied the duplication event. (After AGS 2000.)

- Certain enzymes and other macromolecules required for photosynthesis, such as Rubisco and electron transport proteins.
- Products involved in plant turgor and the responses peculiar to a sessile lifestyle, including phototrophic and gravitropic responses.
- Numerous enzymes and cytochromes involved in the production of the hundreds of thousands of secondary metabolites found in flowering plants.

- A large number of pathogen resistance *R* genes and associated factors. These are generally highly polymorphic, as are mammalian defense-related genes; but unlike the genes of the mammalian immune system, *R* genes are dispersed throughout the genome rather than being localized in a single complex.

Plants share with animals many of the gene families involved in intercellular communication, transcriptional regulation, and signal transduction during development, but there are some notable exceptions. For example, *A. thaliana* lacks homologs of the Ras G-protein family and tyrosine kinase receptors (although there are many serine-threonine kinases), and there are no obvious Rel, forkhead, or nuclear steroid receptor class transcription factors. On the other hand, plant genomes encode variant forms of chromatin remodeling proteins, including histone deacetylases and SNF2 ATPases; several unique families of transcription factors; and novel components that mediate plant hormone function.

As with other model genetic organisms, there is much more to the *Arabidopsis* genome project than the complete genome sequence. The Web site for the The Arabidopsis Information Resource, TAIR (http://www.arabidopsis.org), allows researchers to integrate the genome sequence with an extensive EST database and with the genetic and physical maps; it provides links to functional and molecular genetic data and the literature for specific genes; and it shows an ever expanding list of mutant stocks. An alternative resource for *Arabidopsis* and many other plants, UK CropNet, uses common AceDB or WebAce platforms to coordinate genetic and molecular data (see Box 1.3 for a general discussion of genome database management).

Several sets of recombinant inbred (RI) lines have been particularly useful for quantitative genetic analysis. The molecular maps of these lines are constantly being updated by individual users, which presents a challenge for the ongoing task of aligning physical and genetic map data and means the quality of new data must constantly be assessed and monitored. Gene expression data, structural information, primers for genotyping microsatellites and SNPs, and information about mutant stocks are all immediately accessible.

Grasses and Legumes

Genome initiatives are under way for more than 50 different plant species. From an economic standpoint, the most important of these projects are those for the major feed crops—the grasses maize, rice, wheat, sorghum, and barley; the forage legumes soybean and alfalfa; and the forage rye grasses and fescues. Several of these genomes are so large (as a result of autopolyploidization and the dramatic expansion of repetitive DNA) that whole genome sequencing is impractical, and efforts have instead been focused on comparative genome methods. Both rice (*Oryza sativa*) and maize (*Zea mays*), however, have relatively small genomes and are such key elements of the agricultural economies of the developed world that complete genome sequences have been prioritized. Draft genomes of the two major rice genome cultivars, *japonica* and *indica*, were published by Syngenta (Goff et al. 2002) and the Beijing Genomics Institute (Yu et al. 2002) respectively, and various other consortia are finishing specific chromosomes. Gene identification in maize has been advanced by techniques, such as methylation filtering, that remove intergenic and highly repetitive sequences and effectively reduce the size of the genome for sequencing purposes more than fivefold (Whitelaw et al. 2003). Comparison of genome sequences of rice and *Arabidopsis* suggests that extensive but complex patterns of synteny will be a useful feature of plant genomics (Figure 1.17). *Medicago* (alfalfa) is a true diploid that, along with its crucial role in fixing soil nitrogen, constitutes a major part of forage diets. Thus *Medicago* has been selected as the model legume for whole genome sequencing.

The remaining grasses and legumes are the subjects of extensive EST sequencing and high-resolution genetic map construction in hopes of taking advantage of the expected pervasive synteny within these families. Web sites established by individual research groups integrate research efforts from around the globe. Some useful Web sites include the United Kingdom's CropNet (http://ukcrop.net/), the U.S. Department of Agriculture's Gramene resource (http://www.gramene.org), both of which are designed to facilitate comparative genome mapping, and organism-specific resources such as MaizeGDB (http://www.maizegdb.org/). A common objective of these sites is to link seed stock and real genetic resources to virtual data on linkage and mapping data, QTL localization, EST and sequence data, and biochemical pathway information, among other resources.

The maize, rice, and alfalfa initiatives all focus strongly on quantitative genetic resources, in contrast with the Mendelian focus of many model organism genome projects. Economically important traits include resistance to a broad range of pathogens; flowering time, seed set, grain morphology, and related yield traits; tolerance to drought, salt, heavy metals, and other

Figure 1.17 Rice-*Arabidopsis* synteny. Twenty of 54 genes in a 340-kb stretch of the rice genome (top) retain the same order in five different 80- to 200-kb regions of the *Arabidopsis* genome (below; not drawn to scale). The conserved genes (red and green boxes) are found on both the rice and *Arabidopsis* strands, but are interspersed by a variable number of different genes (yellow boxes) in *Arabidopsis*. Shaded boxes above the rice chromosome indicate that the conserved gene is in the opposite relative orientation on the *Arabidopsis* chromosome(s).

extreme environmental circumstances; and measures of feed quality such as protein and sugar content. Each of these characters can be improved through combinations of genetic engineering and specialized plant breeding techniques, including introgression of germ line from wild ecotypes. More traditional focus on the genetic basis of heterosis, hybrid vigor, and transgressive segregation, as well as efforts to limit inbreeding depression, will also benefit from improved maps and a more general appreciation of the extent and distribution of polymorphism in the genome. Given the role of artificial selection in the recent derivation of grasses in particular, these genome projects promise to reveal much information regarding the evolution of domesticated species. Several genes have been identified in maize that were clearly selected during the transition from wild teosinte to cultivated maize, leading to modification of traits such as glume architecture, ear size, and a change from multiple branches to a single stalk (tiller). Although folklore suggests that cultivated crops are highly inbred, molecular population genetic data contradicts this notion; in fact domesticated maize is highly polymorphic relative to most animal species. As in other crop grasses, diversity in domesticated maize is reduced by a maximum of a mere 30% relative to its presumed wild progenitors (Buckler et al. 2001). However, strongly selected portions of loci, such as the upstream regulatory region of *teosinte branched 1*, (*tb1*) show a marked reduction in diversity that is indicative of a selective sweep that brought one haplotype associated with the selected site close to fixation, purging linked polymorphism in the process (Figure 1.18). The selected site is thought to upregulate expression of the *tb1* gene, which encodes a repressor of lateral branch elongation,

Figure 1.18 *Teosinte branched 1* and the evolution of maize. Modern maize is a derivative of the wild progenitor teosinte, which had multiple tillers. Throughout the coding region of *tb1*, the level of polymorphism (the number of nucleotides that differ between any two alleles in a sliding window along the gene) is substantially the same in a sample of maize and teosinte. However, in the 5′ non-translated region, there is a dramatic reduction in the level of polymorphism in maize relative to that seen in teosinte. (After Wang et al. 1999.)

resulting in the evolution of single-tillered plants (Wang et al. 1999). Crop plants are thus established as excellent model systems for studying the molecular basis of morphological divergence, as well as other types of population genetic analysis.

Other Flowering Plants

Over 90 different angiosperm genome projects around the world are listed on the United States Department of Agriculture Web site (http://www.nal.usda.gov/pgdic/Map_proj). The list includes African projects on beans, corn, and fungal pathogens; Australian projects on cotton, wheat, pine, sugarcane, and nine others; at least 24 European projects that include vegetables such as cabbage, cucumber, and pea, and fruits such as apple, peach, and plum; and over 50 North American projects as diverse as turf grass, chrysanthemum, almond, papaya, and poplar.

The common denominator among all of these projects is the assembly of genetic maps (and in some cases physical maps) and the placement of a common set of plant genes on them. For some species, large EST sequencing projects are also in place, with the twin objectives of enabling comparative genomic analysis (particularly in regions of synteny) and QTL mapping.

Several model organisms in addition to *Arabidopsis* and the grasses are receiving particular attention as a result of a long history of genetic analysis and/or the potential light these organisms may shed on plant evolution. These include the snapdragon (*Antirrhinum majus*), in which numerous classical flowering mutants were initially isolated as a result of transposable element movement; sunflowers (*Helianthus* spp.) and monkey flowers (*Mimulus* spp.), which are of particular interest in studies of hybrid speciation and adaptation in the wild; and the variants of *Brassica oleracea* (cabbage, kale, Brussels sprouts, broccoli, cauliflower, and kohlrabi), which are in the same family (Brassicaceae) as *Arabidopsis* and are a fascinating model for domestication because they all derive from the spontaneous mutation of genes involved in meristem growth (Purugganan et al. 2000).

Forest trees are an example of an area where genomic analysis has the potential for economic impact where classical genetics has been problematic (Figure 1.19). High-density genetic maps of spruce, loblolly, and several other pines, as well as a few species of *Eucalyptus,* have been established using a combination of AFLP and microsatellite markers and applied to the mapping of Mendelian and quantitative trait loci affecting wood quality, growth, and flowering parameters. These maps can be accessed through the Dendrome Web site (http://dendrome.ucdavis.edu). Marker-assisted selection has the promise to improve desired traits dramatically, at least to the extent that phenotypes measurable in seedlings predict mature qualities. Comparative analyses and transcription profiling of genes involved in wood properties including lignins and enzymes that regulate cell wall biosynthesis (many of which can be identified by large EST sequencing efforts) will also have an impact on forest biotechnology throughout the world.

Figure 1.19 Forest genomics.   The Dendrome Web site for forest tree genomics at the University of California/ Davis provides links to over two dozen other forest genomics Web sites around the world, as well as providing access to databases of tree genes, forest genetics, and other resources.

Several staple crop plants are obviously poised to benefit from genomics, including potato and other tubers, tomato, tobacco, beans, and cotton. Analyzing the genome diversity of ecotypes endemic to the original source of these crops—many of which come from tropical regions—has the potential to affect productivity in developing countries as well as to support yield and quality improvements in countries where monocultures are employed.

The political, legal, and sociological implications of proprietary rights to a plant's germ plasm and the polymorphic DNA sequences identified within it are being addressed as much in courts of law as by international agencies. No plant equivalent of the Human Genome Project's ELSI initiative has yet been established.

Microbial Genome Projects

The Minimal Genome

The first complete cellular genomes to be sequenced were of prokaryotes, starting with *Haemophilus influenzae* (Fleischmann et al. 1995) and quickly followed by *Mycoplasma genitalium* (Fraser et al. 1995), three other bacteria, and, in September of 1997, *Escherichia coli* (Blattner et al. 1997). The primary sequences provided immediate information about genome structure (organization of replication, GC content, transposable elements, recombination) and genome content (total number of genes, representation of conserved gene families).

Gene annotation is initially more straightforward for prokaryotes than for eukaryotes, since open reading frames tend to be uninterrupted and genes tend to be closely spaced; however, the assignment of adjacent genes to operons is not trivial. Typically, more than three-quarters of the ORFs in a microbial genome can be assigned a function based on their similarity to genes in other organisms and/or by identifying protein domains. The remaining genes may encode yet-to-be-described functions, they may represent taxon-specific functions, or they may be evolving so fast that their function is conserved despite sequence divergence. Bioinformatic approaches are being employed to complete the functional annotation of microbial genomes. The TIGR server (http://www.tigr.org) provides an excellent resource for tracking the most recent updates.

Similarities in gene content between the three bacterial genomes mentioned above, as well as the first sequenced genome of an Archaea, *Methanococcus jannaschii* (Bult et al. 1996), are presented in Table 1.4. There

TABLE 1.4 *Number of Genes Involved in Defined Cellular Processes*

	M. genitalium	*H. influenzae*	*E. coli*	*M. jannaschii*
Central intermediary metabolism	6 (1.3)*	30 (1.7)	188 (4.4)	18 (1.0)
Energy metabolism	31 (6.6)	112 (6.4)	243 (5.7)	158 (9.1)
Lipid and fatty acid metabolism	6 (1.3)	25 (1.4)	48 (1.1)	9 (0.5)
Cofactor biosynthesis	5 (1.1)	54 (3.1)	103 (2.4)	49 (2.8)
Amino acid biosynthesis	1 (0.2)	68 (3.9)	131 (3.1)	64 (3.7)
Nucleotide metabolism	19 (4.0)	53 (3.0)	58 (1.4)	37 (2.1)
DNA replication and repair	32 (6.8)	87 (5.0)	115 (2.7)	53 (3.0)
Transcription	12 (2.5)	27 (1.5)	55 (1.3)	21 (1.2)
Translation	101 (21.4)	141 (8.1)	182 (4.2)	117 (6.7)
Regulatory functions	7 (1.5)	64 (3.7)	178 (4.2)	18 (1.0)
Transport and binding proteins metabolism	34 (7.2)	123 (7.0)	427 (10.0)	56 (3.2)
Cell structure	17 (3.6)	84 (4.8)	237 (5.5)	25 (1.4)
Cellular processes	21 (4.5)	53 (3.0)	327 (7.6)	26 (1.5)
Other categories	27 (5.7)	93 (5.3)	364 (8.5)	38 (2.2)
Unclassified	152 (32.3)	736 (42.1)	1,632 (38.0)	1,049 (60.4)
Total	**471**	**1,750**	**4,288**	**1,738**

*Numbers in parenthesis represent percentage of all genes.
Sources: Categories adapted according to scheme of Riley (1997).
Data for *M. genitalium* and *H. influenzae* from Fraser et al. (1995). Data for *E. coli* from Blattner et al. (1997).
Data for *M. jannaschii* from http://www.tigr.org/tigrscripts/CMR2/gene_table.spl?db=arg (1/20/01).

are 471 predicted genes in the 0.6 Mb *M. genitalium* genome, 1,750 in the *H. influenzae* genome (1.8 Mb), and 4,288 in *E. coli* strain K12 (4.6 Mb). Thus the average gene length in each species is close to 1.1 kb, indicating that differences in genome size are based on changes in gene number, which in turn reflect duplication and divergence in larger genomes, as well as gene loss in small genomes.

It is usually possible to identify specialized metabolic functions that reflect a microbe's adaptation to a particular ecological niche (such as mammalian genital, respiratory, or enteric tracts) by surveying the predicted enzyme and transporter complement. Indeed, one aim of microbial genomics is to be able to predict metabolic phenotypes on the basis of gene content alone, as discussed in Chapter 6. Comparative analysis of sequences of pathogenic strains of *E. coli* and of various pathogenic species in the *Bacillus* (including anthrax) and *Mycoplasma* genera hint at the genetic basis of virulence and pathogenicity and may suggest novel approaches to antibiotic design.

The concept of the **minimal genome** refers to attempts to define the minimum complement of genes that are necessary and sufficient to maintain a free-living organism—in a sense, to define genetically "What is life"? The possibility that, having defined this complement, genomicists may be able to assemble a species *de novo* has sparked a call for more research into the moral implications of such research (Cho et al. 1999). Nevertheless, two general strategies have been pursued (Figure 1.20). The *bioinformatic* strategy is to identify which genes are present in each and every sequenced genome (Mushegian and Koonin 1998). Since some functions can be performed by nonorthologous genes, the list of completely conserved orthologs must be supplemented with a small number of alternates, but has been estimated at 256 genes. Such a hypothetical organism would be prototrophic (i.e., it would depend on broad-specificity transporters to import essential metabolites) and completely anaerobic, with central metabolism reduced to glycolysis. An *experimental* strategy has been to systematically knock out the function of individual genes: Mutations that cannot be recovered define genes that are likely to be components of the minimal genome (Hutchison et al. 1999). Mutations have been recovered for over 120 of the 470 *M. genitalium* genes, and these are inferred to be dispensable for life.

Similarly, comprehensive mutagenesis of the 4,100 genes of the *B. subtilis* genome has revealed that just 271 genes are indispensable under favorable growth conditions, and most of these are involved in a relatively small number of functions relating to metabolism, cell division and shape, and synthesis of the cellular envelope (Kobayashi et al. 2003). Since **synthetic lethality**—the nonviability in combination of two or more individually viable mutations—must be adjusted for (though has yet to be quantified), as should the likelihood that some genes are required in untested environments, this mutational approach probably underestimates the size of the self-sufficient minimal genome. We can infer that life can be supported by a genome of between 250 and 350 genes, but it is hard to imagine such a minimal organism possessing any competitive advantage outside of a petri dish. Efforts are also under way to build a viable organism from scratch by stitching

(A)

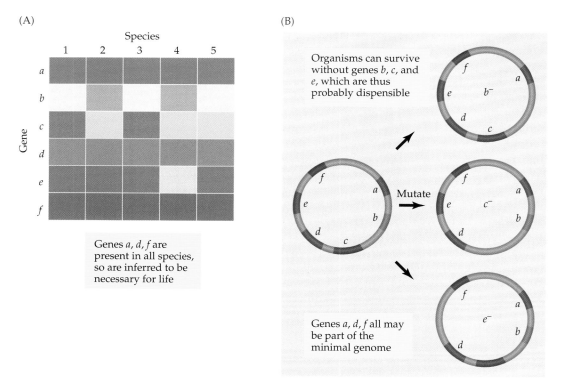

Figure 1.20 Describing the minimal genome. Two strategies have been used in attempts to define the minimal gene content required for microbial life. (A) Comparison of gene content in complete prokaryotic genomes. In the example, deeper color indicates the presence of a given gene, while pale color means the gene is absent in that species. Such comparisons have been compiled for hundreds of diverse prokaryote genomes. (B) Mutagenesis experiments establish which genes are essential by systematically eliminating ("knocking out") each functional gene and seeing whether the organism can survive without it. The overlap of these two approaches may define the minimal genome.

together artificially synthesized genes, building on the resynthesis of a poliovirus from published sequence data (Cello et al. 2002).

One of the first insights provided by genomic analysis has been that bacterial genomes are much more modular and evolutionarily labile than hitherto appreciated. An intriguing feature of the *E. coli* genome is the evidence for genome plasticity in the form of repetitive sequences and insertions of several families of IS class transposable elements, as well as the presence of cryptic prophages and prophage remnants. Another element of bacterial plasticity that has been uncovered by a combination of comparison of phylogenetic distributions and studies of the distribution of GC content and codon biases, is **horizontal gene transfer**—gene exchange between species (Woese 2000). Adaptation and divergence in microbes is now seen to be associated with a dynamic process of loss and gain of genes, notwithstanding the conservation of a common core of othologs that can be revealed by careful phylogenetic analysis of quartets of completed genome sequences (Daubin et al. 2003).

Sequenced Microbial Genomes

The Institute for Genome Research (TIGR) has a Comprehensive Microbial Resource (CMR; http://www.tigr.org/tigrscripts/CMR2/CMRHomePage.spl) that as of the summer of 2004 provides online access to the complete sequences and associated resources for 160 microbial genome sequences. Thirty-nine of these genomes were generated by TIGR and the remainder by consortia in countries from Brazil to Japan; all are collected on a single underlying database, the Omniome. In addition to the CMR resources, each genome is linked to alternate genome pages maintained by NCBI or the sequencing center that generated the data. It is presented with a standardized chromosomal display (Figure 1.21) in which each gene is color-coded according to the predicted molecular function of its product, and includes

(A)

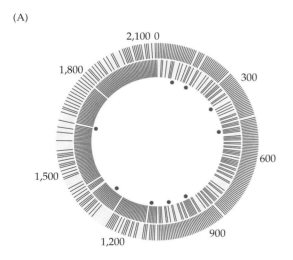

Figure 1.21 TIGR representation of a typical microbial genome. Circular (A) and linear (B) TIGR CMR representations of the 2.1 Mb genome of *Streptococcus pneumoniae*. The outer and inner circles in (A) represent genes encoded on the two strands of the chromosome, with a clear asymmetry that also corresponds with GC content bias. The red dots indicate the locations of nine clusters of genes that are missing in two other strains of the same species, and demarcate potential virulence factors. (B) A close-up view of a 25-kb portion of the chromosome provides evidence for genes on each strand from sequence matches to a gene from an HMM model (blue), BLAST (yellow), or genes in the omniome database of microbial genes (pink). (A after Tettelin et al. 2001.)

(B)

mouse-over links to background information, publications, and data on each individual sequence annotation. A comprehensive set of analytical tools is available through the CMR (Peterson et al. 2001). Search engines allow multiple genomes to be queried on a variety of criteria; whole genomes can be aligned based on exact sequence matches; standard molecular data such as GC skew and restriction maps can be generated; representation of gene families is tabulated; and predicted proteins can be sorted by structural attributes. One tool, MUMMER, is used to align pairs of genomes with selectable minimal alignment lengths from 15 bp to 100 bp (Delcher et al. 1999a). MUMMER can be used online, and its graphical display provides immediate information on synteny and recombinational history. Annotation of microbial genomes can be performed with another tool, GLIMMER (Gene Locator and Interpolated Markov Modeler), that can be downloaded from the main TIGR page (Delcher et al. 1999b). As described in more detail in the next chapter, gene annotation can be performed in a variety of ways, and the CMR utilizes three common types of protein annotation according to the COG, TIGRFAM, and PFAM classification systems. All of these topics are clearly explained through the Help page of the CMR.

EXERCISE 1.4 *Compare two microbial genomes using the CMR*

Use the CMR to find which species of bacterium has the largest number of annotated cell adhesion genes. Compare these genes with the annotated cell adhesion genes in *E. coli* and in *B. anthracis*. Also, estimate the number of genes in each of these two species that are annotated by matches to other genes in genome databases.

ANSWER: Scroll toward the bottom of the CMR homepage until you find the "Multi-Genome Lists by Category" section, then link to the "Genes by Role Category." This page links out to lists of all of the annotated genes in each family of functions, including Cell adhesion (under the "Cellular Processes" heading). Most species have a handful of cell adhesion genes, but *Pseudomonas putida* has 35 such loci, and *Pseudomonas syringae D300* has 21, 15 of which have known roles in pilus biogenesis. By contrast, there are just 4 genes in the cell adhesion list in *E. coli*, all in the curlin surface structural protein (csg) class, and each of these have orthologs in *P. putida* as well as in several other species. *B. anthracis* has two adhesion lipoproteins, but these only show around 25% sequence identity with *Pseudomonas* proteins and are not annotated in the cell adhesion class in this species.

For the second part of the exercise, return to the CMR homepage, and find the "COG/TIGRFAM/PFAM Comparison" tool in the "Multi-Genome Analyses" section. Select representative genomes for the two species and examine the tables that are produced. You should see a table like the one below, which shows that over 40% of the anthrax and nearly 30% of the *E. coli* genes do not match any of the three protein family databases; only 20–25% match TIGRFAM, but around 60% match PFAM classes; and in general, there is more agreement for the *E. coli* than for the anthrax genome.

EXERCISE 1.4 TABLE 1

COG match	TIGR match	PFAM match	*B. anthracis* Number of genes	*B. anthracis* Percent of genes	*E. coli* Number of genes	*E. coli* Percent of genes
X	X	X	85	1	1,233	23
X	X		6	0	85	2
X		X	143	3	1,850	35
X			78	1	472	9
	X	X	868	16	22	1
	X		97	2	14	0
		X	1,958	36	104	2
			2,272	41	1,508	28

Funding for microbial genome research is largely driven by economic and medical motives, but there is a strong evolutionary component to the sampling strategy. Broad sampling increases the power of phylogenetic analysis, which is increasingly seen as a valuable tool for the functional classification of members of gene families. Sampling from a wide variety of habitats, including extreme environments such as thermal vents, is also likely to lead to the identification of novel genes, or gene products adapted for functions that can be utilized by biotechnologists (such as the *Taq* thermally stable DNA polymerase). Table 1.5 provides a survey of the range of sequenced microbial genomes.

Environmental Sequencing

An emerging area of microbial genomics has been dubbed "environmental sequencing," which literally refers to sequencing DNA extracted from an environment such as ocean water, soil, or intestinal flora. The main reason for taking such an approach is that the vast majority of bacteria cannot be cultured in vitro, which implies that our knowledge of microflora is both limited and biased by sampling of a relatively small fraction of the biota. The idea is to circumvent this problem by directly cloning DNA fragments isolated from an appropriate size fraction that largely excludes viruses and multicellular organisms, without any attempt to grow the microbes outside of their environment.

TABLE 1.5 *Sequenced Microbial Genomes*

Archaea	Crenarchaeota	Hyperthermophiles	4 species
	Euryarchaeota	Sulphur metabolizers	1 species
		Halophiles	1 species
		Methanophiles	5 species
		Thermophiles	5 species
Bacteria	Actinobacteria	Corynebacterium (bioengineering)	3 species
		Mycobacterium (animal pathogens)	5 species
		Tropheryma whipplei (reduced genome)	2 isolates
		Streptomyces (bioengineering)	2 species
	Aquificae	*Aquifex* (chemolithoautotroph)	1 species
	Bacteroides	Human gut/oral bacteria	3 species
	Chlamydiae	*Chlamydia pneumoniae*	4 isolates
		Other chlamydia (intracellular pathogens)	3 species
	Cyanobacteria	*Synechococcus, Plochlorococcus*, etc.	6 species
	Deinococcus	*D. radiodurans* (radiation resistance)	1 species
	Firmicutes	*Bacillus* (pathogens, including anthrax)	6 species
		Clostridium (tetanus, other toxins)	4 species
		Listeria (food-borne pathogen)	2 species
		Staphylococcus (including 4 *S. aureus*)	3 species
		Lactobacillae, Streptococci (gut bacteria)	10 species
		Mycoplasmas (small animal pathogens)	8 species
	Proteobacteria	Alpha (*Rhizobium, Rickettsia, Caulobacter*)	11 species
		Beta (*Bordetella, Neisseria*)	8 species
		Delta (*Bdellovibrio* predatory bacterium)	1 species
		Epsilon (*Helicobacter* group)	4 species
		Gamma (diverse *E. coli* group)	25 species
	Spirochaetes	*Borrelia, Treponema* (human parasites)	4 species
	Other Bacteria	Diverse variety	7 species

Source: TIGR Comprehensive Microbial Resource (http://www.tigr.org/tigr-scripts/CMR2/CMRHomePage.spl) and links therein (July, 2004).

The difficulty with this approach is that there is so much microbial diversity that it will take hundreds of human genome equivalents to assemble more than a handful of complete genomes, and most fragments are just gene pieces that cannot easily be assigned to separate species.

Nevertheless, much insight into microbial and genetic diversity has been gained from pilot projects, particularly those that involve the sequencing of ocean water. The identification of novel genes has the potential to change oceanographers' understanding of the mechanisms of photosynthesis and global carbon and nitrogen cycling, as suggested by the discovery of a type of unsuspected rhodopsin in uncultured proteobacteria from Monterey Bay (Béjà et al. 2000). Such proteorhodopsin genes have since been detected in hundreds more samples, suggesting that light harvesting need not be coupled to chlorophyll in cyanobacteria as hitherto thought. By using degenerate gene-specific primers to survey samples from deep sea hydrothermal vents, Campbell et al. (2003) were able to detect a key enzyme involved in carbon fixation in bacteria associated with a polychaete worm, providing insight into how life is sustained in such an extreme environment. Venter et al. (2004) have industrialized environmental sequencing by generating more than a gigabase of sequence derived from about 1,500 liters of seawater from the nutrient-poor Sargasso Sea. In the process, they identified over one million new genes (approximately the equivalent of the number of genes in existing databases) and almost 150 new types of bacteria in an estimated sample of 1,800 total species! The oceanic microbial diversity that remains untapped is more than an order of magnitude greater than this; similarly, protist diversity is virtually unexplored.

Surveys of fecal material suggest that the human gut contains over 500 different species of bacteria, less than 30% of which can be cultured outside the body. Analysis of the dynamics of the intestinal microbial population has utilized diversity in 16S ribosomal RNA sequences, which provide a diagnostic signature for all bacteria. Whole genome sequencing is also employed, for example, of predominate species such as the lactic acid bacteria that serve as probiotics, regulate nutritional uptake, and likely play critical roles in preventing bowel disease (Klaenhammer et al. 2002). Similar strategies are being employed to understand phenomena such as the bacterial contribution to digestion in termites and beetles in the soil, and microbial community structure in the rhizosphere, expanding the reach of genomics from organismal biology to functional ecology.

Yeast

The complete genome sequence of the budding yeast *Saccharomyces cerevisiae* was published in a supplement to the journal *Nature* in May of 1997 (Mewes et al. 1997). The 12 Mb sequence was assembled from roughly 300,000 sequence reads generated in over 100 laboratories and had an initial estimated error rate of 3 in 10,000 bases. Approximately 6,000 predicted genes are spread over 16 chromosomes, which include 53 identified regions of clustered gene duplications. Slightly less than one-half of these genes were functionally annotated (experimentally characterized and/or identified by homology to sequences in other species) prior to completion of the genome sequence.

Chromosome displays and individual gene listings can be accessed through two main *Saccharomyces* sites: the MIPS site of the Munich Information Center for Protein Sequences (http://mips.gsf.de/genre/proj/yeast/index.jsp), and the *Saccharomyces* Genome Database (SGD) at Stanford University (http://www.yeastgenome.org). Sequences of two other well-studied yeasts—the fission yeast *Schizosaccharomyces pombe*, and a yeast commonly associated with opportunistic human infections, *Candida albicans*—are now complete as well. Genomic analysis of numerous other fungi of agricultural and biotechnological interest is under way, including those of the allergenic *Aspergillus* species, various fruit white rots, and the rice blast *Magnaporthe grisea*.

An enormous amount of information is available on the SGD locus pages for all *Saccharomyces* genes, including mutational, structural, biochemical, and transcription data. The SGD site provides links to basic information, as well as "additional information" links that provide extensive detail, including historical notes on the annotation of the gene; several types of functional data ("Function Junction"), and expression data ("Expression Connection"); mapping data; and advanced protein and sequence resources.

EXERCISE 1.5 *Examining a gene in the* Saccharomyces
Genome Database

Answer the following questions about the yeast *GAL4* gene (or a gene of your choice):

a. Write down two aliases (alternate names) for *GAL4*
b. List the three genes on either side of *GAL4* in the *S. cerevisiae* genome
c. What are the three GO annotations for the gene?
d. How many amino acids are there in the *GAL4* protein?
e. What is the phenotype of a *GAL4* mutant?

ANSWER: Each of these questions can be answered directly from the *GAL4* page found by a quick search from the SGD home page. (a) One alias is *GAL8Y* while the systematic name is listed as YPL248C. (b) Clicking on the small gene image at the top right brings up a local chromosomal view, from which you can see that the three genes to the left are ICY2, RPL36B, and GYP5, while those to the right are YPL247C, RBD2, and YPL245W (the "W" and "C" represent Watson and Crick strands for unannotated genes). (c) The three GO annotations (see Box 2.5 on page 124) are listed as transcriptional activator (Molecular Function), galactose metabolism and DNA-dependent regulation of transcription (Biological Process), and nucleus (Cellular Component). (d) You can download the protein sequence of 881 amino acids by linking out to the Entrez Protein page (and examine the protein structure through the PDB site). (e) Null mutants cannot use galactose as a carbon source.

Several complementary approaches to functional characterization of the complete set of yeast genes have been taken. **Systematic mutagenesis** using homologous recombination to knock out open reading frames in heterozygous diploid cells allows complete loss of function to be monitored in haploid cells after induction of sporulation. Winzeler et al. (1999) deleted almost one-third of the yeast ORFs and found that 17% of these were essential for viability in rich medium, while 40% showed quantitative growth defects in rich or minimal medium. Subsequently, over 96% of the yeast genome has been tagged with molecular barcodes associated with gene deletions. A second, complementary approach is a **genetic fingerprinting** strategy, in which natural selection is allowed to sort through a large population of massively mutagenized cells growing in a chemostat under a variety of conditions. These experiments similarly found that about half of the genes (157 of 268) tagged by a transposable element insertion resulted in a detectable reduction in fitness, with many of the genes that showed weak effects not having been characterized previously (Smith et al. 1996). In order to characterize pairwise **genetic interactions** in a systematic manner, Tong et al. (2004) crossed a set of 132 query mutants to a panel of 4,700 viable yeast deletion mutants, and detected over 4,000 functional interactions involving more than 1,000 genes.

A fourth major experimental approach to functional annotation of yeast has been **microarray analysis,** which is discussed in detail in Chapter 4. Using microarrays, changes in the transcriptome have been characterized throughout the cell cycle; during meiosis; in response to environmental shifts, including anoxia and nutrient source; and in a variety of mutant backgrounds. The so-called "compendium" strategy tracks coordinated changes in transcription under a variety of experimental circumstances, assigning genes to functional groups on the basis of co-regulation. Transcriptional data are also being supplemented with protein expression data and proteomic data, as discussed in Chapter 5. Foremost among these approaches is a comprehensive whole genome, two-hybrid screen designed to detect pairs of genes that interact physically within cells. An initial survey of 10% of all possible pairwise interactions has uncovered dozens of novel interactions that can be verified using genetic strategies. Most of these approaches are being pioneered in yeast, but should be transferable to other microbes and even to multicellular eukaryotes.

Parasite Genomics

The World Health Organization (WHO) has singled out for intensive research 10 tropical diseases that affect billions of people worldwide in the hope of eventually eradicating the pathogenic agents responsible (Table 1.6). Mosquito-transmitted viral infection is responsible for two of the target diseases, Dengue hemorrhagic fever and yellow fever; the remaining eight are carried by water- or insect-borne eukaryotic parasites: malaria, leishmaniasis, American and African trypanosomiasis (Chagas disease and sleeping sickness, respectively), lymphatic filariasis (elephantiasis), river blindness (onchocerciasis), schistosomiasis (bilharzia), and dracunculiasis.

TABLE 1.6 *Tropical Diseases Caused and/or Transmitted by Eukaryotic Organisms*

	Vector	Parasite	Number of humans at risk
Dengue hemorrhagic fever	Mosquito	Virus	2.5 billion
Yellow fever	Mosquito	Virus	450 million
Leishmaniasis	Sandfly	Protozoan	350 million
Malaria	Mosquito	Protozoan	2.4 billion
African sleeping sickness	Tsetse fly	Protozoan (*Trypanosoma brucei*)	60 million
Chagas disease (American)	Insect	Protozoan (*Trypanosoma cruzi*)	100 million
Elephantiasis	Mosquito	Filarial nematode	1 billion
River blindness	Blackfly	Nematode	120 million
Schistosomiasis	Water	Platyhelminth (fluke)	500 million
Dracunculiasis	Water-borne crustacean	Nematode (Guinea roundworm)	Near eradication

In addition, numerous parasites carried by invertebrate and animal hosts cause widespread diseases in North America and Europe, including *Cryptosporidium*-induced diarrhea, zoonotic toxoplasmosis, and Lyme disease. Several infectious diseases affecting immunocompromised patients are also emerging. Nematode genomics in particular has direct potential for practical benefits, building on the extensive functional analysis of the *C. elegans* model system. Crop damage caused by parasitic plant nematodes costs billions of dollars and causes untold human suffering around the world each year. In addition, as many as 1 billion people globally may be infected by intestinal and other nematodes that cause diseases including elephantiasis and numerous intestinal disorders such as ascariasis and tricuriasis. These parasites also infect farm animals and pets in the form of hookworm and heartworm.

Current approaches to disease control focus on public health issues such as improving basic living conditions and educating the public about disease transmission, but they have traditionally included chemical interventions that may have deleterious secondary environmental consequences and can lead to increased resistance among the pathogens they are intended to destroy. The complex lifestyles of many of these parasites also limit the effectiveness of chemical intervention, and insect hosts have proven extremely difficult to control. Hence genomic approaches are being supported by WHO, several pharmaceutical giants, and the major genome sequencing centers.

The aims of parasite genomics include:

- Identification of species-specific genes that may be used to generate antigens for vaccination
- Better understanding of the developmental genetics associated with key transitions in the life cycle, suggesting targets for drug design
- Polymorphism surveys that address the population biology of the parasites, informing epidemiological studies.
- In the case of malaria in particular, considerable effort has been devoted to the genomics of the mosquito vector (*Anopheles* spp.), including mapping of loci that affect parasite (*Plasmodium*) transmission.

Web resources providing access to genomic data for these and other projects are available through TIGR, NCBI, and the European Bioinformatics Institute parasite genomics server (http://www.ebi.ac.uk/parasites/parasite-genome.html).

Summary

1. Genome science is a new way of looking at biology that unites genetics, molecular biology, computational biology, and bioinformatics.

2. The core aspects of genome science include the utilization of organismal databases; assembly of physical and genetic maps; genome sequencing; gene annotation; characterization of sequence diversity; expression profiling; proteomics; functional genomics; and comparative genomics.

3. Genetic and physical maps are the foundation upon which most genome projects are built. The two types of maps can be unified both because of the relationship between recombination rate and chromosomal distance, and through cytogenetic maps.

4. "Synteny" refers to the conservation of gene order among groups of several to tens or even hundreds of genes in divergent species. Synteny is the basis for comparative gene mapping.

5. The Human Genome Project (HGP) has assembled a draft sequence of the human genome and is now working to promote use of the sequence information in the domains of general biology, public health, and social policy. Additional research has proceeded apace along lines fostered by a rapidly expanding commercial sector.

6. Mouse, rat, and dog genome sequences have been completed and offer the most immediately accessible vertebrate model organisms for genetic analysis of physiology, development, cell biology, and behavior.

7. A wide range of animal genome projects include cow, horse, pig, sheep, dog, numerous fish species, and the two major invertebrate models, *Drosophila* and *Caenorhabditis*.

8. The first complete plant genome sequence was that of the mustard weed *Arabidopsis*, and significant progress has been made toward completion of the rice and maize genomes. Additional plant genomes being studied include legumes, grasses, fruits, vegetables, forest trees, and ornamentals.

9. Over 200 bacterial genome sequences will soon be available, providing a vast resource for modeling biochemistry, metabolism, and microbial ecology, as well as opening up new avenues of biotechnological research.

10. In addition to the budding yeast *Saccharomyces cerevisiae*, several other fungi of industrial, biomedical, and agricultural importance have genome projects.

11. Genome science promises new strategies for dealing with several of the world's most devastating parasitic diseases using avenues of genetic research that were hitherto unimaginable with classical methods.

Discussion Questions

1. Now that the human genome sequence is nearly complete, how should the excess sequencing capacity at major genome centers be used? Is it good use of public funds to keep sequencing more and more genomes? If so, which genomes should be prioritized?

2. The volume and expense of genome data raises new issues in relation to peer review of research and publication of results. How can editors and reviewers oversee data quality? Should journals demand that companies make complete genome sequences freely available?

3. LIMS (Laboratory Information Management Systems) are an increasingly important part of genomic research. Who should be maintaining them, how should they be integrated with standard laboratory notebooks, and are any potential ethical issues raised by their existence?

4. Propose a genome project for an animal or plant species of particular interest to you. What will be your key aims, in terms of both generating genomic resources and posing scientific questions?

5. Contrast the impact of genomics on the fabric of basic and applied biological research with the impact that molecular biology had in the 1970s. What will be the next revolution?

Literature Cited

Adams, M. D. et al. 1995. Initial assessment of human gene diversity and expression patterns based upon 83 million nucleotides of cDNA sequence. *Nature* 377: 3–174.

Adams, M. D. et al. 2000. The genome sequence of *Drosophila melanogaster*. *Science* 287: 2185–2195.

Arabidopsis Genome Initiative. 2000. Analysis of the genome sequence of the flowering plant *Arabidopsis thaliana*. *Nature* 408: 796–815.

Béjà, O. et al. 2000. Bacterial rhodopsin: evidence for a new type of phototrophy in the sea. *Science* 289: 1902–1906.

Birney, E. et al. 2004. An overview of Ensembl. *Genome Res*. 14: 925–928.

Blattner, F. R. et al. 1997. The complete genome sequence of *Escherichia coli* K-12. *Science* 277: 1453–1462.

Buckler, E. S., J. M. Thornsberry and S. Kresovich. 2001. Molecular diversity, structure, and domestication of grasses. *Genetical Res. Camb*. 77: 213–218.

Bult, C. J. et al. 1996. Complete genome sequence of the methanogenic archaeon *Methanococcus jannaschii*. *Science* 273: 1058–1073.

Campbell, B. J., J. L. Stein and S. C. Cary. 2003. Evidence of chemolithoautotrophy in the bacterial community associated with *Alvinella pompejana*, a hydrothermal vent polychaete. *Appl. Environ. Microbiol*. 69: 5070–5078.

C. elegans Sequencing Consortium. 1998. Genome sequence of the nematode *C. elegans*: A platform for investigating biology. *Science* 282: 2012–2018.

Cello, J., A.V. Paul and E. Wimmer. 2002. Chemical synthesis of poliovirus cDNA: generation of infectious virus in the absence of natural template. *Science* 297: 1016–1018.

Chervitz, S. A. et al. 1998. Comparison of the complete protein sets of worm and yeast: Orthology and divergence. *Science* 282: 2022–2028.

Cho, M., D. Magnus, A. Caplan, D. McGee and the Ethics of Genomics Group. 1999. Ethical considerations in synthesizing a minimal genome. *Science* 286: 2087–2090.

Choy, R. and J. Thomas. 1999. Fluoxetine-resistant mutants in *C. elegans* define a novel family of transmembrane proteins. *Mol. Cell*. 4: 143–152.

Clark, A. G. et al. 2003. Inferring non-neutral evolution from human-chimp-mouse orthologous gene trios. *Science* 302: 1960–1963.

Collins, F. S. et al. 1998. New goals for the U.S. Human Genome Project: 1998–2003. *Science* 282: 682–689

Collins, F. S. et al. 2003. A vision for the future of genomics research. *Nature* 422: 835–847.

Collins, F. S. and D. Galas. 1993. A new 5-year plan for the U.S. Human Genome Project. *Science* 262: 43–50.

Daubin, V., N. A. Moran and H. Ochman. 2003. Phylogenetics and the cohesion of bacterial genomes. *Science* 301: 829–832.

Dehal, P. et al. 2001. Human chromosome 19 and related regions in mouse: Conservative and lineage-specific evolution. *Science* 293: 104–111.

Delcher, A., S. Kasif, R. Fleischmann, J. Peterson, O. White and S. Salzberg. 1999a. Alignment of whole genomes. *Nucleic Acids Res*. 27: 2369–2376.

Delcher, A., D. Harmon, S. Kasif, O. White and S. Salzberg. 1999b. Improved microbial gene identification with Glimmer. *Nucleic Acids Res*. 27: 4636–4641.

Deloukas, P. et al. 1998. A physical map of 30,000 human genes. *Science* 282: 744–746.

Elgar, G. et al. 1999. Generation and analysis of 25 Mb of genomic DNA from the pufferfish *Fugu rubripes* by sequence scanning. *Genome Res*. 9: 960–971.

Fleischmann, R. D. et al. 1995. Whole-genome random sequencing and assembly of *Haemophilus influenzae* Rd. *Science* 269: 496–512.

Fraser, C. M. et al. 1995. The minimal gene complement of *Mycoplasma genitalium*. *Science* 270: 397–403.

Frazer, K. A, X. Chen, D. A. Hinds, P. V. Pant, N. Patil and D. R. Cox. 2003. Genomic DNA insertions and deletions occur frequently between humans and nonhuman primates. *Genome Res*. 13: 341–346.

Frishman. D. et al. 2001. Functional and structural genomics using PEDANT. *Bioinformatics* 17: 44–57.

Goff, S.A. et al. 2002. A draft sequence of the rice genome (*Oryza sativa L.* ssp. *japonica*). *Science* 296: 92–100.

Gopal, S. et al. 2001. Homology-based annotation yields 1,042 new candidate genes in the *Drosophila melanogaster* genome. *Nat. Genetics* 27: 337–340

Gyapay, G. et al. 1994. The 1993–94 Genethon human genetic linkage map. *Nat. Genetics* 7: 246–339.

Hamosh, A., J. Scott, D. Amberger, D. Valle and V. A. McKusick. 2000. Online Mendelian Inheritance in Man (OMIM). *Hum. Mutat*. 15: 57–61

Hattori, M. et al. 2000. The DNA sequence of human chromosome 21. *Nature* 405: 311–319.

Hu, J. et al. 2001. The ARKdb: Genome databases for farmed and other animals. *Nucleic Acids Res*. 29: 106–110.

Hudson, T. J. et al. 1995. An STS-based map of the human genome. *Science* 270: 1945–1954.

Hutchison, C. A. III et al. 1999. Global transposon mutagenesis and a minimal *Mycoplasma* genome. *Science* 286: 2165–2169.

International Human Genome Sequencing Consortium. 2001. Initial sequencing and analysis of the human genome. *Nature* 409: 860–921.

International Human Genome Sequencing Consortium. 2004. Finishing the euchromatic sequence of the human genome. *Nature* 431: 931–945.

International SNP Map Working Group. 2001. A map of human genome sequence variation containing 1.42 million single-nucleotide polymorphisms. *Nature* 409: 928–933.

Jeffreys, A. J., Holloway, J. K., Kauppi, L., May, C. A., Neumann, R., Slingsby, M. T. and Webb, A. J. 2004. Meiotic recombination hot spots and human DNA diversity. *Philos Trans R Soc Lond B Biol Sci*. 359: 141–152.

Klaenhammer, T. et al. 2002. Discovering lactic acid bacteria by genomics. *Antonie Van Leeuwenhoek* 82: 29–58.

Kobayashi, K. et al. 2003. Essential *Bacillus subtilis* genes. *Proc Natl Acad Sci USA* 100: 4678–4683.

Lander, E. S., P. Green, J. Abrahamson, A. Barlow, M. Daly, S. Lincoln and L. Newburg. 1987. MAPMAKER: An interactive computer package for constructing primary genetic linkage maps of experimental and natural populations. *Genomics* 1: 174–181.

Loots, G. G. et al. 2000. Identification of a coordinate regulator of interleukins 4, 13, and 5 by cross-species sequence comparisons. *Science* 288: 136–140.

Mayer, K. et al. 2001. Conservation of microstructure between a sequenced region of the genome of rice and multiple segments of the genome of *Arabidopsis thaliana*. *Genome Res*. 11: 1167–1174.

McKusick, V. A. 1998. *Mendelian Inheritance in Man: Catalogs of Human Genes and Genetic Disorders*, 12th Ed. Johns Hopkins University Press, Baltimore.

Mewes, H. W. et al. 1997. Overview of the yeast genome. *Nature* 387 (Suppl.): 7–9.

Mouse Genome Sequencing Consortium. 2002. Initial sequencing and comparative analysis of the mouse genome. *Nature* 420: 520–562.

Murphy, W., S. Sun, Z. Chen, N. Yuhki, D. Hirschmann, M. Menotti-Raymond and S. J. O'Brien. 2000. A radiation hybrid map of the cat genome: Implications for comparative mapping. *Genome Res.* 10: 691–702.

Mushegian, A. and E. V. Koonin. 1998. A minimal gene set for cellular life derived by comparison of complete bacterial genomes. *Proc. Natl. Acad. Sci. USA* 93: 10268–10273.

Ostrander, E. A. and L. Kruglyak. 2000. Unleashing the canine genome. *Genome Res.* 10: 1271–1274.

Parker, H. G. et al. 2004. Genetic structure of the purebred domestic dog. *Science* 304: 1160–1164.

Pennacchio, L. A. 2003. Insights from human/mouse genome comparisons. *Mammalian Genome* 14: 429–436.

Peterson, J. D., L. Umayam, T. Dickinson, E. Hickey and O. White. 2001. The Comprehensive Microbial Resource. *Nucleic Acids Res.* 29: 123–125.

Purugganan, M. D., A. L. Boyles and J. Suddith. 2000. Variation and selection at the CAULIFLOWER floral homeotic gene accompanying the evolution of domesticated *Brassica oleracea*. *Genetics* 155: 855–862.

Rat Genome Sequencing Project Consortium. 2004. Genome sequence of the Brown Norway rat yields insights into mammalian evolution. *Nature* 428: 493–521

Reboul, J. et al. 2001. Open-reading-frame sequence tags (OSTs) support the existence of at least 17,300 genes in *C. elegans*. *Nat. Genetics* 27: 332–336

Riley, M. 1997. Functions of the gene products of *Escherichia coli*. *Microbiol. Rev.* 57: 862–952

Roest Crollius, H. et al. 2000. Estimate of human gene number provided by genome-wide analysis using *Tetraodon nigroviridis* DNA sequence. *Nat. Genetics* 25: 235–238.

Rubin, G. M. et al. 2000. Comparative genomics of the eukaryotes. *Science* 287: 2204–2215.

Scholz, H., J. Raymond, C. Singh and U. Heberlein. 2000. Functional ethanol tolerance in *Drosophila*. *Neuron* 28: 261–271.

Schwartz, S. et al. 2000. PipMaker: A Web server for aligning two genomic DNA sequences. *Genome Res.* 10: 577–586.

Smith, V., K. N. Chou, D. Lashkari, D. Botstein and P. O. Brown. 1996. Functional analysis of the genes of yeast chromosome V by genetic footprinting. *Science* 274: 2069–2074.

Tettelin, H. et al. 2001. Complete genome sequence of a virulent isolate of *Streptococcus pneumoniae*. *Science* 293: 498–506.

Tong, A. H. et al. 2004. Global mapping of the yeast genetic interaction network. *Science* 303: 808–813.

Venter, J. C. et al. 2001. The sequence of the human genome. *Science* 291: 1304–1351.

Venter, J. C. et al. 2004. Enviromental genome shotgun sequencing of the Sargasso Sea. *Science* 304: 66–74.

Wang, D. G. et al. 1998. Large-scale identification, mapping, and genotyping of single-nucleotide polymorphisms in the human genome. *Science* 280: 1077–1082.

Wang, R., A. Stec, J. Hey, L. Lukens and J. Doebley. 1999. The limits of selection during maize domestication. *Nature* 398: 236–239.

Wheeler, D. L. et al. 2000. Database resources of the National Center for Biotechnology Information. *Nucleic Acids Res.* 28: 10–14.

Whitelaw, C. A. et al. 2003. Enrichment of gene-coding sequences in maize by genome filtration. *Science* 302: 2118–2120.

Winzeler, E. A. et al. 1999. Functional characterization of the *S. cerevisiae* genome by gene deletion and parallel analysis. *Science* 285: 901–906.

Woese, C. 2000. Interpreting the universal phylogenetic tree. *Proc. Natl. Acad. Sci. USA.* 97: 8392–8396.

Yu, J. et al. 2002. A draft sequence of the rice genome (*Oryza sativa L.* ssp. *indica*). *Science* 296: 79–92.

2 *Genome Sequencing and Annotation*

The first objective of most genome projects is to determine the DNA sequence either of the genome or of a large number of transcripts. This endeavor leads to the identification of all or most genes, and to the characterization of various structural features of the genome. In this chapter we explain the basic principles of how blocks of DNA sequence are obtained, and how these blocks are serially assembled first into contiguous stretches of sequence (contigs) and ultimately into a whole genome sequence. We will outline the essence of the bioinformatic strategies for sequence alignment (since alignment is the basis of sequence assembly), comparison of cDNA/EST and genomic sequences, and annotation of open reading frames. In addition to identifying individual genes, DNA sequences reveal information about other features of the genome, including repetitive elements; centromeres and telomeres; variable distribution of GC content; and evolutionarily conserved elements of yet-to-be-determined function. This chapter concludes with a discussion of how genes are annotated by comparison with, and evolutionary analysis of, similar predicted protein sequences from other organisms.

Automated DNA Sequencing

The Principle of Sanger Sequencing

Almost all of the genome-scale sequencing performed, up to and including the Human Genome Project, has made use of the basic chain termination method developed in 1974 by Frederick Sanger. The idea behind Sanger's method is to generate all possible single-stranded DNA molecules differing in length by one nucleotide that are complementary to a template. The template starts at a common 5' base and extends anywhere up to 1 kilobase or more in the 3' direction. These single strands of DNA are labeled in such a way as to allow us to infer the identity of the 3'-most base in each molecule. Separation of the molecules according to size by electrophoresis results in a ladder of bands, with each adjacent band corresponding to a class of molecule that differs by the addition of one particular base. The sequence is then "read" from this ladder, as shown in Figure 2.1.

The sequencing reaction is primed by annealing an oligonucleotide of 20 or so bases to a denatured template. The template is usually a plasmid that contains a cloned piece of DNA. The same "universal" primer can be used in all reactions, since it recognizes a plasmid sequence adjacent to the inserted DNA clone. For some applications, genomic DNA is sequenced directly, in which case the primer is specific to the part of the gene being sequenced. The enzyme DNA polymerase is then used to catalyze synthesis of the complementary strand in the presence of all four deoxynucleotides (dNTPs).

The trick used to generate molecules differing in length by one nucleotide is to randomly terminate polymerization of the growing single strand sequence by the incorporation of a dideoxynucleotide (ddNTP). Whenever a ddNTP incorporates, the absence of a hydroxyl group on the sugar-phosphate backbone means that there is nothing for the next dNTP to attach to, and polymerization is effectively terminated. Once the sequencing reaction terminates, the DNA is again denatured, and the fragments are separated on a gel, by virtue of the differential retardation of the migration of molecules of different length and electric charge through a semiporous matrix such as that produced by an acrylamide polymer.

Throughout the 1980s, most sequencing was done manually. The process was labor-intensive in terms of performing the reactions, preparing and running slab polyacrylamide gels for electrophoretic separation, and reading the reactions by the human eye. A radioactive label, typically ^{33}P or ^{35}S, was incorporated into the sequencing product as part of one of the dNTPs, and four separate reactions had to be set up for each sequence—one for each dideoxy terminator. The four reactions were run side by side on a large slab gel, and were stopped after an appropriate time so that all products within the desired size range could be visualized by exposing the gel to X-ray film. With a separation of 0.5 mm between bands corresponding to each molecule that differs in length by one nucleotide, generally only up to 500 bases could be read from a single set of four lanes on a 30-cm gel. At least three runs of different durations had to be performed to sequence any fragment longer than 1,000 bp. Manual sequencing was also limited by problems associated with poor resolution of short stretches of sequence, notably GC-rich regions that tend to compress, which led to ambiguities that could not always be resolved by sequencing the opposite strand.

High-Throughput Sequencing

In the 1990s, several advances were made that enabled the automation of Sanger sequencing, without which genome-scale sequencing would be impossible. The new techniques and equipment include:

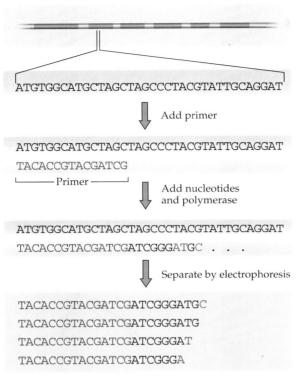

Figure 2.1 The principle of dideoxy (Sanger) sequencing.
Single-strand molecules are synthesized from a template and randomly terminated by the addition of a labeled dideoxynucleotide (ddA, ddT, ddC, or ddG), then separated by electrophoresis. The sequence is visualized either by radioactivity (manual sequencing; one lane per base on a fixed gel) or fluorescence (automated sequencing; each base is identified by computer as it emerges from a single lane).

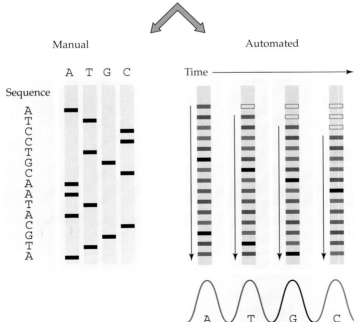

- Four-color fluorescent dyes have replaced the radioactive label. Attachment of these dyes to the ddNTPs results in a fluorescent tag that directly marks only the terminated DNA molecule; consequently, a single sequencing reaction spiked with all four ddNTPs is sufficient to sequence any template. (For some applications, the dyes are attached to the primer, in which case four reactions are still performed but may be pooled in a single lane before electrophoresis.)

- Rather than stopping the electrophoresis at a particular time, the products are scanned for laser-induced fluorescence just before they run off the end of the electrophoresis medium. The sequence is collected as a set of four "trace files" that indicate the intensity of the four colors; a peak in the trace distribution implies that the particular base was the last one incorporated at the position. Such traces can be read automatically (as described in the next section) resulting in enormous savings in time and reducing the scoring errors that inevitably creep into manual readings.

- Improvements in the chemistry of template purification and the sequencing reaction (which is now usually performed in a thermocycler with multiple priming cycles, rather than as a less robust single extension reaction), including use of bioengineered thermostable polymerases that can read through secondary structure with high fidelity, has extended the length of high quality sequence. Reads greater than 1,200 bp are possible with current technology, though the 500–900 bp range is more common.

- Slab gel electrophoresis gave way to capillary electrophoresis with the introduction in 1999 of Applied Biosystem's ABI Prism™ 3700 automated sequencers, which in turn were updated with ABI Prism™ 3730 DNA analyzers in 2003. These sequencers give extremely high quality, long reads; save time and money by abolishing the laborious and often frustrating step of gel pouring; and add a new level of automation in that the capillaries are loaded by robot from 96-well plates, rather than by hand. Each machine can handle up to twenty 96-well plates per day, or approximately 2 Mb of sequence—which is two orders of magnitude greater than the output of a single investigator just a decade earlier. Other automated DNA sequencing systems are also available, including Amersham's MegaBACE 1000™ and Li-Cor's 4300 DNA Analysis systems.

Reading Sequence Traces

The reading of raw sequence traces, or **base-calling,** is now routinely performed using automated software that reads bases, aligns similar sequences, and provides an intuitive platform for editing. Genome-scale sequencing requires only minimal human input—which is a good thing, because if it took just a quarter of an hour for a single person to properly edit any given read, it would take seven people a very boring week to process the output from one day's operation of a single ABI-3700. Automated software such as the freely available **phred** program developed at the University of Washington (http://www.phrap.org; Ewing et al.1998; Ewing and Green 1998), or commercial equivalents, convert traces into sequences that can be deposited in a database within seconds after the completion of a sequencing run. These programs assign probability scores to the accuracy of each base call as the trace is read, and this information is utilized in subsequent alignment steps.

Whereas human observers integrate multiple pieces of information in calling a base from a sequence trace, software uses an algorithm in which the process is broken down into a series of steps, as outlined in Figure 2.2. First, the four traces corresponding to the four fluorescence spectra are merged into a single file, maintaining the register of the peaks. Next the computer calculates where it expects to find a peak, using an averaging process based on the mean distance between peaks over some stretch of the sequence. This ensures that an N is called in place of an A, C, G, or T where no base is seen, and that only a single base is called where two appear to be present. The most efficient algorithms can adjust for the increase in spacing between peaks that occurs as the run proceeds, as well as for variation due to changes in local GC content.

Subsequently, the algorithm detects local maxima for each of the four channels and checks to see that each peak occurs with the predicted spacing relative to the adjacent peaks. Since not all peaks are the same height, and runs of the same base can show reduced peak resolution, algorithms typically employ a threshold that computes the relative magnitude of a local minimum and maximum to decide whether a peak is real. Similarly, on occasion the tail of one peak can be higher than the maximum of the next peak of a different nucleotide, so the program must be able to call the correct peak rather than simply detecting the highest signal. Where two peaks are called

Figure 2.2 The phred base-calling algorithm.
Programs such as phred convert computer-generated traces into base sequences and assess the probable accuracy of each base call.

but there is only space for one, or vice-versa, peaks can be either omitted or split based on probability measures. Uncalled peaks are inserted if there is a local maximum of one dye that was not initially called as a peak. Automated base calling usually takes less than a half a second per 1-kb trace and results in a string of letters corresponding to the 5′-to-3′ order of nucleotides.

Good software should also be able to identify instances where two different bases are present at the same site, either in heterozygotes or as a result of the mixing of two samples. Alignment with other sequences greatly assists in the identification of such single nucleotide polymorphisms (SNPs), but in the first stage of trace reading, these may simply be designated as unassigned or ambiguous bases and labeled N. Probability scores can also be used to flag sites as potential SNPs.

Several common problems can lead to errors in sequencing, some of which are shown in Figure 2.3. For example, the first 50 or so bases of a read are typically "noisy" due to the anomalous migration of short DNA fragments that contain bulky dyes (Figure 2.3A). Similarly, traces become progressively less uniform as a run proceeds, and the effects of diffusion are amplified while the relative mass differences between successive fragments decreases (Figure 2.3C). Figure 2.3B shows how alignment of the sequence traces of two different alleles reveals SNPs and insertion-deletion (indel)

Figure 2.3 Automated sequence chromatograms. Each trace is accompanied by two lines of sequence: the automated call below the trace, and a manually edited call above it. (A) This sequence shows "noisiness" typical of the first 30 or so bases of a run. (B) The middle two rows show a segment of two sequences that are polymorphic for both SNPs and an indel. (C) A decline in sequence quality typically occurs after about 800 bp.

polymorphisms. Dye-terminator chemistry deals quite well with compression problems, but anomalies such as reduced signals of G following A must be accounted for. Not all reactions work as well as they should, resulting in low signal-to-noise ratios and effects such as a large peak of one dye that overwhelms the true signal; these must also be recognized either by the software or by manual editors.

EXERCISE 2.1 *Reading a sequence trace*

Read the following DNA sequence obtained by direct sequencing of a single individual organism, assuming that on the trace green is A, red is T, blue is C, and black is G. Remark on any ambiguities in the sequence.

ANSWER: The sequence reads:

GCTCATTTGAATAACCTGAAATGCANCAA (C/T) AACACACATTCATTTATC

Note that the base labeled N is impossible to call due to poor sequence quality. Four nucleotides further downstream, two peaks of the same height are observed at the same location, suggesting that the individual is heterozygous at this site for a C and T single nucleotide polymorphism (SNP).

All automated base-calling algorithms make errors, and some may deal with particular types of ambiguity better than others. For this reason, it has become standard to assign error probability estimates to each base. Error probabilities generated by phred are calculated based on four parameters:

1. The variance in peak spacing over a seven-peak window centered on the called base;

2. The ratio of the largest uncalled to the smallest called peak in that window;

3. The same ratio over a three-peak window;

4. The number of bases between the current base and the nearest unresolved one.

The error probability, p, is then converted to a phred score, q, which is 10 times the negative logarithm of p. Thus, a phred score of less than 13 means that there is a greater than 0.05 probability that the base is incorrectly called, while a score of 30 or more means that the associated error probability is 0.001. In general, scores greater than 20 are regarded with high confidence. The distribution of phred scores shown for data from a set of 26,000 ABI 3700 sequence reads in Figure 2.4 shows that the predicted error rate increases for longer fragments, but that automated sequence quality up to

Figure 2.4 Phred quality value distributions. The distribution of phred scores for 26,000 dye-terminator sequence reads. Dark blue bars represent only bases 100–400 in each sequence, showing generally better scores than for all bases in the readable portion of each sequence trace (in light blue). (Data from Ewing and Green 1998.)

at least 500 bases is generally expected to be very good. The length of high-quality reads continues to increase as sequencing technology improves.

Contig Assembly

The "finishing" stage in sequencing a stretch of DNA longer than a single clone entails alignment, editing, and error correction. All of these steps are customarily done with sequence editing software, an example of which is the combination of the **phrap** assembler and **consed** graphic editor (Gordon et al. 1998), which complement the phred base-caller and are widely used in academic settings. High-level commercial editors also aim to link assembled sequences immediately to relational databases and annotation applications. The key features of successful editors are:

- The use of color to illustrate key features such as the different bases, quality scores, regions of sequence conservation, and contrasts between automated and manual base calls.
- The ability to view and navigate along the actual traces of the sequences being compared, and to tag ambiguities and features of interest with notes.
- Easy display of the complementary strand.
- Tools for manual sequence editing, including inserting and deleting bases without disrupting the original trace files, yet propagating the edits throughout the assembly as appropriate, including adjustments to linked output files as requested by the editor.
- A flexible alignment algorithm that implements user-defined alignment parameters.
- Computation of probability scores associated with a calculated consensus sequence.
- The ability to identify potentially polymorphic sites.
- Provision of tools to guide error correction.

Editing starts with the chromatogram files (for example, .abi or .scf standard chromatogram format) that contain the fluorescence trace profiles for each of the four channels of a read. These files are generally left untouched by editors. They are converted to base call files by programs such as phred (which creates .phd files) that include the base calls, quality scores, peak positions, and any tagged information provided by human annotators. If a separate assembly program such as phrap is used, it creates a third input file (.ace) that includes information on the alignment, the consensus contig sequence, and quality scores for each base in the consensus. After manual editing, multiple versions of .phd and .ace files may be stored, each of which can be called up later for subsequent reediting, perhaps following addition of newly obtained sequences.

Consed provides a graphic interface that allows the user to call up and interactively edit individual reads and/or contigs assembled by phrap. The aligned reads window shown in Figure 2.5 presents the consensus sequence as well as each individual sequence that contributes to an alignment. Gray shading indicates the quality of each base call, with white indicating the highest quality. The user can manipulate features of the display, including the use of color to indicate quality or match, and the inclusion of tags. Nav-

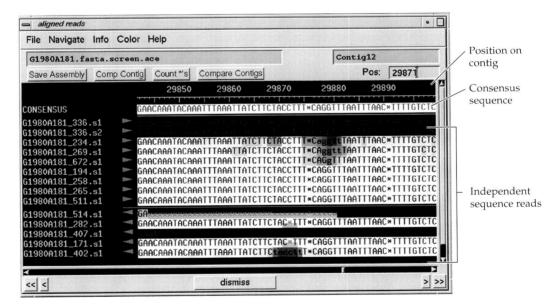

Figure 2.5 An aligned-reads window in consed. The consensus sequence is shown across the top, as calculated by phrap from the individual files included in the alignment below. Gray shading and lowercase lettering indicate reduced quality of a particular base call. Asterisks show the location of a potential insertion or deletion, one example of which may be a true polymorphism, indicated by the yellow bar highlighting a region of discrepancy between two high-quality regions. The pink arrowheads indicate which strand was sequenced, in this case suggesting that the discrepancy may be an artifact of which strand is represented. The light blue bar marks a region of misalignment. Scroll bars and clickable options allow the user to navigate and edit the entire assembly. (From Gordon et al. 1998.)

BOX 2.1 Pairwise Sequence Alignment

The single most important class of bioinformatics tools deals with pairwise alignment of DNA and amino acid sequences. We shall see that pairwise sequence alignment and database searching are essentially the same computational problem, and that the underlying algorithms for pairwise alignment are found in a variety of other bioinformatics applications.

Consider the following two simple sequence alignments:

	Alignment 1	**Alignment 2**
Sequence 1	ACGCTGA	ACGCTGA
Sequence 2	A--CTGT	ACTGT--

Intuitively, we recognize that the first alignment is "superior" to the second. Why? Because the residues in the same column are mostly identical (four exact matches), and there is no obvious way to improve the alignment quality. In contrast, the second alignment has less sequence identity (three exact matches), and we can recognize the potential improvement offered by "sliding" the CTG subsequence of sequence 2 to match the CTG in sequence 1.

The goal of pairwise sequence alignment is to formalize this intuitive procedure. We seek alignments that contain high levels of sequence identity, few mismatches and gaps, and little opportunity for improvement. We will work under the assumption that the observed similarity in sequences to be aligned is the result of either random chance (in which case the similarity is likely of no interest) or of a shared evolutionary origin.

By placing residues from two sequences in the same column of an alignment, we are implying that they are **homologous,** meaning that the two residues are descendants of a common ancestral residue. Homology is not to be confused with sequence identity; two residues may be identical without being homologous. Two positions in an alignment column can have any of the relationships shown in the figure below. Parts C and D of the figure point out that it is possible to have the same true alignment arise through different evolutionary events; when we are dealing with pairwise alignments, the data do not allow us to distinguish between insertion and deletion events, thus explaining the origin of the term "**indel.**"

Figure A Common evolutionary events and their effects on alignment. The alignments are shown beneath diagrams of how they might occur. Two positions in an alignment may contain identical residues (A); they may hold different residues as the result of one or more substitution events (B); or one of the sequences can hold a gap or indel, the result of insertion (C) or deletion (D) events.

Of course, sequences do not come to us in aligned form. We must infer the location of past indel and substitution events. Several computational approaches for performing this task exist, but the most widely used are score-based methods in which inferred events in an alignment are penalized by amounts proportional to their rarity. For example, we may assign substitutions a penalty of –1 unit, while less common indels receive a penalty of –5 units. A match (identity) might receive a score of +3 units. The score for a candidate alignment is then the sum of the penalties at each of the positions in the alignment. Using this scoring scheme, the four alignments in Figure A have scores of 9, 5, 4, and 4, respectively. If we search through all possible alignments and compute the score for each, we can identify the alignment (or alignments) with the optimal score and use it as our best guess for the true alignment.

It is worthwhile to understand the computational approach for finding the optimal pairwise alignment of two sequences. Understanding the algorithms enhances the ability to work with database searching tools, and the computational approach is widely used in many other bioinformatics tools. The essence of the alignment problem is to be able to take two input sequences, and by inserting gaps into one or both of them, maximize the total alignment score. Needleman and Wunsch (1970) described the first computationally tractable algorithm for finding optimal pairwise alignments. The algorithm is an example of **dynamic programming,** and it guarantees that we find all optimal alignments in a manner that avoids exhaustive searching of all possible alignments of two sequences of lengths m and n.

Consider the example given at the beginning of this box, where sequence 1 has seven residues (ACGCTGA) and sequence 2 has 5 residues (ACTGT). One way to find the optimum alignment would be to compute every possible alignment score, which is not so difficult in this example but would be computationally intensive if the two sequences were ten times longer, or if one sequence was being aligned with a database of thousands of sequences. Instead, the Needleman-Wunsch approach uses a recursive algorithm based on the addition of one position at a time to the best alignment of residues up to that point. That is, if we already know the best alignment of the first $(i - 1)$ residues of sequence 1 with the first $(j - 1)$ residues of sequence 2, then we can quickly compute the best alignment of residues i against j. This is because there are only three ways to terminate the longer alignment, illustrated as follows for alignment of the first four residues ($i = j = 4$):

1. Having already aligned residues 1 through $(i - 1)$ of sequence 1 and 1 through $(j - 1)$ of sequence 2, when we add both residues i and j, we add the match score or penalty that they generate (in this case, a penalty due to a C–G mismatch):

```
ACG                                    ACGC
|| .     (Score = 3 + 3 – 1 = 5)  →    || . .    (score = 5 – 1 = 4)
ACT                                    ACTG
```

2. Having already aligned residues 1 through $(i - 1)$ of sequence 1 and 1 through j of sequence 2, when we add residue i we incur the penalty for aligning residue i with a gap:

```
AC-G                                   AC-GC
|| _ |   (Score = 3 + 3 – 5 + 3 = 4) → || _ | _   (score = 4 – 5 = –1)
ACTG                                   ACTG-
```

3. Having already aligned residues 1 through i of sequence 1 and 1 through $(j - 1)$ of sequence 2, when we add residue j, we incur the penalty for aligning a gap with residue j:

```
ACGC                                   ACGC-
|| . _   (Score = 3 + 3 – 1 – 5 = 0) → || . _ _   (score = 0 – 5 = –5)
ACT-                                   ACT-G
```

(*Continued on next page*)

BOX 2.1   *(continued)*

Clearly, the best option was adding both residues without a gap. Formally, the score for alignment of *i* residues of sequence 1 against *j* residues of sequence 2 is given by:

$$S(i,j) = \mathbf{max} \begin{cases} S(i-1, j-1) + c(i,j) \\ S(i-1, j) + c(i, -) \\ S(i, j-1) + c(-, j) \end{cases}$$

where $c(i,j)$ is the score for alignment of residues *i* and *j* and takes the value 3 for a match or −1 for a mismatch, and $c(-,j)$ is the penalty for aligning a residue with a gap, which takes the value −5 given the default match and penalty scores. (Note that assigning different scores may give different optimal alignments as shown in Exercise 2.2.)

We can display this algorithm using a two-dimensional matrix. The (i,j) entry in the matrix is the best score for aligning the first *i* residues of sequence 1 (on the *y*-axis) with the first *j* residues of sequence 2 (the *x*-axis). The optimal alignment score of two sequences of length *m* and *n* can then be found in the upper right of the matrix, the (m,n) element.

We begin by filling in the first row and first column of the matrix, which corresponds to initial gaps (e.g., aligning the A in sequence 2 against a gap in sequence 1 costs −5; aligning the first three positions of sequence 1 against initial gaps in sequence 2 costs −15).

i = 7	A	−35					
i = 6	G	−30					
i = 5	T	−25					
i = 4	C	−20					
i = 3	G	−15					
i = 2	C	−10					
i = 1	A	−5	3				
i = 0	−	0	−5	−10	−15	−20	−25
	−		A	C	T	G	T
j =		0	1	2	3	4	5

From this starting point, we can fill in each cell in the table using the algorithm. The entry for S(1,1), is the maximum of the following three events:

$$S(0,0) + c(A,A) = 0 + 3 \quad = \quad 3$$
$$S(0,1) + c(A,-) = (-5) + (-5) = -10$$
$$S(1,0) + c(-,A) = (-5) + (-5) = -10$$

In a similar fashion, we find S(2,1) as the maximum of three values: $(-5) - 1 = -6$; $(3) - 5 = -2$; and $(-10) - 5 = -15$. The best entry is the addition of the C indel to the A–A match, for a score of −2. If we fill in the entire matrix, we get the following:

A	−35	−27	−19	−9	−3	1
G	−30	−22	−14	−4	2	2
T	−25	−17	−9	−1	−1	7
C	−20	−12	−4	0	4	3
G	−15	−7	1	5	4	−1

C	−10	−2	6	1	−4	−9
A	−5	3	−2	−7	−12	−17
−	0	−5	−10	−15	−20	−25
	−	A	C	T	G	T

If you are having trouble following the algorithm, just imagine filling in the cells of the matrix by exhaustively computing the best score for each alignment of i by j residues. The first column shows what would happen if you aligned each of the i residues of sequence 1 with gaps; the second column if you aligned them against A and a series of gaps, and so on. The entry for row i and column j indicates the optimal score for aligning i residues of sequence 1 against j residues of sequence 2. These scores were computed by iterative algorithm rather than exhaustive computations for each possibility. The score for the optimal alignment of the whole sequence is indicated in the top right hand corner, the (m,n) cell. In this example it is the alignment of 7 against 5 residues, having a score of 1.

In order to see what the actual alignment associated with this score looks like, we perform a **traceback** step. Beginning at the upper right (m,n) element, we determine which of the three possible "directions" was taken to reach that cell. For example, the 1 in the (7,5) cell could only be reached by the addition of the mismatch A–T, the value of 1 being obtained by adding a mismatch penalty of −1 to the 2 found in the (6,4) cell. Had the (7,5) cell been reached from the (7,4) cell, the score would have been (−3) + (−5) = −8; had it been reached via adding an indel from the (6,5) cell, the entry would have been 2 + (−5) = −3. We will indicate with arrows the directions taken, and step our way back through the entire matrix to the (0,0) cell.

A	−35	−27	−19	−9	−3	1
G	−30	−22	−14	−4	2	2
T	−25	−17	−9	−1	−1	7
C	−20	−12	−4	0	4	3
G	−15	7	1	5	4	−1
C	−10	−2	6	1	−4	−9
A	−5	3	−2	−7	−12	−17
−	0	−5	−10	−15	−20	−25
	−	A	C	T	G	T

A cell with two arrows indicates two equally good alignments, and any continuous path from the upper right cell to the lower left cell represents an alignment with the optimal score. For example, the value of −4 in the (4,2) position could be obtained from either of two directions—either by adding a penalty of −5 to the 1 in cell (3,2), or by adding a match of 3 to the −7 in cell (3,1). Every such split doubles the number of best-scoring alignments. In this example, this means that there is a pair of optimal alignments, each having score 1, for these two sequences, and these are:

```
ACGCTGA          ACGCTGA
A--CTGT          AC--TGT
```

Note that each alignment includes four matches, one mismatch, and two indels. The ambiguity has to do with which C in sequence 1 aligns with the C in sequence 2.

The original Needleman-Wunsch algorithm has undergone many modifications for both efficiency and improved biological reality. The outstanding computational biology book by Waterman (1995) provides examples.

igation along a sequence is facilitated either by scroll bars or, for contigs that exceed several kilobases in length, by manual entry of numerical positions. It can be performed using criteria that automatically identify features such as ambiguous bases, regions of poor alignment, high-quality reads that disagree with the consensus, and homopolymeric stretches. Editing of individual bases is facilitated by calling up the trace file, which allows visual comparison of two or more traces, again using a series of interactive tools.

The alignment algorithms used by various assemblers are generally modifications of methods first introduced by Needleman and Wunsch in 1970, and by Smith and Waterman in 1981; these methods are described in Box 2.1. Not all alignments of the same set of sequences are identical, primarily because they assign different default parameters—and use them differently—to weight the effects of insertions and deletions relative to single-base mismatches. Algorithms that also use quality scores from the contributing sequence reads add further complexity, as does weighting attached to the frequency of mismatches. Although default settings are the easiest to use, they are not necessarily the settings most likely to yield the correct alignment. Manual inspection remains the most efficient way to resolve ambiguities, but is too time-consuming to be performed on a genome scale.

EXERCISE 2.2 *Computing an optimal sequence alignment*

Compute the best possible alignment for the following two sequences, assuming a gap penalty of –5, a mismatch penalty of –1, and a match score of +3. Would your answer be any different if the gap penalty was also –1 (rather than –5)?

AGCGTAT   and   ACGGTAT

ANSWER: Three possible high-quality alignments are:

```
AGCGTAT              AGC-GTAT              AGCG-TAT
| · · | | | |   and   | - | - | | | |   and   | - | | - | | |
ACGGTAT              A-CGGTAT              A-CGGTAT
```

The second and third alignments will produce the same score with these penalties, namely $(6 \times 3) - (2 \times 5) = 8$ with a gap penalty of –5; or $(6 \times 3) - (2 \times 1) = 16$ with a gap penalty of –1. By contrast, the first alignment gives a score of $(5 \times 3) - (2 \times 1) = 13$. Consequently, the first alignment is best with a large gap penalty, but either of the other two alignments would be better with a small gap penalty.

The final task of a sequence editor is to help in the resolution of gaps and ambiguities. To this end, the Autofinish function within consed (Gordon et al. 2001) designs finishing reads by suggesting primers and identifying templates that will help to bridge gaps between contigs. Consed also facilitates the comparison of two or more contigs for consistency—for example, by identifying regions that may be erroneously inverted in one assembly or may contain an insertion or site of recombination that may have generated a chimeric clone. The program will also identify local regions of poor sequence and design primers that can be used for a new round of PCR amplification and sequencing across the ambiguous region. This is particularly important in a genome project where a pre-set, genome-wide error acceptance rate has been established, as design of further experiments to improve the sequence quality can be automated.

Once a multiple sequence alignment has been produced, a consensus sequence is derived on the basis of the quality of the reads at each position. If one or two of the reads in a contig disagree with the others, and they are of low quality, the consensus will disregard them and instead reflect the high-quality sequences. Furthermore, by combining the probability scores of multiple reads, the confidence in the actual consensus is greatly increased. *It is this averaging of multiple reads that provides high confidence in automated DNA sequence determination and circumvents the impossibility of manually verifying every sequence in a genome.* That is to say, in high-throughput sequencing applications, it is much more efficient to sequence every stretch of DNA multiple times and derive a consensus than to manually curate a single trace for each sequence. Nevertheless, most genome sequencing projects make the original trace files available to individual investigators to examine online.

Emerging Sequencing Methods

Although dideoxy-based sequencing has supported the vast majority of genome research to date, a number of novel sequencing technologies are on the horizon (Meldrum 2000). Since even in the mid-2000s the sequencing of a mammalian genome still cost in the range of $30–50 million dollars, the NHGRI has called for proposals to develop technologies for dramatic reduction in sequencing costs, with a target of $1,000 per human genome by the year 2010. The J. Craig Venter Foundation has posted a $500,000 reward for the first person to achieve this feat. Several of the emerging technologies have been shown to work in principle for short sequences in the range of 20–50 bases. They may prove most useful for resequencing applications in which the goal is to determine whether different individuals have the same DNA sequence as that observed in a reference sample.

- *Sequencing by hybridization* (SBH) makes use of the complementarity of the two strands of DNA molecules to detect whether an exact match to an oligonucleotide is present in a sample of DNA. As described in Chapter 3, this principle is already being used for resequencing applications, in which a variant detector array (VDA) is manufactured that contains all possible oligonucleotides in a molecule several kilobases in length, as well as mismatches at each site. The company Perlegen has used this technology to resequence entire primate chromosomes and to scan individual human genomes for 1.5 million SNPs for clinical applications.

- *Mass spectrophotometric techniques* have been demonstrated for the sequencing of fragmented DNA molecules up to 50 bases in length. The identity of fragmented oligonucleotides can be determined by reference to the time of flight of a set of standards through a vacuum chamber. In principle, utilizing advances in information theory, it should be possible to determine the full sequence of a molecule that is divided into all possible oligonucleotides, supporting de novo sequencing. MS methods are rapid and likely to become cost-effective. Sequenom utilizes the approach in massively parallel genotyping based on the differential mass of polymorphic nucleotides.

- Ultrafast and relatively inexpensive sequencing of long DNA fragments might be achieved using *nanopore sequencing strategies* (Deamer and Akeson 2000). One approach monitors changes in electrical current as the nucleotide base blocks a tiny membrane pore as a single strand of DNA or RNA passes through it. Another approach, being pursued by US Genomics, channels the nucleic acid through a single-molecule fluorescence reader. Such methods promise to allow the sequencing of molecules hundreds of kilobases long in just a few minutes.

- Several other single-molecule approaches still employ the basic Sanger approach, but avoid the reagent and liquid-handling costs associated with capillary-based automated sequencing. Solexa, Visigen, and Genovoxx focus on technologies for detecting fluorophore incorporation into single molecules on an array, with various modifications designed to overcome technical obstacles associated with incorporation and imaging. The 454 Corporation has instead opted to amplify single molecules, using a massively parallel picoliter scale amplification and detection system.

- An intriguing method for small-laboratory-scale genome analysis that is illustrated in Figure 2.6 is known as **polony sequencing** (Mitra et al. 2003). Polonies are PCR colonies: rather than cloning a DNA molecule into a plasmid and growing it in cells, or amplifying it in a plastic tube, the PCR reaction is performed in the acrylamide matrix on a glass microscope slide. The DNA to be sequenced is greatly diluted so that single molecules form PCR templates a few microns apart from one another. This results in tens of thousands of tiny colonies of identical DNA molecules. Cycles of fluorescent nucleotide incorporation, followed by scanning of the slide in a microarray reader, leads to the sequencing of several kilobases of DNA per slide.

It is possible that ultrahigh-throughput de novo sequencing technologies currently under development may increase sequencing capacities by a further order of magnitude. In addition, a not-implausible extension of these emerging techniques will be the ability to obtain a thorough read-out of gene expression at the cellular level by lysing cells and directly reading the sequences of thousands of complete mRNA molecules.

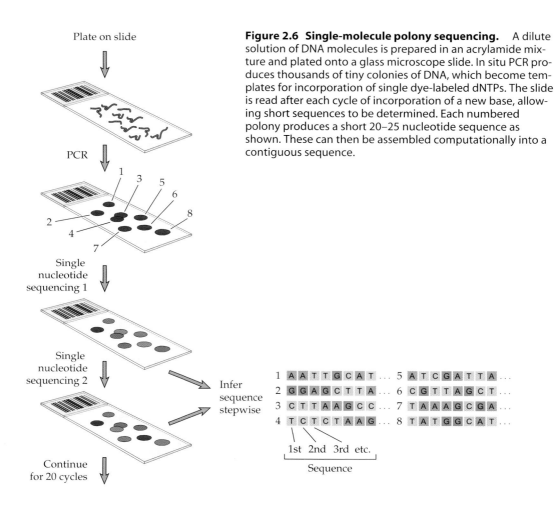

Figure 2.6 Single-molecule polony sequencing. A dilute solution of DNA molecules is prepared in an acrylamide mixture and plated onto a glass microscope slide. In situ PCR produces thousands of tiny colonies of DNA, which become templates for incorporation of single dye-labeled dNTPs. The slide is read after each cycle of incorporation of a new base, allowing short sequences to be determined. Each numbered polony produces a short 20–25 nucleotide sequence as shown. These can then be assembled computationally into a contiguous sequence.

Genome Sequencing

Whole chromosome sequences are reassembled from the sequences of hundreds of thousands of fragments, each typically between 500 and 1,000 bp in length. Two general strategies for fragmentation and reassembly are used, known as **hierarchical sequencing** and **shotgun sequencing.*** The distinc-

---

*For an historical overview of genome sequencing, starting with the first recombinant DNA molecule and moving through early sequencing developments, the first genome sequence (SV40), map-making, and proposals for the human genome project, see **http://www.sciencemag.org/feature/plus/sfg/human/timeline1.shtml** at Science Online.

Figure 2.7 Hierarchical versus shotgun sequencing. In hierarchical sequencing, the chromosomes are first cloned as large BAC or P1 fragments up to 200 kb. These are physically ordered, and a subset that gives a minimal overlap for complete genome coverage is chosen for shotgun sequencing. In the whole genome shotgun approach, no attempt is made to order the clones in advance. Instead, the whole genome is assembled using computer algorithms that order contigs based on their overlapping sequences.

tion between them is that in the hierarchical approach, the first step is to develop a low resolution physical alignment that is used to ensure the sequence is obtained in large, ordered pieces, whereas the shotgun approach simply breaks the genome into small, sequenceable units and relies on computer algorithms to assemble them. The difference is summarized in Figure 2.7. Most new genome sequencing projects adopt the whole genome shotgun approach, bypassing the use of hierarchical strategies to guide assembly.

Hierarchical Sequencing

Hierarchical strategies, which are also known as *top-down, map-based,* or *clone-by-clone strategies,* were developed in the late 1980s, at a time when the cost of chemicals was exorbitant, high-volume automated protocols had not yet been developed, and computers were not yet powerful enough to handle whole genome shotgun sequencing. The general approach of breaking a genome into smaller and smaller units, the relative locations of which are known before sequencing commences, had two further advantages: it fostered assembly of high-resolution physical and genetic maps and it allowed groups working around the globe to form consortia and work together without risking repetition, since each group could concentrate on a particular chromosome.

The technology for cloning large fragments of genomes and maintaining them in a stable state progressed rapidly throughout the 1990s, with the result that the *E. coli, S. cerevisiae, C. elegans,* and *A. thaliana* genome projects were

Figure 2.8 Cloning vectors used in genome sequencing. Different size inserts are cloned in different types of vectors. Phagemids (A) are derivatives of M13 that can shuttle between double-stranded plasmid and single-stranded bacteriophage forms. Like standard cloning plasmids (B), the multiple cloning site (mcs) into which the foreign DNA is inserted is flanked by primer binding sites for sequencing. P1 clones (C) are assembled as plasmids, but are converted into phage using the *cosN* or *LoxP* recognition sites for Ter and Cre recombinases. As with BAC clones (D), copy number within *E. coli* cells is kept low by use of the F plasmid origin of replication.

the beneficiaries of ever more efficient techniques of hierarchical sequencing. The basic strategy of hierarchical sequencing remains applicable to any organism, and in fact was the basis of the publicly funded human genome project.

The first step in top-down sequencing is to clone the genome as manageable units of some 50–200 kb in length. Large insert cloning vectors include bacterial artificial chromosomes (BACs, which carry up to 300 kb of DNA), P1 phage-derived PAC clones (~100 kb), and smaller phage-derived cosmids or fosmids (50 kb). As shown in Figure 2.8, BAC and PAC clones are similar in organization to ordinary plasmid cloning vectors, with the major difference that vector-encoded sequences help keep the copy number close to one per cell, which limits the potential for recombination and maintains the integrity of the large clones. Special features of PAC clones allow the initiation of phage packaging during the cloning step, excision of the high-copy-number plasmid sequences after bacterial infection, and induction of replication immediately prior to harvesting of DNA.

DNA libraries are constructed by partial digestion or shearing of genomic DNA by sonication, after which the fragments are ligated into a multiple cloning site (mcs) in the vector of choice using standard recombinant DNA procedures. The source of genomic DNA can be the whole genome, or a chromosome isolated by pulsed-field gel electrophoresis or some other method. Researchers typically aim for 5- to 10-fold redundancy, meaning that every portion of the genome should be represented at least five times in the library of clones. Since each clone will have different ends, it should in principle be possible to select a scaffold of clones that forms a contiguous sequence covering a whole chromosome—a **tiling path**—by aligning the regions of overlap (Figure 2.9). The tiling path can be assembled using a combination of three methods: hybridization, fingerprinting, and end-sequencing.

Figure 2.9 Hierarchical assembly of a sequence-contig scaffold (supercontig).
A minimal tiling path through a library of aligned BAC clones that ensures complete coverage of the chromosome is chosen. After sequencing independent shotgun libraries for each BAC, small gaps in the sequenced clone contigs remain. These are closed as far as possible by merging the two BAC sequences, as well as by the addition of mate-pair information (yellow) and cDNA structural information (red), which establishes the orientation and distance between cloned segments.

(A)

(B)

(C)

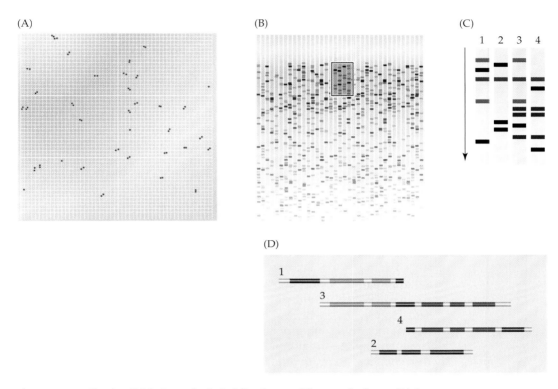

(D)

Figure 2.10 Aligning BAC clones by hybridization and fingerprinting. (A) A macroarray of BAC clones is probed with a short, radioactive fragment to identify all BACs that carry a specific fragment. (B) These clones are digested with a restriction enzyme, end-labeled, and separated by gel electrophoresis. (C) Software converts the bands to a virtual profile, shown hypothetically for a small portion of four bands (highlighted box in part B). Shared bands (red or blue) imply that the two clones share the same sequence. Green indicates the vector band common to all clones. (D) The fingerprint profile is then converted into a BAC alignment. In this example, clone 2 does not share any bands with the others and so is placed into a separate BAC contig, while the other three clones form a tiling path.

Hybridization. All of the clones in a library that carry a particular sequence can be identified rapidly by hybridizing a small radioactively or chemically labeled probe containing the sequence to a filter on which is printed an array of tens of thousands of clones (Figure 2.10A). Robots are used to pick colonies and replica-plate them on nitrocellulose membranes in high-density grids. If the sequence is unique in the genome, it should "light up" only a small number of clones, which can be compared with one another. Subsequently, the end of one of these clones can be used as a probe in a fresh screen for adjacent clones, in a procedure known as chromosome walking (see Figure 1.2). Alternatively, the library can be interrogated with thousands of random probes from throughout the genome, leading to the generation of islands of overlapping clones that eventually will be linked by some other method.

Fingerprinting. The most cost-effective way to assemble contigs of large insert clones is to compare and align them according to restriction digest profiles (Figure 2.10B,C). Restriction enzymes that cleave DNA at 6-bp recognition sites cut on average just under once every 4 kb, and thus liberate between 20 and 30 fragments when used to cleave a 100-kb BAC clone. These fragments can be separated either by agarose or acrylamide gel electrophoresis; they are then placed into size bins (following appropriate statistical normalization that accounts for variability among gels). Software has been developed that contrasts fragment profiles and aligns the clones according to regions of overlap (Figure 2.10D). Given sufficient redundancy, fingerprinting procedures regularly result in the assembly of Mb-length contigs, and suggest a minimum set of clones that are then individually subject to shotgun sequencing.

End-sequencing. A common way to identify clones spanning the gaps that remain after fingerprinting is to sequence both ends of the collection of BAC clones. Once a critical threshold of assembled sequence has been achieved, there is a high probability that at least one end of a BAC will lie within an assembled region, implying that the other end of the clone either extends into the gap, or even links to an adjacent contig. For example, end-sequencing 10,000 BAC clones will provide a sequence tag every 5 kb along a 10-Mb genome; this density will close most gaps of less than 50 kb. End-sequencing is also an important component of the techniques used to verify sequence assemblies.

Once a tiling path has been chosen, the individual BAC clones are sheared into small fragments that are subcloned for automated sequencing. A common cloning vector is an M13-derived phagemid, which exists as both double-stranded plasmid DNA and single-stranded phage, facilitating rapid and efficient template preparation. With this vector, only short fragments package efficiently, and reads are most readily obtained in a single direction, up to 1 kb. An alternative method is to clone 2- to 3-kb fragments into a plasmid vector and sequence from both ends. Shearing the BAC into small pieces is done by sonication, which ensures that the ends of each fragment are unique. Enough sequencing reactions are performed to approach tenfold coverage of each sequence, ensuring that the majority of any given sequence is covered at least once. This in turn allows computational reassembly of the complete sequence, and also increases sequence accuracy as regions of low quality are compensated for in other reactions. Inevitably, however, small gaps remain, and these must be filled in a sequence finishing phase.

Shotgun Sequencing

In shotgun sequencing, computer algorithms are used to assemble contigs derived from hundreds of thousands of overlapping sequences. As in hierarchical sequencing, the aim is to achieve 5- to 10-fold redundancy for each fragment of the sequence, but the contigs are generated from a plasmid library that has been constructed from a single whole genome rather than from individual BAC clones. Multiple sequences are aligned using algorithms that successively screen out repetitive sequences, overlap reads of the same sequence, generate unitigs and scaffolds, and resolve repeats. At that point, a genome sequence may be more than 90% assembled, but the finishing phase—closing gaps, cleaning up ambiguities, and improving the overall quality—can take as much time and effort as the initial shotgun phase. In general, the details of sequence assembly are only superficially described in publications and users are asked to trust the assemblies produced by the genome sequencing centers without actually examining the raw data themselves. The following description refers to software used by Celera Genomics, but similar algorithms underly all sequence assemblies.*

The **Screener** is employed to mask (that is, mark and hide) sequences that contain repetitive DNA (including microsatellites with a less than 6-bp repeat) and known families of interspersed repeats such as LINE elements, Alu repeats, retrotransposons, and ribosomal DNA. These sequences are not removed, but they are screened out of the alignment algorithms so that they do not contribute to determining overlap. Once contigs have been assembled, however, the length and location of repetitive sequences is taken into account in subsequent stages of genome assembly. Cloning vector sequences are also trimmed, and attempts are made to recognize contaminating bacterial, viral, or other extraneous sequences.

The **Overlapper** compares every unscreened read against every other unscreened read, searching for overlaps of a predetermined length and identity. This screening process is essentially the same as performing a BLAST search (Box 2.2) of each sequence read against each other sequence read. In the case of the human genome, overlaps of at least 40 bp with no more than 6 differences were required. The differences allow for misidentification and other sequencing errors, as well as for the presence of polymorphisms present in heterozygotes or libraries assembled from multiple individuals. Overlaps of this specificity have a probability of appearing once in every 10^{17} comparisons, and so are exceedingly unlikely to occur twice in a single genome (unless the sequence is recently duplicated). Parallel processing on 40 supercomputers, each with 4 gigabytes of RAM, allowed the 27 million screened human sequence reads to be overlapped in less than five days.

Repeat-induced overlaps due to the presence of a low copy number (greater than one) of a sequence, including tandem and dispersed duplicates, are resolved using the **Unitigger,** as diagrammed in Figure 2.11. A **unitig** is a contig formed from a series of overlapping, unambiguously unique sequences. These sequences are identified first. Instances in which

---

*The initial proposal for shotgun sequencing of large genomes, including calculations of the necessary clone and sequence coverage, can be found in Weber and Meyers (1997), along with a rebuttal by Green (1997). See also Venter et al. (1998) for the announcement of Celera's intention to proceed with sequencing of the human genome.

BOX 2.2 Searching Sequence Databases Using BLAST

Without a doubt, the single most extensively utilized bioinformatics tool is the Basic Local Alignment Search Tool, or BLAST. The BLAST program by Altschul et al. (1990) is used to search large databases of molecular sequences, returning sequences that have regions of similarity to a **query sequence** provided by the user.

In Box 2.1, we described the Needleman-Wunsch algorithm for global alignment of a pair of sequences, in which all residues from both sequences are included. In order to understand the workings of BLAST and similar search tools (notably FASTA), it is necessary to introduce the concept of **local alignment.** Local alignment algorithms attempt to find *isolated regions* in sequence pairs that have high levels of similarity. It is this property that makes local alignment ideal for database searches. The user provides a query sequence, which is then compared to the entire database. One can envision the concatenation of all sequences in the database as the **target sequence.** BLAST then searches for regions of the target sequence with similarity to the user's query sequence. Thus, we see the fundamental importance of pairwise alignment methods in bioinformatics.

Interpreting BLAST Output

Algorithms guaranteed to find the best local alignment of two sequences have been developed (Smith and Waterman 1981). However, as is the case with many bioinformatics applications, in practice these methods are often too slow to use. BLAST is an effective heuristic search method that is not guaranteed to find the best local alignment, but has been especially effective in practice. In this box we will use an example that pays particular attention to interpreting the output of the widely used NCBI BLAST program.

DeSalle et al. (1992) extracted 16S ribosomal DNA from an insect preserved in 25-million-year-old fossilized amber. BLAST can be used to verify that the amplified DNA was not a laboratory contaminant, and to identify the closest living relative of the ancient organism. Figure A shows some results from BLAST after probing GenBank using a 92 bp fragment from the fossilized insect (GenBank accession number S45649).

The BLAST report (graphical portion in Figure A, textual portion in Figure B) ranks **hits** in order of a measure of statistical significance called the **E-value,** with the most significant hits listed first. Formally, the E-value is the number of hits with the same level of similarity that you would expect by chance if there were no true matches in the

FIGURE A A BLAST output.   The graphical overview of BLAST clearly indicates which local region of our query sequence has similarity to a sequence in the database. The NCBI implementation of BLAST also has many features available through use of your mouse, including the actual alignment, along with a variety of useful summary statistics.

database. Thus, a hit with an E-value of 0.01 would be expected to occur once every 100 searches even when there is no true match in the database; a hit with E-value 1.0 is expected every time you search.

The E-value is similar in spirit to the traditional *p*-value of statistical hypothesis tests. The definition of a *p*-value is the probability of obtaining a result as extreme as the observed one, if in fact there is truly no effect. In the context of database searching, the desired *p*-value would be the probability of finding a sequence similarity as similar as the observed match if there were really no true matches in the database. Clearly, the E-value doesn't satisfy this definition, since it can take any positive value. Conveniently, though, the E-value approximates the *p*-value when it is small, say less than 0.1. Since we are primarily interested in "unusual" hits, it is typically safe to interchange E-values and *p*-values.

For the results of the example search, the E-values for the top hits are all very small—even the largest is a mere 3×10^{-13}, strongly suggesting that the similarities are not the result of random chance.

If we click on the value in the "Score" column, a display of the actual local alignment appears.

Scoring Matrices

Although BLAST has many settings that can be adjusted by the user, it is wise to leave most adjustments to experts. The exception to that statement is the choice of a scoring matrix. In Box 2.1, we saw a very simple scheme for scoring sequence matches and mismatches: all mismatches received the same penalty. A scoring matrix extends this idea to allow some mismatches to be penalized less than others. For instance, a leucine-isoleucine mismatch might be penalized less than a more biochemically radical leucine-tryptophan substitution.

Some of the best scoring matrices derive their scores from empirical collections of sequence alignments. The PAM family of scoring matrices is based on the pioneering work of Margaret Dayhoff and her colleagues (Dayhoff et al. 1972). The penalties were derived from closely related species to avoid the complications of unobserved multiple substitutions at a single position. More recently, it was recognized that the goal of most database searches was to identify very ancient homologies. A scoring matrix based on data from distantly related sequences was the logical step, resulting in the BLOSUM family of scoring matrices (Henikoff and Henikoff 1992).

When possible, the selection of a scoring matrix should depend on the level of similarity that one hopes or expects to find when performing a database search. Practical experience suggests that the BLOSUM62 matrix is quite suitable for general use.

```
                                                              Score    E
Sequences producing significant alignments:                  (bits)  Value

gi|256517|gb|S45649.1|S45649   16S rRNA [Mastotermes electrod...  139   5e-33
gi|12005612|gb|AF246514.1|AF246514   Drosophila ornatipennis ...   86   7e-17
gi|11119031|gb|AF304735.1|AF304735   Sphyracephala bipunctipe...   84   3e-16
gi|3552018|gb|AF086859.1|AF086859   Mystacinobia zealandica l...   84   3e-16
gi|256518|gb|S45650.1|S45650   16S rRNA [Mastotermes darwinie...   84   3e-16
gi|15341487|gb|AF403473.1|AF403473   This canus 16S ribosomal...   82   1e-15
gi|15341483|gb|AF403469.1|AF403469   Icaridion debile 16S rib...   82   1e-15
gi|13435200|ref|NC_002697.1|   Chrysomya chloropyga mitochond...   82   1e-15
gi|13384216|gb|AF352790.1|AF352790   Chrysomya chloropyga mit...   82   1e-15
gi|3552016|gb|AF086857.1|AF086857   Calliphora quadrimaculata...   82   1e-15
gi|15341485|gb|AF403471.1|AF403471   Malacomyia sciomyzina 16...   80   4e-15
gi|15341463|gb|AF403449.1|AF403449   Helcomyza mirabilis 16S ...   80   4e-15
```

FIGURE B A local alignment. Here we see the representation of several of the reported matches. The primary section of the BLAST output is a list of the best matches to our query sequence. Following the exact match to our query sequence itself, we see that the top matches are to known insects, such as *Drosphila ornatipennis*, providing confirming evidence that the extracted DNA was, in fact, from an insect.

(A)

ACGTTGTCGACTAGGATCGCTCGTGAGGG   ATGCAGCAGCAGCAGTTGTGAGAATCCAC

       ATCGCTCGTGAGGGGCT . . . . . ATAATGCAGCAGCAGCA

(B)

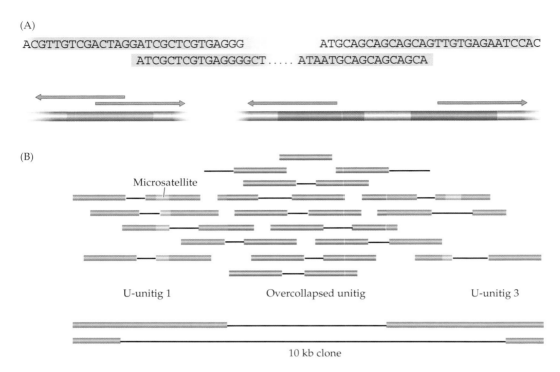

Figure 2.11 U-unitigs and repeat resolution.   (A) Sequence alignment between two or more shotgun clones can arise between unique sequences (left) or repetitive sequences (right). (B) The Overlapper aligns unitigs, which are identified as unique sequence alignments (U-unitigs) or overcollapsed repeats (blue). Two unitigs can be aligned and oriented by using mate-pair sequence information from the ends of longer (10- or 50-kb) clones, as shown at the bottom, while mate-pairs from 2-kb fragments allow assembly of scaffolds despite the presence of simple repeats such as microsatellites (light blue) that are masked before performing alignments.

two different sequences have been overcollapsed into a unitig are then identified with a discriminator that computes the likelihood that the number of times each sequence is represented in the overlap is greater than a statistical threshold. Simulations indicate that the set of certain and highly likely unitigs (the U-unitigs), formed from 6× shotgun coverage can cover up to 98% of nonrepetitive euchromatin in stretches of unique DNA greater than 2 kb in length—which is approximately 75% of the human genome.

The **Scaffolder** uses mate-pair information to link U-unitigs into scaffold contigs. For both the human and *Drosophila*, the vast majority of sequence reads were obtained from both ends of 2- or 10-kb clones, and these were correctly paired more than 98% of the time. (The remaining 2% represented annotation errors that arise inevitably in high-throughput work.) Wherever two or more mate-pairs link two U-unitigs with the same predicted spacing, it can be assumed that they are accurately linked. After iteration of this process, 50-kb mate-pairs can be used to verify the contiguity of the assembled scaffold.

Most of the remaining gaps are due to repeats, and are resolved by more computationally aggressive and error-prone steps, recognizing that it is better to produce large scaffolds that can be verified than to leave the genome in thousands of disconnected pieces. Figure 2.12 shows the estimated coverage of the fly and human whole genomes after initial assembly: in both cases, 84% or more of the genome was covered by scaffolds at least 100 kb in length, while most scaffolds were in the Mb range. Increasing sequence coverage from 5× to 10× results in a 10% increase in the proportion of scaffolds of lengths up to 1 Mb. The choice of level of coverage must balance genome size, availability of funds, and the needs of the project, but obviously the more sequences, the more complete the assembly will be. Where a closely related species has already been sequenced, it may be possible to hang a 3× coverage assembly on the scaffold of that species, but there are risks in doing so and most projects are carried out independently.

At least three different approaches can be taken to closing any remaining gaps. Where it is clear that a gap is solely due to a repetitive sequence that was screened out, this repeat sequence can be reinserted into the assembly, with the proviso that it may be a generic repeat sequence rather than the actual sequence (including unique polymorphisms) at that position.

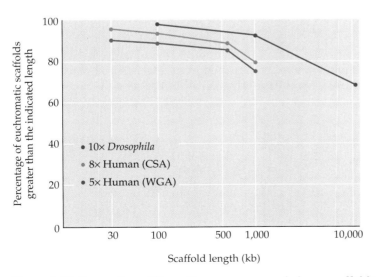

Figure 2.12 Proportion of fly and human genomes in large scaffolds.   The plot shows the percentage of scaffolds that have a length greater than that indicated for the *Drosophila* 10×, human 8× (compartmentalized shotgun assembly, CSA) and human 5× (whole genome assembly, WGA) sequences generated by Celera. The fly and CSA assemblies include shredded sequences generated from BAC clones by public genome sequencing efforts.

Where it is not clear what the gap sequence may be, it will often be possible to identify a clone that spans the gap (because the mate-paired ends of the clone derive from two different scaffolds) and then to carefully sequence just that clone. For this reason, shotgun sequencing projects include some fraction of 50-kb clones in addition to 2-kb and 10-kb clones. Alternatively, gaps may be bridged by sequences from different projects, including cDNA sequences (two exons may be separated by a large intron that falls in the gap) and comparative alignment with other species in which two adjacent scaffolds always map to the same locus. This information at least links the scaffolds together, and directs finishing efforts to design PCR primers to bridge the gap.

The scaffolds are then assigned to chromosomal locations. This is achieved most simply by matching the assembly to previously cloned gene sequences that have been placed on genetic maps by linkage. Representative clones can also be placed on cytological maps by fluorescent in situ hybridization (FISH).

Sequence Verification

The veracity of any whole genome sequence must be assessed at three levels: its completeness, the accuracy of the base sequence, and the validity of its assembly.

- *Completeness.* Microbial genomes tend to be sequenced in their entirety, but may contain small gaps generally of the order of one kilobase or less that prove difficult to close. It should be recognized, however, that *any* sequence is that of a single isolate and that among isolates, some length polymorphism, including insertion and deletion of whole genes, is to be expected. Most higher eukaryotic genomes contain large stretches of heterochromatin that are simply excluded from analysis and may never be sequenced except for small islands such as the *Raf* serine kinase gene that is dispersed in heterochromatin in the vicinity of the centromere of the second chromosome of *Drosophila*.

- *Accuracy* is assessed by probability scores, as described in earlier in this chapter (see the section titled "Reading Sequence Traces"), and can always be increased simply by sequencing more clones to cover a specific region, possibly using novel chemistry to resolve ambiguities.

- *Validity of assembly* is not trivial to assess, and can be approached either by measuring internal consistency, or by contrasting the assembly with genetic or pre-existing physical maps. Two excellent measures of internal consistency are the error rate in the alignment of predicted restriction profiles with observed fingerprints, and the correct spacing of paired end-sequences from clones of different sizes. The spacing between mate-pair reads from 50-kb clones is particularly useful for confirming the orientation and extent of repeat content. Disagreements are called breakpoints.

In the case of the human genome, at the time of publication of the draft assemblies in February 2001, hundreds of inconsistencies and gaps were

Chromosome 5

Chromosome 22

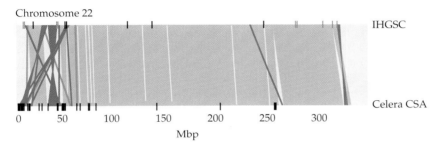

Figure 2.13 Alignment of two draft human genome assemblies. First-draft genome assemblies are far from complete, as indicated by the discrepancies between the public (IHGSC) and Celera compartmentalized shotgun assembly alignments for chromosomes 5 and 22. Green links indicate matched alignment; orange lines indicate sequence blocks greater than 50 kb that are out of order, and blue lines blocks that are also in the opposite orientation. Small black notches indicate breakpoints, and light blue notches are runs of 10 kb of Ns. (Redrawn from data available at Science Online: **www.sciencemag.org/cgi/content/full/291/5507/1304/DC1**.)

reported for each chromosome. As shown in Figure 2.13, some of these differentiated the public and private drafts by more than half the length of a chromosome. Since the two "finished" chromosomes, 21 and 22, showed only a handful of breakpoints, it is likely that most of these represent discrepancies that arose during the assembly process. Finishing is a very expensive process, and for some species the draft genome is sufficient as individual investigators who have an interest in a particular region will generally resolve ambiguities in the course of their research. However, proper ordering of clones is essential where the genome sequence is used to guide linkage and association studies described in the next chapter, and draft genome sequences can be deficient in this regard.

Genome Annotation

EST Sequencing

The most direct way to identify a gene is to document the transcription of a fragment of the genome. This can be done using microarrays, as will be described in Chapter 4, but elucidation of the structure of a transcript requires sequencing of one or more cDNAs. These are reverse-transcribed (complementary DNA, or cDNA) copies of mRNAs, and a collection of cloned cDNAs is known as a **cDNA library.** Full-length cDNAs are usually too long to be sequenced in a single reaction, so the term **expressed sequence tag (EST)** is used to describe a single sequence that represents a partial sequence of a cDNA. Since most genes are represented by multiple transcripts produced by alternative splicing, and since each transcript can be represented by several ESTs covering different stretches of sequence, single genes can be represented by multiple ESTs.

The first step in generation of a cDNA library is isolation of the polyA fraction of RNA from a tissue. After the mRNA strand is converted to DNA, generally using a polyT oligonucleotide to prime reverse transcription, the double-stranded DNA fragments are cloned into a plasmid. The vast majority of expressed sequences will be present at a frequency of less than 1 clone in 10,000; each of several hundred sequences will constitute between 0.01% and 1% of the total transcripts; and there will be a handful of sequences that are present at even higher frequencies. Consequently, sequencing random clones results in a high level of redundancy, in the sense that at least half of the transcripts will be represented two or more times well before all the different transcripts are identified. The efficiency of identifying new genes can be improved using several strategies that help normalize the library based on subtractive screening prior to sequencing of the cDNA fragments (Carninci et al. 2000).

Because a large proportion of genes are differentially expressed in time and space, as well as under a variety of environmental conditions or between the two sexes, sampling a wide array of tissues is necessary in order to identify as many genes as possible. Both the public and private sectors have developed atlases of human gene expression, drawing on hundreds of different cDNA libraries, many of which are derived from tumors or transformed cell lines. Clone sequences can be downloaded from the NCBI's GenBank dbEST database (Boguski et al. 1993) at http://www.ncbi.nlm.nih.gov/dbEST, or from ENSEMBL at http://www.ensembl.org. Similarly, the vast majority of genome projects are developing EST collections that are an essential resource in confirming the correspondence between open reading frames and expressed genes, and these are available both through the above URLs and at the home Web sites for each organism.

Prior to the completion of the Human Genome Project, estimates of the number of genes based on the number of observed ESTs exceeded 100,000—three to four times what we now believe to be the true number of human genes. The two main reasons for such overestimates are the prevalence of alternative splicing, and artifacts due to incompleteness of cDNA fragments, as diagrammed in Figure 2.14.

Alternative splicing refers to the phenomenon whereby the same gene gives rise to multiple different transcripts, based on which combination of exons are incorporated in the mature mRNA transcript. The average human gene contains between 10 and 15 exons and encodes three or more different proteins as a result of alternative splicing. When different exons of a single

(A) Gene structure

(B) Complete cDNAs

(C) ESTs

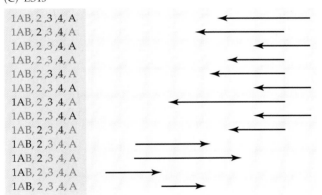

Figure 2.14 Relationship between gene structure, cDNA, and EST sequences.   (A) Alternative splicing and the use of alternate 5' start sites or alternate polyA signals results in (B) the generation of multiple cDNA transcripts from individual genes. (C) EST sequences can be derived from either the 5' or 3' end of a cDNA clone and are generally incomplete, resulting in the representation of different exons in different clones from the same gene.

gene are included along with the 3'-most exon in a collection of sequence tags, they will initially be predicted as multiple genes. Complementary DNA synthesis is a notoriously error-prone process, and internal priming can occur, also giving rise to different ESTs from one gene. Further, the probability that the 5' end of a clone corresponds to a true start site of transcription decreases as a function of the length of the transcript. As a result, EST sequences derived from the 5' end often commence at internal sequences, and will tend to identify different sequences that may be interpreted as belonging to different genes. Errors can also occur close to the 3' end as a result of internal deletions, or artifacts of mis-splicing, while some genes have alternate natural 3' ends.

Comparing EST and genomic DNA sequences readily resolves the issue of single genes that are represented as independent ESTs. Proper definition of gene structure, however, requires isolation of full-length cDNAs from high-quality libraries, followed by sequencing of the complete cDNA sequences of multiple isolates. The position of the 5′ end is usually confirmed with a technique such as RNA extension or RNAse protection. The representative major transcript for a gene is known as a RefSeq (reference sequence), but genome browsers typically display evidence for the full complement of alternate transcripts by default. In sequence files, the predicted coding sequence of an mRNA—that is, excluding the 5′ and 3′ untranslated regions—is annotated as the CDS. Many organisms now have public databases of full-length cDNA sequences, notably including the H-InvDB of over 21,000 full-length human cDNA sequences at http://www.h-invitational.jp (Imanishi et al. 2004), the *Drosophila* GOLD collection, and the mouse FANTOM and the *Arabidopsis* RIKEN full-length cDNA projects.

Ab Initio *Gene Discovery*

Protein coding sequences within a whole genome sequence can be identified by a process known as *ab initio* **gene discovery,** using software that recognizes features common to protein coding transcripts. These features include the existence of long open reading frames, particularly ones for which the codon bias is typical of that observed for the species being studied; proximity of transcriptional and translational initiation motifs and 3′ polyadenylation sites; and splicing consensus sequences at putative intron-exon boundaries.

In bacteria and in eukaryotes with small, compact genomes, programs such as GeneFinder and Grail have been found to predict accurately in excess of 90% of all true genes. The genomes of higher eukaryotes, however, tend to have numerous introns and extensive noncoding intergenic regions—not to mention quirks such as *trans*-splicing and genes located within the introns of other genes—that make *ab initio* gene discovery considerably more difficult. A new generation of programs, including Genie, Genscan, HMMgene, and FGENES, incorporate statistical approaches into **hidden Markov models (HMMs)** to help overcome some of these difficulties (Box 2.3; reviewed by Rogic et al. 2001).

Irrespective of which discovery algorithm is used, all computationally identified putative genes must be confirmed by a second line of evidence before being elevated to gene status in the genome annotation. The six standard types of direct evidence are:

1. Identity to a previously annotated cDNA reference sequence.

2. A match to one or more EST sequences from the same organism.

3. Similarity of the nucleotide or conceptually translated protein sequence to such sequences from other organisms in GenBank or other databases.

4. Protein structure prediction that matches a domain in the PFAM database.

5. Association with predicted promoter sequences, including a TATA box consensus sequence, proximity to clusters of recognized transcription factor binding sites, and often (in mammals) proximity to a CpG island.

6. Proof that a mutation in the sequence produces an observable phenotype.

BOX 2.3 Hidden Markov Models and Gene Finding

Many of the bioinformatics tools used for predicting features or functions of sequences belong to the family of methods that rely on **hidden Markov models,** or **HMMs.** HMMs allow for local characteristics of molecular sequences to be modeled and predicted within a rigorous statistical framework, and also allow the knowledge from prior investigations to be incorporated into analyses.

An Example of the HMM

It is perhaps simplest to introduce HMMs with an example. Let us suppose that every nucleotide in a DNA sequence belongs to either a "normal" region (N) or to a GC-rich region (R). As the name implies, the bases G and C are more prevalent in the GC-rich categories. Furthermore, let us assume that the normal and GC-rich categories are not randomly interspersed with one another, but instead have a patchiness that tends to create GC-rich islands located within larger regions of normal sequence. A short representation of such a sequence might be:

NNNNNNNNNRRRRRNNNNNNNNNNNNNNNNNNNNRRRRRRRNNNN

The **states** of the HMM are the two categories, either N or R. The two states emit nucleotides with their own characteristic frequencies. The word "hidden" in the name "hidden Markov model" refers to the fact that the true status of the states is unobserved, or hidden. Keeping in mind that sites in GC-rich regions are more likely to be either G or C, a possible DNA sequence having this underlying collection of categories is:

TTACTTGACGCCAGAAATCTATATTTGGTAACCCGACGCTAA

At first glance, this may appear to be a typical random collection of nucleotides; the observed 60% AT and 40% GC is not too far from the expected counts from a "random" sequence. However, if we focus on the red GC-rich regions, we see that the frequency of GC in them is 83% (10/12), compared to a GC frequency of 23% (7/30) in the other sequence regions. The main use of HMMs is to identify these types of features in sequences.

HMMs have the ability to capture both the patchiness of the two classes and the different compositional frequencies within the categories. They have proven invaluable in applications such as gene finding, motif identification, and prediction of tRNAs. In general, if we have sequence features that we can divide into spatially localized classes, with each class having distinct compositions, HMMs are a good candidate for analyzing or finding new examples of the feature.

Let us now become more specific. Suppose that studies of other DNA sequences reveal that normal and GC-rich regions have the following base compositions (the compositional biases are exaggerated for the purpose of illustration):

	A	T	G	C	Mean length
Normal (N)	0.3	0.3	0.2	0.2	10
GC-rich (R)	0.1	0.1	0.4	0.4	5

The process of obtaining these frequency estimates from empirical data is called *training* the HMM. An HMM describing these data is represented as in Figure A.

The workings of an HMM can be broken into two steps: the assignment of the hidden states, and then the emission of the observed nucleotides *conditional* on the hidden states. To illustrate, consider the simple sequence TGCC. One possible way that this sequence could arise is from the set of hidden states NNNN. In the case that we know (or assume)

(*Continued on next page*)

BOX 2.3   *(continued)*

Normal   GC-rich

FIGURE A Training the HMM.   The *states* of the HMM are the two categories, N or R. *Transition probabilities* govern the assignment of states from one position to the next. In the current example, if the present state is N, the following position will be N with probability 0.9, and R with probability 0.1. The four nucleotides in a sequence will appear in each state in accordance to the corresponding *emission probabilities.*

the hidden states, the probability of the observed sequence is simply a product of the appropriate emission probabilities: $\Pr(\text{TGCC} \mid \text{NNNN}) = 0.3 \times 0.2 \times 0.2 \times 0.2 = 0.0024$, where, for example, the notation $\Pr(T \mid N)$ is the *conditional probability* of observing a T at a site given that the hidden state is N.

The HMM provides the probability of this occurrence. In general, this probability is computed as the sum over all hidden states as:

$$\Pr(\text{sequence}) = \sum \Pr(\text{sequence} \mid \text{hidden states})\Pr(\text{hidden states})$$

using a simple application of the rules of conditional probability (see, for example, Moore and McCabe 1999). The description of the hidden state of the first residue in a sequence introduces a technical detail beyond the scope of this discussion, so we simplify by assuming that the first nucleotide is in a Normal region. With that simplification in hand, there are $2 \times 2 \times 2 = 8$ possible sets of hidden states, or **paths** through the Markov model, and the TGCC sequence could have been produced by any of these paths. We apply the law of total probability, in conjunction with conditional probability rules, to obtain:

$$\Pr(\text{TGCC}) = \Pr(\text{TGCC} \mid \text{NNNN})\Pr(\text{NNNN}) + \Pr(\text{TGCC} \mid \text{NNNR})\Pr(\text{NNNR}) +$$
$$\Pr(\text{TGCC} \mid \text{NNRN})\Pr(\text{NNRN}) + \Pr(\text{TGCC} \mid \text{NRNN})\Pr(\text{NRNN}) +$$
$$\Pr(\text{TGCC} \mid \text{NNRR})\Pr(\text{NNRR}) + \Pr(\text{TGCC} \mid \text{NRNR})\Pr(\text{NRNR}) +$$
$$\Pr(\text{TGCC} \mid \text{NRRN})\Pr(\text{NRRN}) + \Pr(\text{TGCC} \mid \text{NRRR})\Pr(\text{NRRR})$$

To illustrate the calculation, (recall that we assumed the first position was in a normal state):

$\Pr(\text{TGCC} \mid \text{NNNN})\Pr(\text{NNNN}) =$
$\Pr(T \mid N)\Pr(G \mid N)\Pr(C \mid N)\Pr(C \mid N) \times \Pr(N - N)\Pr(N - N)\Pr(N - N) =$
$(0.3 \times 0.2 \times 0.2 \times 0.2) \times (0.9 \times 0.9 \times 0.9) = 0.00175$

Similarly,

$\Pr(\text{TGCC} \mid \text{NNRR})\Pr(\text{NNRR}) =$
$\Pr(T \mid N)\Pr(G \mid N)\Pr(C \mid R)\Pr(C \mid R) \times \Pr(N - N)\Pr(N - R)\Pr(R - R) =$
$(0.3 \times 0.2 \times 0.4 \times 0.4) \times (0.9 \times 0.1 \times 0.8) = 0.000691$

Notice that the observed sequence is slightly more likely for the first path, NNNN. If we extend this notion and compute the probability of the sequence for all possible paths, we can use the path that contributes the maximum probability as our best estimate of the unknown hidden states. For the sample sequence, one finds that the most probable path is in fact NNNN, which is slightly higher than the path NRRR (.00123), pointing out the role of spatial clustering. If the fifth nucleotide in the series were also a G or C, the path NRRRR

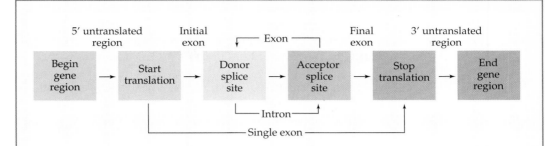

FIGURE B Schematic of the hidden states included in an HMM. Boxes denote *signal sensors* for regulatory elements, coding region start sites, intron donor and acceptor sites, and translation stop sites; arrows indicate *content sensors* for intergenic regions, exons, and introns. Each of these regions emits nucleotides with frequencies characteristic of that region, with these frequencies being obtained by training the HMM on data sets of many known genes.

would be more likely than NNNNN, providing evidence for the existence of a GC-rich island. This approach is one of several for predicting the hidden states (see Sonnhammer et al. 1998 for a thorough discussion).

Gene Finding

We conclude our treatment of HMMs with a more complex example, that of predicting genes from large unannotated DNA sequences. Figure B displays a version of the HMM implemented in the GENSCAN program (Burge and Karlin 1997). Rather than go into extensive details of the intricacies of the model, the figure simply points out the main features.

   We close by pointing out that gene prediction is an inexact science. Different prediction methods will give different predictions. Most of the time the predictions are quite similar, however. We illustrate this fact with a region of gene predictions from the UCSC Human Genome Browser (http://genome.ucsc.edu), shown in Figure C.

Figure C Predicting genes. Three different prediction methods (Ensembl, Fgenesh, and Genscan) were used on a region of chromosome 17 that includes the well-annotated GOSR2 gene. The black images below indicate the location of matching cDNA/EST sequences.

Each of these procedures can be automated to some extent, but human oversight appears to be indispensable for sifting through discrepancies and noise. For most genome projects, users in the scientific community at large are encouraged to submit corrections and updates to annotations as they identify them.

The three biggest deficiencies of computational gene discovery methods are imprecise or incomplete characterization of gene structure, characterization of false positive genes, and failure to identify true genes. Comparison of predicted genes with previously characterized genes indicates that *ab initio* methods correctly identify all of the intron-exon boundaries for at most only three-quarters of the genes in higher eukaryotic genomes, and in some cases considerably fewer. Given the complexity of alternative splicing in most of these cases, thorough characterization of gene structure by sequencing of complete cDNAs should be regarded as a routine component of thorough annotation. Comparison of the genome sequences of related species is also likely to considerably improve the annotation of gene structure, particularly as multiple sequence alignments become available.

Perhaps the more important fact is that probably 80–90% of all true genes *are* identified, with a false-positive rate of less than 10%. False positives include some putative genes for which there is simply no evidence either of transcription or similarity to other genes in the database, as well as unfiltered pseudogenes and/or genes associated with transposons that are unlikely to have a function in the organism. False negatives can be detected using a different gene finding program, or by adjusting the stringency of the original program's search parameters or even the properties of the database that is searched (Gopal et al. 2001).

Confirmation that an annotated gene corresponds to a true gene ultimately depends on functional data. Many genes were first identified following characterization of a mutation or polymorphic variant of the gene associated with an abnormal or extreme phenotype. As will be described in Chapter 5, reverse genetic **gene knockout** strategies—modifying a gene so that it is nonfunctional and seeing what effect this has on the phenotype—are perhaps the most direct method for confirming not just that a gene is expressed, but also that it is functional.

The expression of genes for which no EST or cDNA counterpart has been identified can be established by specifically searching for transcription from the predicted locus. An efficient way to do this is to identify open-reading-frame sequence tags (OSTs; Reboul et al. 2001) by amplifying predicted genes from a cDNA library using primers designed to anneal to both ends of putative open reading frames. Alternatively, fluorescently labeled cDNA can be hybridized to microarrays constructed with genomic DNA probes from the candidate genes (Penn et al. 2000). Transcription from loci that are not predicted to encode proteins is also commonly observed when cDNA is hybridized to so-called tiling arrays of short oligonucleotides that represent all genomic sequences (Shoemaker et al. 2001). Some of this may be attributed to non-functional priming of transcription, some to production of regulatory RNAs, and some to genes that just are not predicted by bioinformatic algorithms. Since over one-quarter of the genes in most newly sequenced genomes do not show similarity to genes in other organisms, the development of novel tools for the characterization of predicted but not yet described genes is a major goal of functional genomics.

Annotation of alternative splicing is generally imprecise, as it relies on computational prediction and the sequences of a relatively small number of cDNA clones. A possible solution to this deficit is **digital polony exon profiling,** a modification of the polony technique (referred to earlier in the context of novel methods for genome sequencing; see Figure 2.6) that allows annotation of the combinatorial diversity of exon usage in different tissues. Reverse amplification of cDNA ends (5′-RACE) is used to specifically amplify transcripts from a single locus in the acrylamide matrix, but instead of sequencing these, oligonucleotide probes corresponding to each predicted exon are hybridized to the slide. By using multiple fluorescent dies, and reprobing the slide with different probes as necessary, it is possible to determine which of the thousands of polonies have each exon, and hence to infer the prevalence of each transcript type (Zhu et al. 2003).

Regulatory Sequences

Comparative genomics is used increasingly to annotate the regulatory sequences of genes, which are generally recognized as conserved non-coding sequences. These tend to be located 5′ to the transcription start site, but may also be intronic or located at the 3′ end of the gene, as much as several tens of kilobases from the transcript sequence. The general approach known as **phylogenetic footprinting** was first demonstrated through detection of regulatory elements in a globin gene by comparison of mouse and human genomic DNA, and has since become a standard method for directing molecular biologists to candidate regulatory regions. Multiple sequence alignments provide the most information for detecting conserved elements, particularly when phylogenetic information is used with a statistical model of the rates of evolution of conserved and non-conserved sequences (Blanchette and Tompa 2002). Two popular open source programs for visualizing pairwise sequence alignments are PipMaker (http://bio.cse.psu.edu; see Figure 1.12) and VISTA (http://www-gsd.lbl.gov/vista).

As it has become apparent that as many as 30% of the functionally defined binding sites for transcription factors in human regulatory regions are not conserved in rodents (Dermitzakis and Clark 2003), novel computational methods for finding regulatory elements are being developed. The approach shown in Figure 2.15 is called **phylogenetic shadowing,** and is conceptually similar to phylogenetic footprinting, with the difference that several closely related species are compared rather than pairs of distantly related ones. Boffelli et al. (2003) showed that by applying a model of sequence evolution to an alignment of five or six primates, they could discern 10 candidate binding sites, each 50–100 base pairs long, in the *apo(A)* promoter, most of which were confirmed to bind to a nuclear protein extract and to be necessary for normal levels of transcription. The technique has been used to

Figure 2.15 Phylogenetic shadowing. A multiple alignment of sequences from at least five closely related species allows fitting of a statistical model that formulates the ratio of likelihoods that each window of 50 nucleotides is evolving at a relatively fast or slow rate. For the *apo(A)* promoter, this led to the identification of 10 conserved elements (C) and 7 non-conserved elements (N). Oligonucleotides of several of the C elements were found to bind nuclear extracts, while disruption of most of these and several other C elements resulted in reduced transcription from a reporter gene in cell culture. Hypothetical examples of conserved and non-conserved samples are illustrated.

compare fungal and plant sequences as well, and eShadow software (Ovcharenko et al. 2004) is available online at http://eshadow.dcode.org.

Description of computational methods for detecting short regulatory motifs is deferred to Box 4.4. Many of these methods are based on searching for clusters of motifs that have been shown: (1) experimentally to bind to specific transcription factors; (2) to be required for transcription of reporter genes; or (3) to be overrepresented in potentially regulatory genomic DNA (Rajewsky et al. 2002). Gibbs sampling approaches can be used to find single short motifs, but there is typically a high false-positive rate, both because of the potential for error in finding short sequences and the confounding effects of other positive and negative regulatory elements in promoters. In the case of *Drosophila* embryonic patterning, more success has been seen where clusters of sites that bind to interacting proteins are detected, and software such as Ahab has been developed for automating cluster-based motif detection. An emerging challenge is to use comparative sequence data simultaneously to identify conserved elements and to study how their use evolves within a lineage.

Non-Protein Coding Genes

The identification of genes whose product is not a protein but a functional RNA is complicated by several features that distinguish these genes. First, for the most part the transcripts are not polyadenylated, and so are not represented in standard cDNA libraries. Second, the constraint on sequence divergence is at the level of secondary structure rather than codon sequence, so that sequence

TABLE 2.1 *Classes of Noncoding RNA in the Human Genome*

Class	Function	Number	Localization
tRNA	Protein synthesis	~500	Dispersed large clusters
rRNA	Protein synthesis	~200 each	Tandem arrays
U snRNAs	Splicing	<20 each	Dispersed in clusters
snoRNAs	rRNA modification	~100	Dispersed single copy
miRNAs	Gene regulation	~175	Dispersed single copy
Others	Various	~20 ??	Single copy

Source: As reported by IHGSC (2001).

divergence between species is often too great to identify the genes purely by sequence similarity. And finally, relatively little is known about the function and distributions of non-protein-coding RNAs (ncRNAs; Table 2.1) other than those involved in transcriptional processing and translation.

Nevertheless, experience from the analysis of the content of mitochondrial genomes, as well as knowledge gained from detailed biochemical analysis of ribosomal complexes, has led to the development of algorithms that are thought to identify a substantial fraction of functional RNAs. As with protein-coding genes, first-pass annotation is unavoidably incomplete and imprecise, but intrinsic interest in the biology of RNA-mediated catalysis ensures the ongoing characterization of these gene classes.

Transfer RNAs (tRNAs) fold into a characteristic cloverleaf structure by virtue of the assembly of short stretches of base-pairing in which the local complementarity of sequence is more conserved than the actual sequences involved. This structure is identified by using software such as tRNAscan-SE (Lowe and Eddy 1997), which has an advanced search grammar based on sequence similarity as well as base-pairing potential to sort true tRNAs from likely pseudogenes. Although there are 61 codons to be decoded (excluding the three stop codons), only 45 anticodons are required because "wobble" pairing allows third-position U or C nucleotides to be recognized by the same tRNA species. As shown in Figure 2.16A, there are actually 48 classes of identified tRNA in the human genome, but, as in all eukaryotes, most of these are present in multiple copies, yielding a total of 497 canonical human tRNAs. The frequency of each tRNA type correlates with codon usage in mRNAs (Figure 2.16B), and is thus thought to be responsible in part for the evolution of codon biases in genomes. Transfer RNAs are dispersed throughout genomes, but tend to occur in clusters containing multiple different classes.

Ribosomes contain four types of ncRNA: the 28S, 5.8S, and 5S rRNAs found in the large subunit, and the 18S small subunit rRNA. Each of these are represented by between 150 and 300 copies in higher eukaryotic genomes, in which they are found in single arrays of tandemly duplicated genes, although related single copy sequences can also be identified in dispersed locations. The highly repetitive nature of these clusters makes thorough structural characterization of rRNA complexes difficult, due both to

Figure 2.16 Transfer RNA content in the human genome.   (A) The table indicates the percentage of codons for each color-coded amino acid below the diagonal, and the number of tRNAs with the associated anticodon above the diagonal in each box. (B) This plot shows that codon bias (expressed as the percentage of the particular amino acid that is encoded by each codon) is only partially correlated with the number of tRNAs with the complementary anticodon. Thirteen possible codons are not associated with exactly complementary tRNAs, and so must be decoded by wobble anticodons.

(A)

	NUN		NGN		NAN		NCN	
UNU	46	0	18	10	44	1	45	0
UNC	54	14	22	0	56	11	55	30
UNA	8	8	15	5				
UNG	13	6	6	4			100	7
CNU	13	13	29	11	41	0	9	9
CNC	19	0	32	0	59	12	19	0
CNA	7	2	28	10	26	11	11	7
CNG	40	6	11	4	74	21	21	5
ANU	36	13	24	8	47	1	15	0
ANC	48	1	36	0	53	33	24	7
ANA	16	5	28	10	43	16	20	5
ANG	100	17	12	7	57	22	20	4
GNU	18	20	26	25	47	0	17	0
GNC	24	0	40	0	53	10	34	11
GNA	11	5	23	10	43	14	25	5
GNG	47	19	11	5	57	8	24	8

(B)

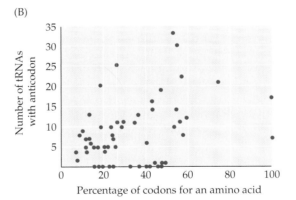

biases against sequencing repetitive DNA in genome projects and problems associated with the assembly of such regions. As with the tRNAs, many dispersed copies of the genes are likely to be pseudogenes, relatively recently generated insertions that are accumulating mutations and perform no physiological function.

Eukaryotic rRNA is heavily modified in the nucleolus by protein-RNA complexes, including a family of small nucleolar RNAs (snoRNAs). Two families of these, the C/D box and H/CA families, guide site-specific methylation and pseudouridylation, and 84 single-copy snoRNA genes have been found dispersed throughout the human genome.

The splicing of primary transcripts into mature mRNA is mediated by protein-RNA complexes that include spliceosomal RNAs designated as U1 through U12. The genes for these molecules also occur in loosely structured tandem arrays of up to 20 copies, though some are dispersed as single copies. Identification of these genes currently requires sequence-matching to previously characterized RNAs, and these genes are likely to be both underrepresented and improperly characterized in shotgun genome sequences. Small RNAs that mediate the editing of mRNAs remain to be annotated using genomic methods.

Several other classes of ncRNA are known to exist, including RNA components of enzymes such as telomerase and RNase P. These genes have been identified through sequence identity to purified products. In addition, a small number of non-protein-coding regulatory genes has been identified by cloning of genetic lesions that affect an RNA that does not include an ORF. For example, the *Xist* gene has no open reading frame, but is required for dosage compensation (X chromosome inactivation) in mammals. Searches are also under way for transcripts from portions of the genome that are not predicted to encode a protein, yet may produce an important product. The possibility exists that eukaryotic genomes encode numerous genes whose functional product is a cryptic RNA.

A large family of such genes came to light only in the early 2000s with the discovery of microRNAs (miRNAs), which are short ~22-nucleotide hair-pin RNAs that bind to 3' UTRs of mRNAs and function in gene regulation either by repressing translation or promoting mRNA degradation. An online miRNA Registry (http://www.sanger.ac.uk/Software/Rfam/mirna) catalogs over 1,000 known miRNAs from numerous organisms as of July 2004, along with their predicted structure and genomic locations. The founding members of this class of gene are *let-4* and *lin-7,* from the nematode *C. elegans*, which were first identified as regulators of developmental timing. There are over 100 known miRNAs in this organism, many of which appear to be conserved in other animals and even in plants.

Structural Features of Genome Sequences

There are numerous structural features of genomes that are of biological interest. These include the distribution of repetitive elements; variation in GC content; simple sequence repeats (SSRs, including microsatellites and minisatellites); degree and arrangement of segmental duplications; and the structure of the centromeres and telomeres. Analysis of such sequences is relevant to the basic principles of genome evolution, as well as to understanding the role of chromatin structure in gene regulation and chromosome replication.

Repetitive sequences.   Interspersed repetitive sequences account for just a few percent of some eukaryotic genomes, such as yeast and *Drosophila*, but constitute more than 50% of the human genome, and over 90% of the genomes of species as diverse as certain crickets, lilies, and amoebae. As a consequence, genome size varies over several orders of magnitude, whereas gene content in multicellular organisms varies over less than one order of magnitude. This **C-value paradox** is now hypothesized to arise as a result of lineage-specific differences in the relative rates of expansion of repetitive sequences and deletion of nearly neutral sequences. A small shift in the balance of these two processes might lead to a rapid expansion or contraction of genome size, but it is not yet known whether or how natural selection shapes the balance. Analysis of the age and frequency distribution of repetitive sequences, as well as the deletion and substitutional mutation rates in pseudogenes in whole genomes, promises to shed light on the mechanisms and causes of the evolution of genome size (Petrov et al. 2000).

At least five classes of repetitive elements can be recognized:

1. Transposon-derived repetitive elements
2. Inactive mRNA-derived copies of cellular genes (pseudogenes)
3. Simple sequence repeats (microsatellites and short tandem repeats/ VNTRs)
4. Segmental duplications of up to 300 kb
5. Blocks of non-interspersed repeats, including ribosomal gene clusters and heterochromatin.

All of these are described in detail in most standard genetics textbooks. Transposable elements (transposons) can be grouped into a large number of families by phylogenetic analysis, but are generally grouped into the four categories shown in Figure 2.17. One category transposes by way of a DNA intermediate, using a transposase enzyme to catalyze excision and reinsertion, and includes the famed *Drosophila* P-elements, bacterial IS and yeast Ty elements, and the widespread mariner family of transposons. The other three categories transpose by way of an RNA intermediate that is converted to DNA by a reverse transcriptase enzyme. These include autonomously transposing LINEs of up to 8 kb; non-autonomous derivative SINES (which use the LINE enzyme to transpose) of just a few hundred bases in length; and retrovirus-like long terminal repeat (LTR) containing elements.

GC content.   Genomes show wide variation in their overall GC content. Factors contributing to this variation are thought to include environmental temperature of microbial niches, levels of methylation (particularly of cytosine, which mutates to thymidine at an elevated rate when methylated), and recent transposon activity (the nucleotide biases of transposons often differ from those of the remainder of the genome).

Over stretches of hundreds of kilobases, GC content should vary by less than 1% as a result of random sampling, but most genomes show a mosaic of nucleotide bias ranging over as much as 30% (Figure 2.18). Boundaries between regions with different GC content are not sharp, but nevertheless may be identified computationally, and some authors divide the genome into **isochores** that may have structural and biological significance (Bernardi 2000). Karyotypic bands revealed by nuclear dyes such as Giemsa tend to correlate with GC content (dark bands being more AT-rich), possibly reflect-

(A)

(B)

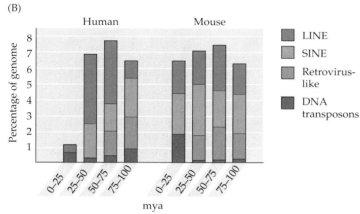

Figure 2.17 Classes of repetitive sequences and their frequency in mammalian genomes. (A) The structure of the four major classes of repetitive elements (LINEs and derivative SINEs, retrovirus-like LTR elements, and DNA transposon families). The average length of each type of sequence is shown at the right. (B) The percentage of the human and mouse genomes each class of element constitutes. Age bins were estimated from the degree of sequence divergence from the consensus for each family of element. The human genome has relatively more LINEs than the mouse, but the rate of transposition slowed around 25 million years ago (mya). (After IHGSC 2001.)

ing their propensity to coil into superstructure, but clearly other features of the DNA contribute to chromatin assembly. CpG dinucleotides are underrepresented in mammalian genomes overall, but cluster as CpG islands between 0.5 and 2 kb in length that are significantly enriched just upstream of genes (Larsen et al. 1992).

Simple sequence repeats. **Microsatellites** are defined as simple sequence repeats (SSRs) with a repeat length of up to 13 bases, whereas longer repeats give rise to **minisatellites.** The most common classes are dinucleotide, trinucleotide, and tetranucleotide repeats, which occur collec-

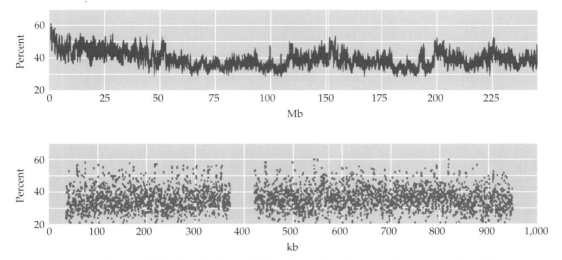

Figure 2.18 Distribution of GC content along human chromosome 1.   GC content varies between 20% and 65% at several different levels of resolution, including for the entire 220 Mb of chromosome 1 averaged over 1-Mb windows (top) and within just 1 Mb for 200-bp windows (bottom). A gap in the IHGSC sequence can be seen at the 400-kb mark on the 1-Mb scale. (After Venter et al. 2001, and IHGSC 2001).

tively at a rate of at least 1 every 10 kb in a wide range of eukaryotic genomes. SSRs may arise by a variety of mechanisms, probably most commonly replication slippage, but transposon-associated microduplications have also been documented, and a fraction of SSRs could be conceived by substitutional mutations as well. Once established, unequal recombination between repeats can generate stepwise changes in repeat number greater than the standard single repeat increase or decrease (Calabrese et al. 2001).

Microsatellites have a much higher mutation rate than standard sequences—up to 0.001 gametes per generation—and have a high probability of back mutation. They are extremely useful in estimating evolutionary relationships between populations within a species, but generally evolve too rapidly to be phylogenetically informative between species (for example, in *Drosophila*, many microsatellites found in one species are not even present in sibling species). Human microsatellites average at least 10 alleles with heterozygosity per locus over 80%, and hence they have become a crucial tool in forensic analysis and paternity testing. The FBI considers a set of 13 SSRs to uniquely describe each American, although issues relating to the inference of individual identification in the courtroom are still under debate. SSRs are also important in human genetic studies both as markers for pedigree analysis and because a family of diseases (including Huntington's disease and some forms of muscular dystrophy and ataxia) can be traced in part to trinucleotide expansion (Table 2.2; Cummings and Zoghbi 2000).

TABLE 2.2 *Microsatellite-Associated Human Diseases*

Disease	Gene	Type	Number of repeats		Location
			Normal	Disease	
Fragile X syndrome	FMR1	CGG	6–53	60–230	5′ UTR
	FMR2	GCC	6–35	60–200	5′ UTR
Myotonic dystrophy	DMPK	CTG	5–37	50–'000s	3′ UTR
Friedreich ataxia	X25	GAA	7–34	34–100	Intron 1
Kennedy disease	AR	CAG	9–36	38–62	Coding
Huntington disease	HD	CAG	6–35	36–120	Coding
Haw River syndrome	DRPLA	CAG	6–35	49–88	Coding
Spinocerebellar ataxia	SCA1	CAG	6–44	39–82	Coding
	SCA2	CAG	15–31	36–83	Coding
	SCA3	CAG	12–40	55–84	Coding
	SCA6	CAG	4–18	21–33	Coding
	SCA7	CAG	4–35	37–300	Coding
	SCA8	CTG	16–37	110–250	Coding
	SCA12	CAG	7–28	66–78	5′ UTR

Source: From Cummings and Zoghbi (2000).

Segmental duplication.   Segmental duplications are remarkably common in vertebrate and plant genomes. Three percent of the human genome, for example, shows a match to another sequence at 90% or more identity over at least a kilobase. Duplications are almost as likely between chromosomes as within them. Some duplications cover more than 50 Mb, and up to one-quarter of the genes in higher eukaryotic genomes may have a paralogous pair. This level of duplication is one of the major surprises to emerge from genome sequences, and the consequent potential for redundancy is just beginning to be explored in relation to homeostasis of developmental and physiological processes, potential for evolutionary divergence, impact on genetic heterogeneity and the etiology of disease, as well as the technical demands it places on gene mapping. In invertebrates, a substantial fraction of genes are tandemly duplicated, and it is well known that this phenomenon constitutes a considerable impediment to functional genetic dissection.

Structure of centromeres and telomeres.   The evolutionary dynamics of centromeric and telomeric regions are markedly different from those of general euchromatin. As a rule, heterochromatin consists of long, highly repetitive stretches of DNA and includes junkyards of transposable elements, interchromosomal duplications, and even insertions of large chunks of mitochondrial DNA (most notable in the *Arabidopsis* genome, which is polymorphic for one such insertion; Lin et al. 1999). For this reason, centromeric chromatin is not typically sequenced in its entirety. The structural properties of centromeres that mediate mitosis and meiosis remain to be characterized in most species, but are yielding to comparative sequence analysis (Hall et al. 2004). It is also clear that heterochromatin contains interspersed unique genes, a striking example being relatively gene-rich regions of otherwise heterochoromatic Y chromosomes in several species.

Telomeric DNA similarly remains to be thoroughly characterized, in part because it is often refractory to traditional cloning, in part because it consists of long tandem repeats. These repeats are actually generated during replication by a telomerase enzyme that uses an RNA template to prime lagging-strand DNA synthesis. Subtelomeric regions in the human genome contain relatively short stretches of interchromosomal duplications that may be a remnant of the process by which new telomeres are created during karyotype evolution.

The ENCODE Project.   As much as 5% of mammalian genome sequence is highly conserved even in marsupials, whereas only at most 2% of a typical mammalian genome encodes transcripts. It is estimated that the human genome contains about 60,000 conserved **non-genic sequences (CNGs),** each generally at least 100 nucleotides in length and over 85% identical across the Mammalia (Dermitzakis et al. 2003). The distribution of nucleotide substitutions in many of these elements is distinct from those typically associated with ncRNAs and regulatory motifs. Observations such as this suggest that some of the so-called "junk" in genomic DNA will eventually be assigned a function.

To facilitate mapping of all of the functional elements of the human genome, the NHGRI has established the ENCODE (ENCyclopedia Of DNA Elements) Project. One percent of the genome has been targeted for intensive computational and experimental study designed to develop and evaluate methods for detecting genes, regulatory regions, replication origins, structural features of the chromatin, and whatever other elements may exist.

Functional Annotation and Gene Family Clusters

One of the first goals of any genome sequencing project is to classify as many genes as possible into putative functional families. Answers to the questions "Are there any genes that are conspicuous by their absence?" and "Which genes are overrepresented relative to other genomes?" fuel hypotheses about gene function that may lead biotechnologists to probe new ways of attacking pathogens or harnessing microbes for useful purposes.

More generally, functional annotation is an essential step toward understanding how genes and gene products interact, using gene expression and proteomic information as discussed in Chapters 4 and 5. As with the structural annotation of genes, however, it is essential to realize that any functional annotation that depends solely on alignments and comparison is incomplete and prone to error. Classical genetic, biochemical, and cell biological methods must all be used in order to dissect the true function of genes.

First-pass classification is achieved by sequence-similarity searches, using software such as BLAST. Often protein similarities can be identified by conceptual translation of similar nucleotide sequences detected with BLASTn, but due to the redundancy of the genetic code, amino acid sequences tend to be more highly conserved than nucleotide sequences. BLASTp can be used to detect amino acid similarities, or BLASTx, which translates a nucleotide sequence into all possible reading frames and scans these against a protein database. The basis for such searches is described in Box 2.2, which includes a description of the E-value ("Expectation") score that is used to evaluate the significance of a match. A common result of BLAST procedures is that between one-third and one-half of all of the predicted proteins do not match a protein for which any functional data are available; hence these genes are classified as "unknown function" or "orphans." This is as true of multicellular eukaryotic genomes (human, fly, worm, weed) as it is of yeast and prokaryotes. Protein structure determination and other proteomic methods (discussed in Chapter 5) will gradually bring this number down.

EXERCISE 2.3 *Perform a BLAST search*

Using the gene that you identified in Exercise 1.1, perform a BLASTn search for homologs in other species. What conclusions can you reach regarding whether this gene is part of a gene family, or regarding its phylogenetic distribution?

ANSWER: Continuing our example of IL13 from Chapter 1, copy the transcript sequence either from the Ensembl or GENBANK page. (For Ensembl, click on the "Transcript Structure" box, and the sequence appears; for LocusLink, click on the "RefSeq" link, and then go to the mRNA annotation NM 002188). Return to the NCBI home page, click on "BLAST" and then bring up the "Nucleotide-nucleotide BLAST (blastn)" page and paste your sequence into the search box. Using the default parameters, perform a search against the nr sequence database. Depending on the number of queries in the queue, this could take a minute or two. In this case, over 150 hits are putative matches to the 1250 nt search, at least 10 of which are different representations of the same human IL13 sequence in the database. Very high match scores are also seen for several primate IL13 loci, and weaker matches over smaller extents of the sequence are seen for other mammals. The closest match in *Drosophila* has an E-value of just 0.17, due to a 23-bp perfect match element that is probably not indicative of a homologous gene. Consequently, IL13 appears to be unique to vertebrates (and possibly even to mammals, since no fish or bird sequences have significant E-values), and to be present in a single copy per genome.

It is believed there is a finite number of structural protein domains in the combined proteome of all organisms, and that once these domains have been identified, it ought in principle to be possible to place all proteins into a plausible cluster. Many genes evolve at such a fast rate that alignment based on sequence similarity alone is unlikely to be successful, and in these cases clustering will depend on structural predictions.

Clustering of Genes by Sequence Similarity

Another common result obtained with protein BLAST searches is that each query sequence matches multiple proteins from one or more species. This can happen either because one domain in the query is present in a family of proteins, or because multiple domains match different proteins. These possibilities are readily distinguished because the alignment software reports a series of keywords associated with each positive hit and indicates the location of the stretch of similarity, as seen in Figure 2.19. When comparing closely related species, a query sequence containing multiple domains will typically identify a protein or proteins with the same domain structure.

However, over a period of divergence of tens of millions of years, domains shuffle, and over hundreds of millions of years different domains are found clustered in different combinations. There is some biological logic to domain clustering in that multiple domain matches tend to classify the gene product in the same broad category, such as transcription factor or receptor. That is, two different DNA-binding domains may be combined on the same protein, or a transmembrane region may be linked to a range of different intracellular and extracellular domains.

The output of a protein BLAST search is thus not so much a match to another gene as a match to one or more protein domains. Numerous databases have been established that classify protein domains according to criteria agreed upon by groups of experts. One of the first to emerge was the Enzyme Commission (EC) hierarchical classification of enzymes, in which each enzyme is assigned a number that reflects the subclassification of function. For example, EC1.1.1.1 is alcohol dehydrogenase. Similar classification schemes for nonenzymatic proteins are not so obvious, since most of these proteins perform functions whose biochemical or biological consequences are dependent on the context in which the protein is used. Nevertheless, a general scheme devised for the classification of bacterial proteins has attained wide acceptance.

Structural biologists have assembled protein databases such as PFAM that allow researchers to immediately access data concerning biochemical properties of a set of predicted proteins. The InterPro classification system goes a step further in classifying individual protein domains. Gene annotations now typically link directly to InterPro classifications. Table 2.3 lists the ten most common InterPro domains in the human genome and their frequencies in other multicellular eukaryotic genomes.

It follows that it is important to draw a distinction between protein function prediction and classification of genes as members of the same family. The latter involves distinguishing between paralogs and orthologs, and is addressed in the next section. The former allows us to make some very gen-

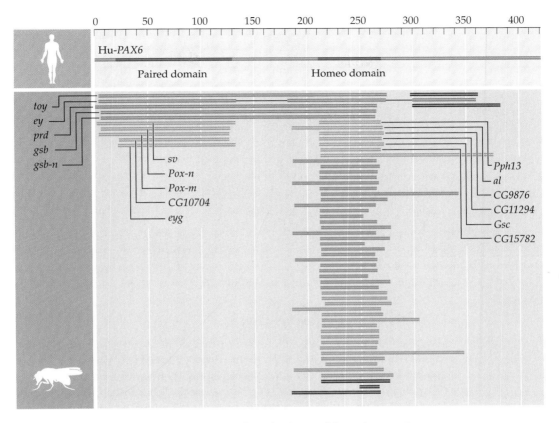

Figure 2.19 Alignment of human *PAX6* against the *Drosophila melanogaster* genome. The basic protein BLAST search identifies 103 proteins in the fly proteome, with varying degrees of similarity to two portions—the Paired domain and the Homeo domain—of human *PAX6*. Sixty of the matches are shown graphically; mouse-over provides the identity of each match as shown for the 16 closest matches. Note that *toy, ey, prd, gsb,* and *gsb-n* match to both domains.

eral conclusions about the content of genomes based on the association between protein domains and function.

The major classes of protein molecular function are enzyme, signal transduction (including receptors and kinases), nucleic acid binding (including transcription factors and nucleic acid enzymes), structural (including cytoskeletal, extracellular matrix, and motor proteins), and channel (voltage and chemically gated). Several other important categories include immunoglobulins, calcium-binding proteins, and transporters. Subclasses within each group vary widely in frequency among genomes, as do the absolute numbers and relative proportions of each major category. For example, the human genome shows a marked increase by comparison with the fly and worm in proteins involved in neural development, signaling during development, hemostasis, and apoptosis.

TABLE 2.3 *Number of InterPro Protein Domains in Some Eukaryotic Genomes*

Domain type	InterPro ID	Number of domains				
		Human	Fly	Worm	Yeast	Weed
Immunoglobulin	IPR003006	765	140	64	0	0
C₂H₂ zinc finger	IPR000822	706	357	151	48	115
Protein kinase	IPR000719	575	319	437	121	1,049
Rhodopsin-like GPCR	IPR000276	569	97	358	0	16
P-loop motif	IPR001687	433	198	183	97	331
Reverse transcriptase	IPR000477	350	10	50	6	80
Rrm domain	IPR000504	300	157	96	54	255
G-protein WD-40 repeat	IPR001680	277	162	102	91	210
Ankyrin repeat	IPR002110	276	105	107	19	120
Homeodomain	IPR001356	267	148	109	9	118

Source: From IHGSC 2001, Table 25.

Clusters of Orthologous Genes

Sequence alignment serves the useful purpose of identifying which genes to include in a phylogenetic analysis. Prior to genome sequencing, many genes were cloned by hybridization to cloned genes from another species, or by degenerate PCR. These methods were not guaranteed to identify all the potential relatives of a particular gene, and in fact a typical procedure was to focus on the one or two clones that either gave the strongest hybridization signal or happened to be identified first.

The postgenomic equivalent of this bias is to simply assume that your gene is the same as the one that shows the closest sequence match in another genome. If similar genes from a number of different species all have the closest match, all the better. The problem arises when the query gene matches multiple members of a family of genes that have discrete functions. In this case, the closest match may well be very misleading in terms of function.

Consequently, classification of molecular function starts with the identification of as large a set of potential family members as possible. The PSI-BLAST (Altschul et al. 1997) and PHI-BLAST (Zhang et al. 1998) algorithms and related procedures have been developed for this purpose. The basic idea of PSI-BLAST is to align the sequences obtained in an initial protein database search and use this to construct a profile (see Box 5.1), which is then used to initiate a fresh search. The process is iterated until no further matches are identified. A true family of genes ought to be bounded by a significance cut-off, so that there is a limit as to which proteins will be included in a family, as long as realistic parameters are chosen. The iterative procedure has the effect of reducing the degree of sequence similarity required for a significant match.

A slightly different approach to protein classification has been to assemble **clusters of orthologous genes,** or **COGs** (Tatusov et al. 2001). COGs are created by identifying the best hit for each gene in complete pairwise comparisons of a set of genomes. Comparison of the 185,500 proteins encoded by the genomes of 66 microbes led to the identification of 4,873 COGs as of September 2003. These represent about 75% of all predicted microbial proteins and can be accessed at http://www.ncbi.nlm.nih.gov/COG. Similarly, just over 50% of 110,000 eukaryotic proteins from fly, nematode, human, *Arabidopsis*, two yeasts, and a microsporidian, assemble into 4,852 eukaryotic KOGs (Tatusov et al. 2003), 20% of which form a conserved core of proteins.

As shown in Figure 2.20, triangles are drawn linking genes that have the same best hit in at least one direction (for example, the genes that are the best hit in both species B and C for a gene in species A are also the best hit for at least one of the comparisons of B against C). These triangles are then merged together to assemble a cluster of putative orthologs.

COGs will typically consist of both orthologs and paralogs. A **paralog** is a duplicate copy of a gene that arose subsequent to the split between the two lineages that are being compared. An **ortholog** is a gene in another lineage that is derived from the same ancestral gene that was present prior to the lineage split (Mindell and Meyer 2001). Thus, using Figure 2.21A as a guide, if there are two copies of a human gene *HuA*, namely *HuA* and *HuA'*, these two copies are paralogs, and they are both orthologs of a mouse gene *MmA*. Distinguishing which copy has retained the ancestral function is not necessarily possible simply by comparing the levels of sequence similarity. To see why, suppose (trivially) that *HuA* and *MmA* differ at five amino acids, none of which affect the function of the protein, while *HuA'* and *MmA* differ at just four amino acids, one of which changes a critical residue involved in substrate binding. In this case, clustering by similarity would erroneously identify *HuA'* and *MmA* as the most likely to be functionally equivalent. The situation is additionally complicated by the fact that such a high proportion of genes have duplicates that arose prior to the divergence of the two species being compared.

It cannot be emphasized strongly enough that COGs provide at best only a hint as to gene function. Errors are increasingly likely to arise at more and more refined levels of a clustering hierarchy. This is because, particularly in animals and plants, genes tend to duplicate and diverge in function in lineage-specific manners, but clustering by overall sequence divergence does not reflect key events involving single amino acid changes that shift the functional specificity of enzymes, kinases, and transcription factors. Phylogenetic methods are considered to provide a much more reliable indicator of functional subclassification.

Phylogenetic Classification of Genes

Molecular evolutionists have developed sophisticated tools for the assembly of gene phylogenies, as described in Box 2.4. The methods are the sub-

BOX 2.4 Phylogenetics

A concept that has arisen several times in this discussion is that of homology, or *shared descent from a common ancestral sequence*. There are many reasons why we may want to use an aligned set of DNA or amino acid sequences to propose an **evolutionary tree** or **phylogeny** that displays the hierarchical ancestral relationships among the sequences. We may simply wish to understand the pattern of relatedness of a group of species. There are also less obvious objectives. Many proteins are members of multigene families, and we may want to classify a newly discovered gene into the appropriate subfamily. We may want to test predictions about the phylogenetic placement of one or more sequences, perhaps inferring the source of a viral infection. In any case, there is a large and growing literature on the inference of phylogenies using sequence data. We will provide only a brief overview of some of the more common methods. For further details, see the excellent presentation by Swofford et al. (1996) or the handbook by Hall (2001).

General Principles

Before describing the leading phylogenetic reconstruction methods, a few technical issues must be introduced. First is the distinction between rooted and unrooted tree topologies (Figure A), which is important because most phylogenetic methods can only produce unrooted trees. Additional information regarding evolutionary rates or the most ancient relationships is needed to root the inferred trees.

A second important concept is that of branch length, which in molecular terms is defined as the average number of nucleotide substitutions per site. For example, along a branch of length 0.2, the average nucleotide site has undergone 0.2 changes. Clearly, that number is not achievable in nature; it represents the average value over sites, most of which will have changed either 0 or 1 times.

The primary methods of phylogeny reconstruction are **parsimony, distance,** and **likelihood.** There are many variants within each of the three broad classifications. The shared thread among all of the methods is an attempt to *identify the topology that is most congruent with the observed data*.

The methods differ in their mechanisms for measuring this congruence. Some methods (parsimony, likelihood, and some distance methods) define a metric between topology and data and require (in principle) an exhaustive search through all possible tree topologies. Unfortunately, the number of topologies increases explosively. The number of unrooted tree topologies for n sequences is $3 \times 5 \times 7 \times \ldots \times (2n - 5)$—a value that exceeds 2 million when n reaches 10. In practice, heuristic search methods are used to search through a likely subset of the vast collection of possible topologies. The commercially available software PAUP* (Swofford 2002; http://paup.csit.fsu.edu) incorporates many of the algorithms and is widely used to construct phylogenies.

Parsimony Methods

Parsimony methods are probably the most widely used and most intuitive methods of phylogenetic inference. The

FIGURE A Rooted and unrooted trees. For any four taxa (the tips of the branches) there are three distinct unrooted trees. Each unrooted tree can be rooted on any of its five branches; two of the five possible rooted trees for the center unrooted tree are shown.

underlying idea is a simple one: The best topology is the one requiring the smallest number of changes to explain the observed data. In Figure B we see a short alignment of four DNA sequences, along with the three possible unrooted topologies for the four sequences. Consider position 1 in the alignment. If the true topology were topology 3, a bit of thought reveals that at least two nucleotide substitutions (for example, one on the branch leading to sequence A, and a second on the branch leading to sequence B) would be required. At position 2, the data can be explained with a single C→T (or T→C) change on the interior branch of topology 3.

Likelihood Methods

In its mechanics, likelihood reconstruction of phylogeny is similar to parsimony. A search through tree topologies is necessary, with the congruency measure being the maximum probability of the data given the topology. The **maximum likelihood estimate** of the true tree is the topology providing the highest such probability. In order to compute the probability, one must first select a probabilistic model describing the change of sequences over time. Many factors might be included in such a model, including base frequency biases and unequal probabilities for different types of nu-

cleotide substitutions. This feature should be contrasted with most parsimony methods, where all changes are treated equivalently, regardless of the type of substitution or the amount of time on a given branch. Likelihood methods tend to be very powerful tools for phylogeny reconstruction, but are typically limited to relatively small data sets (say, less than 50 sequences) because of their computational expense.

Distance Methods

Distance methods are the oldest family of phylogenetic reconstruction methods. As do likelihood methods, they rely on a probabilistic model of sequence evolution. However, distance methods use the model only to calculate pairwise *evolutionary distances* for all pairs of sequences—a task that is much faster than the likelihood calculations. Given a matrix of pairwise evolutionary distances, there are many distance methods to choose from. While many require topology searching, some of the more popular ones do not, relying instead on a predetermined algorithm for obtaining the estimated phylogeny.

One of the most effective distance methods is the **neighbor joining (NJ)** method, which is very fast and tends to have a high probability of reconstructing the correct tree (when used with a proper choice of pairwise distance).

Evaluating Reconstructed Trees

As with any statistical inference, an estimate without some notion of its reliability is essentially useless. For the phylogeny problem, we would like a measure of data support for particular groupings, or **clades,** in an estimated tree. The most common approach for measuring support is through the use of **bootstrapping,** as introduced by Felsenstein (1985). Numerical resampling techniques are used to compute bootstrap support levels for every node in the tree topology. Bootstrap values near 100% indicate clades that are strongly supported by the data, while lower levels indicate reduced support. What the precise bootstrap level to indicate a statistically significant clade should be is the source of tremendous controversy, but values greater than 70–80% are often taken to indicate fairly strong support for the clade.

FIGURE B Maximum parsimony. The table within the figure shows the minimum number of changes at each site required by each topology, along with the total number of changes obtained by summing across sites. Applying the maximum parsimony principle to these data, tree 3 would be our best estimate of the "true" phylogeny because it requires the smallest number of nucleotide changes (five) to explain the aligned data.

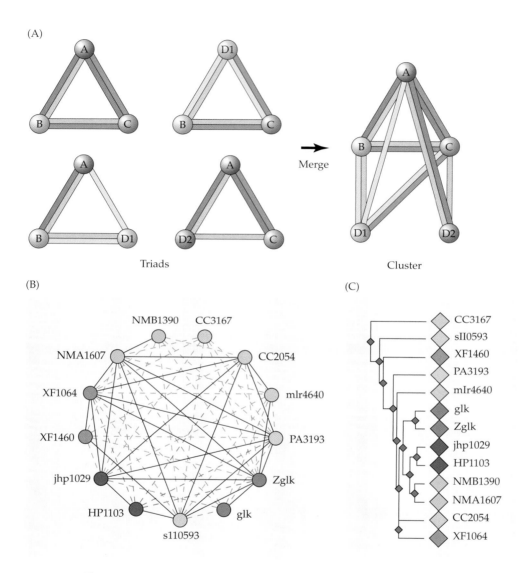

(A)

Triads

Merge

Cluster

(B)

(C)

Figure 2.20 Clusters of orthologous genes. COGs are assembled by merging triads of genes from different species that are the best match to one another. (A) Each triad shows one gene from each of four species (genes A, B, C, and D) linked to the gene that is the best match in the other species. Thus genes A, B, and C are the best matches to one another in the respective species' genomes, and although A is the best match for both D1 and D2, D1 is not the best match for A, so these genes are joined by a single line. (B) COG0837, identified as glucokinase, as viewed at the NCBI Web site. In the merger of the triads on the left, dashed lines indicate examples where the best match occurs only in one direction; solid lines imply that each gene is the best match in the genome of the other. The tree at the right (C) shows a more standard representation of the relationship among the genes.

(A)

(B)

(C)

Figure 2.21 Orthologs and paralogs. When any three sequences are contrasted, usually two will cluster together. (A) In this example, the two human sequences differ from the mouse sequence at three common sites, and *HuA* has two further changes compared to a single unique change in *HuA'*. However, the difference in charge introduced by changing an aspartate [D] to an asparagine [N] may confer a different function to the *HuA'* gene, highlighting the fact that degree of sequence similarity does not always indicate which proteins are most similar functionally. (B) Duplicate genes may be identified as pairs of orthologs if the sequences between species (*HuA* and *MmA*; *HuA'* and *MmA'*) are more related than within species. (C) If *HuA* and *HuA'* are more similar to one another than to *MmA* and *MmA'*, they are likely to be recently duplicated paralogs.

ject of numerous complete textbooks (e.g., Graur and Li 2000; Hall 2001). Different procedures for phylogenetic analysis can yield very different answers, but procedures exist for attempting to ascertain the likelihood that a particular phylogenetic hypothesis is correct. Whichever method is used, one of the goals of phylogenetic analysis is to ascertain how groups of similar genes isolated from a set of species are related by descent. In the preceding example, assume now that the mouse gene is also duplicated. Then if *HuA* and *MmA* are found to cluster in a phylogenetic analysis, as do *HuA'* and *MmA'*, then it is reasonable to infer that genes *A* and *A'* duplicated and diverged in function prior to the divergence of the mouse and human lineages (Figure 2.21B). However, if *HuA* and *HuA'* cluster more closely together, as do *MmA* and *MmA'* (Figure 2.21C), then it is likely that two independent duplications have occurred, and hence neither pair of genes are orthologs and functional assignments should be treated with more caution.

The above caveats notwithstanding, the most likely function of unknown genes can be inferred by phylogenetic analysis so long as the functions of a subset of similar genes or proteins are known, as shown in Figure 2.22. Application of these methods to large families of genes in multiple taxa is sometimes referred to as phylogenomics (Eisen 1998).

EXERCISE 2.4 *A simple phylogenetic analysis*

A few years ago, a pressing question in the area of human evolution was the exact phylogenetic relationship between humans, chimpanzees, and gorillas. Fossil, morphological, and early molecular data provided conflicting and inconclusive results as to which pair of organisms were most closely related. Suppose the following four short DNA sequences are available, including a sequence from the outgroup organism, orangutan:

Human	CGAAATGCAT
Chimpanzee	CGGACTGTAT
Gorilla	CAAGCTGTAC
Orangutan	TAAGACGTAG

Carry out a phylogenetic analysis to infer the rooted evolutionary tree of human, chimpanzee, and gorilla.

ANSWER: There is only one possible unrooted tree for three species, so we cannot answer the question using only sequences from human, chimpanzee, and gorilla. By adding the outgroup sequence, we can infer the unrooted tree for the four sequences and then root the tree on the branch leading to orangutan. The result of this rooting will also provide a root for the human-chimpanzee-gorilla subtree.

Although distance or likelihood methods would also be acceptable, we will use the parsimony method as discussed in Box 2.4. The three possible unrooted trees for the four species are:

Following the method in Box 2.4, the required number of changes for each tree are:

	1234567890	Total
Tree A	1111210102	10
Tree B	1212210102	12
Tree C	1212110102	11

Thus, Tree A is chosen as the best estimate. Since we know that the orangutan is the outgroup, we can place the root of the tree on the branch leading to the orangutan, as follows:

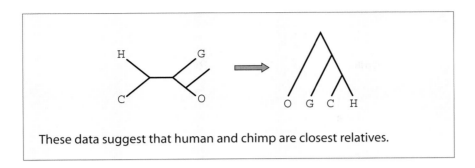

These data suggest that human and chimp are closest relatives.

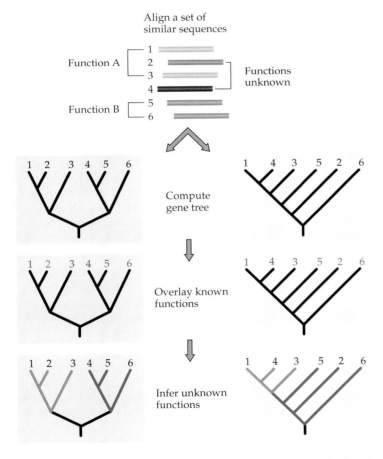

Figure 2.22 Inferring gene function from phylogenetic analysis.   In the example on the left, an ancient duplication event is inferred to have resulted in the evolution of two types of function, and the unknown function of protein 2 is most likely the same as that of proteins 1 and 3. By contrast, in the example on the right, no deep branch separating two clades is observed, but the function of the unknown protein 2 is most likely the same as that of the ancestral function (protein 6), since the more recently derived protein 5 retains the same function. (After Eisen 1998.)

Gene Ontology

It has become clear that annotation on the basis of molecular function alone is insufficient to describe or predict biological function. Three examples illustrate the general point. Perhaps the best known example of the evolution of a novel function is the reuse of enzymes such as lactate dehydrogenase as lens crystallins, which has occurred multiple times in vertebrate lineages. In these cases, an enzyme is converted into a structural protein. Just as dramatically, the *Drosophila* ortholog of the mammalian dioxin receptor is encoded by the *spineless-aristapedia* gene, which is one of the key regulatory genes that control antenna differentiation. The only clear functional similarity here is the involvement of the protein product in the olfactory system. In both of these cases, knowledge from one species is not directly transferable to another.

Less dramatic, but no less important, are examples where the *molecular* function has remained the same but the *physiological* function has evolved. In some cases, apparently divergent functions (such as those of the *Hox* genes in patterning both the cranial nerves of vertebrates and the appendages of invertebrates) can be traced to a shared common function in patterning along the body axis. In others, one or more biological functions are clearly derivative, such as the use of the *Toll-dorsal* pathway in the embryonic patterning of flies as opposed to innate immunity in both vertebrates and invertebrates. In still other cases, functions performed by a single gene in one taxon are split between two genes in another.

Despite these reservations, there is considerable conservation of gene function across the metazoa, much of which extends to the fungi and plants. Just over 1,500 genes have unambiguous orthologs among the fly, worm, and human genomes—meaning that there is a single closest match in each genome. Although plants have novel families of transcription factors and seem to use different mechanisms for intercellular communication, most of their core physiological functions are the same as those of animals. Consequently, while developmental roles may diverge, it is well accepted that knowing the cell biological and molecular functions and/or the subcellular localization of each protein is a useful aspect of gene annotation.

To this end, the Gene Ontology (GO) consortium (Ashburner et al. 2000) has established a project to annotate all these features for the complete set of genes identified by genome projects. Several thousand genes from each of the major model organisms are already annotated in GO, which is accessible on the Web at http://www.geneontology.org, or through links on the main genome project pages.

The GO site classifies protein function in three major categories, as shown in Figure 2.23 and described in more detail in Box 2.5. *Cell biological function* refers to the nature of the process that is regulated or affected, such as cell growth and division, respiration, signal transduction, and cAMP biosynthesis. *Molecular function* corresponds to the biochemical activities discussed in a previous section, such as enzyme, nucleic acid binding, DNA helicase, or tyrosine kinase. *Cellular component* refers to the place in a cell where the gene product is active, such as the cell surface, Golgi apparatus, or spliceo-

(A)

Biological process

▶ Behavior
▶ Cell communication
▶ Cell growth and maintenance
▶ Death
▶ Developmental processes
▶ Perception of external stimulus
▶ Physiological processes
▶ Viral life cycle

Molecular function

▶ Antitoxin
▶ Anticoagulant
▶ Antioxidant
▶ Apoptosis regulator
▶ Cell cycle regulator
▶ Cytoskeletal regulator
▶ Defense/immunity protein
▶ Vitamin transporter
▶ and many more

Cellular component

▶ Cell fraction
▶ Cell wall
▶ Extracellular
▶ Intracellular
▶ Membrane
▶ Unlocalized

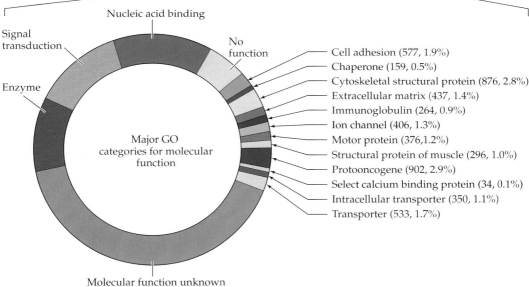

Nucleic acid binding

Signal transduction

Enzyme

No function

Major GO categories for molecular function

Cell adhesion (577, 1.9%)
Chaperone (159, 0.5%)
Cytoskeletal structural protein (876, 2.8%)
Extracellular matrix (437, 1.4%)
Immunoglobulin (264, 0.9%)
Ion channel (406, 1.3%)
Motor protein (376,1.2%)
Structural protein of muscle (296, 1.0%)
Protooncogene (902, 2.9%)
Select calcium binding protein (34, 0.1%)
Intracellular transporter (350, 1.1%)
Transporter (533, 1.7%)

Molecular function unknown

(B)

Evidence codes

▶ IMP, inferred from mutant phenotype
▶ IGI, inferred from genetic interaction
▶ IPI, inferred from physical interaction
▶ ISS, inferred from sequence/structural similarity
▶ IDA, inferred from direct assay
▶ IEP, inferred from expression pattern
▶ IEA, inferred from electronic annotation
▶ TAS, traceable author statement
▶ NAS, non-traceable author statement
▶ NR, not recorded

Figure 2.23 Gene Ontology categories. The Gene Ontology (GO) Web site categorizes genes by their probable function. (A) The three main functional rubrics are shown, with a partial list of the specific functions within each category. The graph shows the number and proportion of genes in each GO molecular function category in the human genome. (Graph after Venter et al. 2001, Figure 15.) (B) The evidence codes of the GO site indicate how functional information was obtained.

BOX 2.5 Gene Ontologies

In the not-so-distant past, a *Drosophila* geneticist might have queried a sequence database using the partial sequence of a gene named *mip*, and found very strong matches to the genes GAP, gap1, rI533, sextra, CG6721, and Ras-GAP. Upon brief inspection of the gene sequences, she would have reached the puzzling conclusion that all of these genes were identical in sequence. Might this have been an interesting example of sequence conservation? No, it was merely an example of the same gene being named multiple times in different locations in the literature. To make matters more confusing, once she solved the nomenclature issue she would most likely have found a bewildering array of functions assigned to this gene by various researchers. This state of affairs made it all but impossible to carry out automated annotation of genes, and it was a major obstacle to creating unified databases that span multiple organisms and use varying vocabularies.

The Gene Ontology™(GO) consortium (www.geneontology.org) is the most well-known attempt to provide a standard vocabulary for describing the functions of genes. GO assigns a single name to each gene product, keeps track of known synonyms to that name, and provides a hierarchical annotation of the gene's functions. In the GO system, a gene product is considered to have one or more molecular functions, be involved in at least one biological process, and be associated with any number of cellular components. For each of these three major aspects, the GO team is developing a controlled vocabulary for use in gene annotation. For instance, the *sextra* gene is found to be a synonym of the Gap1 gene product, which is involved in the biological processes "mitosis" and "sister chromatid cohesion," among others. Gap1/sextra is associated with the cellular component "plasma membrane," and has the molecular function "Ras GTPase activator activity." Each of these terms is itself a member of a hierarchical system. "Plasma

membrane" is found to be in the class "membrane," which in turn is a member of the class "cell." If the cellular component were localized even further or limited to structures of specific organisms, finer categories would have been provided (e.g., "apical plasma membrane" or "caveolar membrane").

The figure illustrates the structure of the GO classification scheme for an *Arabidopsis* gene named *INNER NO OUTER*. Notice that the central hub corresponds to a single gene product. Splitting from this hub are three separate branches, which correspond to the three GO ontologies: biological process, cellular component, and molecular function.

The use of GO is rapidly becoming a universal standard for gene annotation. It eliminates confusion arising from nonstandard nomenclature, and enables databases to interact more effectively. Consequently, for example, if a student discovers that a previously uncharacterized gene in her organism is homologous to an annotated gene in another species, the GO categories allow her to formulate hypotheses about what the function of the gene may be. Her collaborator might then initiate a bioinformatic search for potential interacting partners, starting by scanning through the lists of genes in other organisms with similar gene ontologies. Both the hierachical structure of GO and the uniformity of the vocabulary facilitate rapid hypothesis formulation.

The GO project arose in 1998 as a solution to barriers that inhibited communication between FlyBase (*Drosophila*), the Mouse Genome Database, and the *Saccharomyces* database (Ashburner et al. 2000). Since that time, many other groups have joined the consortium, including the databases at TIGR, Gramene (Monocots), WormBase (*C. elegans*), and TAIR (*Arabidopsis*). NCBI's Entrez Gene tool allows users to query for GO terms. GO consortium members and GO users have developed tools to encourage the widespread

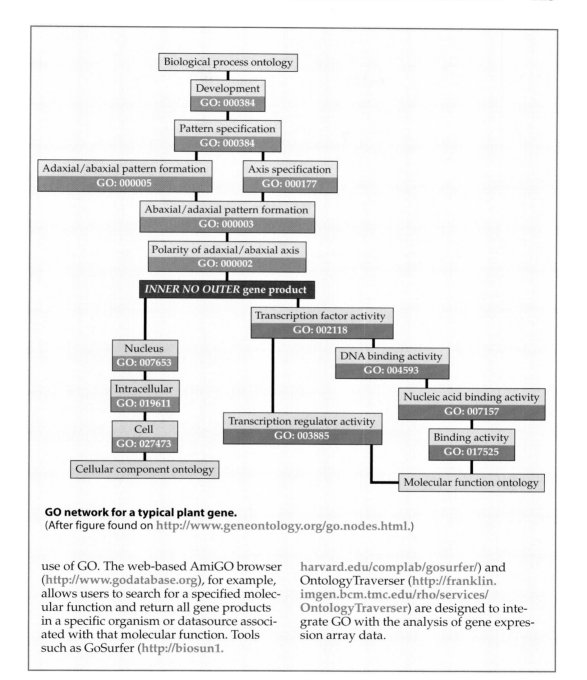

GO network for a typical plant gene.
(After figure found on http://www.geneontology.org/go.nodes.html.)

use of GO. The web-based AmiGO browser (http://www.godatabase.org), for example, allows users to search for a specified molecular function and return all gene products in a specific organism or datasource associated with that molecular function. Tools such as GoSurfer (http://biosun1. harvard.edu/complab/gosurfer/) and OntologyTraverser (http://franklin. imgen.bcm.tmc.edu/rho/services/ OntologyTraverser) are designed to integrate GO with the analysis of gene expression array data.

some. Each of these terms is arranged hierarchically, from general to specific function. The organization of the GO site is dynamic in the sense that attributes are continually updated as new information becomes available, and in the sense that the structure allows the researcher to scroll through biological associations to identify likely genetic and biochemical partners. The current organization assumes function in an idealized general eukaryotic cell type, so does not attempt to convey information as to which tissue or cell type is affected, nor does it capture the pleiotropy of gene function. However, such information is typically linked to online literature resources.

In GO, there are over 7,000 terms describing molecular function and 5,000 describing biological process, with some annotations having as many as 12 levels within a hierarchy of terms. This is too deep for efficient computational searches and development of statistical models of protein families. Consequently, an allied PANTHER database has been developed that has a simplified ontology schema built on HMMs and manual curation of multiple sequence alignments (http://panther.appliedbiosystems.com). This database can be used to analyze the conservation of amino acid sequence segments in subfamilies of proteins, to rank the likelihood that individual replacements affect protein function, and to annotate predicted proteins from novel genome sequences.

Other modes of functional classification are conceivable. In reality, the Web resources for each organism portray different attributes in their annotation. WormBase, for example, allows a nematologist to pull up images of which particular cells express a gene product, while FlyBase offers comprehensive links to allelic information and genetic interactions, among other data. TIGR provides gene indices online for numerous animals, plants, protists and fungi, using their own algorithm for defining protein families, at http://www.tigr.org/tdb/tgi/. The essential focus of emerging annotation methods is in linking gene expression and regulatory information to the static properties that can be inferred from a gene sequence. Dynamic modeling of gene function will depend on successfully integrating as much knowledge of genes as databases can reasonably and sensibly portray.

Summary

1. Almost all genome sequencing is currently performed using automated fluorescence-based dideoxy chain-termination (Sanger) sequencing on machines that can generate in excess of a megabase of raw sequence per day.

2. Sequence traces are converted to raw sequences using algorithms that assign confidence scores to each base according to the spacing, height, and quality of the peaks that represent each nucleotide. Phred quality scores are then used in the automated generation of sequence assemblies, or contigs, based on alignment of overlapping sequence reads.

3. Two main strategies are used to assemble whole genome sequences: hierarchical sequencing and shotgun sequencing. The former relies on prior ordering of clones to build the genome sequence in megabase-

sized pieces, while the latter uses computational power to assemble the entire sequence simultaneously.

4. Repetitive sequences are a major obstacle to sequence assembly, but can be overcome by masking the repeats from initial alignment and using mate pairs to provide information on spacing and orientation of sequences that are interrupted by repeats.

5. First-draft genome sequences assembled from between 5- and 10-fold redundant sequences have an error rate greater than 1 in 10,000 bases and short sequence gaps every few hundred kilobases.

6. "Finishing" refers to the process of filling the gaps, improving the accuracy, and resolving ambiguities. It is labor-intensive and can be more time-consuming than preparation of the draft.

7. Sequence alignment uses an algorithm that imposes penalties for mismatches and indels and allows the most likely alignment to be inferred from the scores of each possible alignment. This is a computationally intensive process, so tools such as BLAST and FASTA use a heuristic method to focus searches only on the subset of plausible matches.

8. Annotation of genes using software that searches for open reading frames, splice sites, and promoters using hidden Markov models (HMMs) is efficient overall, but can miss genes and often fails to identify intron-exon boundaries with accuracy.

9. Estimates of the number of genes in a genome based on the number of distinct ESTs in a set of libraries tend to greatly overstate the true number of genes, due to alternative splicing of complex transcripts and to errors that occur during cDNA cloning. Alignment of EST with genomic sequence can collapse families of ESTs into single genes.

10. Conservative annotation of gene locations utilizes two or more pieces of information, generally including de novo prediction, cDNA evidence, and similarity to sequences in other genomes. These approaches are likely to underestimate the true gene content, since many transcripts are tissue-specific or novel.

11. Comparative analysis of genome sequences is expected to assist in the identification of conserved regulatory sequences required for control of transcription and possibly of functions of the chromatin.

12. Numerous structural features of genomes such as the distribution of repetitive DNA, sequence content of centromeres and telomeres, and variation in GC content, are of biological interest in addition to the identification of genes.

13. COGs are clusters of orthologous genes identified by connecting best-hit matches of each gene in the complete genomes of other organisms.

They provide a draft picture of the distribution of gene families in each completed genome.

14. Phylogenetic methods should be used to investigate the likely functional relationships among a group of aligned gene sequences. Orthologs are sequences in two lineages that are derived from a common ancestor, while paralogs are duplicate sequences in the same lineage.

15. The Gene Ontology and PANTHER projects seek to categorize genes not only according to predicted function, but also according to the location of the gene product in a cell, and the biological process or processes that it regulates.

16. There are numerous ways of grouping and categorizing proteins, so it is generally advisable to contrast the predictions made by several methods and to integrate predictions made directly from sequence comparisons with whatever molecular biological information is available in the species of interest. The World Wide Web facilitates online comparisons across diverse species.

Discussion Questions

1. By what criteria can we judge the accuracy of a draft (or complete) genome sequence? If 50% of the base pairs are apart by more than 100 kb in two different drafts, what areas of genome research are most likely to be adversely affected?

2. Should the fact that both of the draft human genome sequences were derived predominantly from individual males be cause for concern?

3. Do you think that the gene counts derived from first-pass annotation of complete genome sequences are more likely to be underestimates or overestimates of the true number of genes? What are the criteria used for gene discovery?

4. Different tools for predicting gene function often give very different predictions. How useful are bioinformatic methods for functional annotation, and what can be done to improve their accuracy?

5. What are some of the applications for emerging DNA sequencing methods such as single molecule sequencing and sequencing by hybridization?

Web Site Exercises

The Web site linked to this book at http://www.sinauer.com/genomics **provides exercises in various techniques described in this chapter.**

1. Edit two sets of overlapping sequence trace files with phred, then merge them into a contig using phrap and consed.

2. Perform a BLAST search with the consensus sequence derived from Exercise 1.

3. Use Clustal to align the sequences obtained in Exercise 2.

4. Run GenScan on a scaffold sequence containing the full-length gene.

5. Perform a phylogenetic analysis of the aligned sequences in Exercise 3.

Literature Cited

Altschul, S., W. Gish, W. Miller, E. Myers and D. J. Lipman. 1990. Basic local alignment search tool. *J. Mol. Biol.* 215: 403–410.

Altschul, S., T. Madden, A. Schaffer, J. Zhang, Z. Zhang, W. Miller and D. Lipman. 1997. Gapped BLAST and Psi-BLAST: a new generation of protein database search programs. *Nucl. Acids Res.* 25: 3389–3402.

Ashburner, M. et al. 2000. Gene ontology: Tool for the unification of biology. The Gene Ontology Consortium. *Nat. Genetics.* 25: 25–29.

Bernardi, G. 2000. Isochores and the evolutionary genomics of vertebrates. *Gene* 241: 3–17.

Blanchette, M. and M. Tompa. 2002. Discovery of regulatory elements by a computational method for phylogenetic footprinting. *Genome Res.* 12: 739–748.

Boffelli, D., J. McAuliffe, D. Ovcharenko, K. Lewis, I. Ovcharenko, L. Pachter and E. M. Rubin. 2003. Phylogenetic shadowing of primate sequences to find functional regions of the human genome. *Science* 299: 1391–1394.

Boguski, M., T. Lowe and C. Tolstoshev. 1993. dbEST: Database for "expressed sequence tags." *Nat. Genet.* 4: 332–333.

Burge, C. and S. Karlin. 1997. Prediction of complete gene structures in human genomic DNA. *J. Mol. Biol.* 268: 78–94.

Calabrese, P. P., R. T. Durrett and C. F. Aquadro. 2001. Dynamics of microsatellite divergence under stepwise mutation and proportional slippage/point mutation models. *Genetics* 159: 839–852.

Carninci, P. et al. 2000 Normalization and subtraction of cap-trapper-selected cDNAs to prepare full-length cDNA libraries for rapid discovery of new genes. *Genome Res.* 10: 1617–1630.

Cummings C. and H. Zoghbi. 2000. Trinucleotide repeats: Mechanisms and pathophysiology. *Annu. Rev. Genomics. Hum. Genet.* 1: 281–328.

Dayhoff, M. O. 1972. *Atlas of Protein Sequence and Structure*, Vol. 5. National Biomedical Research Foundation, Silver Springs, MD.

Deamer, D. and M. Akeson. 2000. Nanopores and nucleic acids: Prospects for ultrarapid sequencing. *Trends Biotech.* 18: 147–151.

Dermitzakis, E. T. and A. G. Clark. 2003. Evolution of transcription factor binding sites in Mammalian gene regulatory regions: conservation and turnover. *Mol. Biol. Evol.* 19: 1114–1121.

Dermitzakis, E. T., A. Reymond, N. Scamuffa, C. Ucla, E. Kirkness, C. Rossier and S. E. Antonarakis. 2003. Evolutionary discrimination of mammalian conserved non-genic sequences (CNGs). *Science* 302: 1033–1035.

DeSalle, R., J. Gatesy, W. Wheeler and D. Grimaldi. 1992. DNA sequences from a fossil termite in Oligo-Miocene amber and their phylogenetic implications. *Science* 257: 1933–1936.

Eisen, J. A. 1998. Phylogenomics: Improving functional predictions for uncharacterized genes by evolutionary analysis. *Genome Res*. 8: 163–167.

Ewing, B. and P. Green. 1998. Base-calling of automated sequencer traces using phred. II. Error probabilities. *Genome Res.* 8: 186–194.

Ewing, B., L. Hillier, M. Wendl and P. Green 1998. Base-calling of automated sequencer traces using phred. I. Accuracy assessment. *Genome Res.* 8: 175–185.

Felsenstein, J. 1985. Confidence limits on phylogenies: An approach using the bootstrap. *Evolution* 39: 783–791.

Gopal, S. et al. 2001. Homology-based annotation yields 1,042 new candidate genes in the *Drosophila melanogaster* genome. *Nat. Genetics* 27: 337–340.

Gordon, D., C. Abajian and P. Green. 1998. Consed: A graphical tool for sequence finishing. *Genome Res.* 8: 195–202.

Gordon, D., C. Desmarais and P. Green. 2001. Automated finishing with autofinish. *Genome Res.* 11: 614–625.

Graur, D. and W.-H. Li. 2000. *Fundamentals of Molecular Evolution.* Sinauer Associates, Sunderland, MA.

Green, P. 1997. Against a whole-genome shotgun. *Genome Res.* 7: 410–417.

Hall, A. E., K. C. Keith, S. E. Hall, G. P. Copenhaver and D. Preuss. 2004. The rapidly evolving field of plant centromeres. *Curr. Opin. Plant Biol.* 7: 108–114.

Hall, B. G. 2001. *Phylogenetic Trees Made Easy: A How-To Manual for Molecular Biologists.* Sinauer Associates, Sunderland, MA.

Henikoff, S. and J. Henikoff. 1992. Amino acid substitution matrices from protein blocks. *Proc. Natl. Acad. Sci. (USA)* 89: 10915–10919.

Imanishi, T. et al. 2004. Integrative annotation of 21,037 human genes validated by full-length cDNA clones. *PLoS Biology* 2: 856–875.

International Human Genome Sequencing Consortium. 2001. Initial sequencing and analysis of the human genome. *Nature* 409: 860–921.

Larsen, F., G. Gundersen, R. Lopez and H. Prydz. 1992. CpG islands as gene markers in the human genome. *Genomics* 13: 1095–1107.

Lin, X. et al. 1999. Sequence and analysis of chromosome 2 of the plant *Arabidopsis thaliana*. *Nature* 402: 761–768.

Lowe, T. and S. Eddy. 1997. tRNAscan-SE: A program for improved detection of transfer RNA genes in genomic sequence. *Nucl. Acids Res*. 25: 955–964.

Meldrum, D. 2000. Automation for genomics. II. Sequencers, microarrays, and future trends. *Genome Res.* 10: 1288–1303.

Mindell, D. and A. Meyer. 2001. Homology evolving. *Trends Ecol. Evol.* 16: 434–440.

Mitra, R. D., V. L. Butty, J. Shendure, B. R. Williams, D. E. Housman and G. M. Church. 2003. Digital genotyping and haplotyping with polymerase colonies. *Proc. Natl. Acad. Sci. (USA).* 100: 5926–5931.

Moore, D. S. and G. P. McCabe. 2002. *Introduction to the Practice of Statistics*, 4th Ed. W. H. Freeman, New York.

Needleman, S. and C. Wunsch. 1970. A general method applicable to the search for similarities in the amino acid sequence of two proteins. *J. Mol. Biol.* 48: 443–453.

Ovcharenko, I., D. Boffelli and G.G. Loots. 2004. eShadow: a tool for comparing closely related sequences. *Genome Res.* 14: 1191–1198.

Penn, S., D. Rank, D. Hanzel and D. Barker. 2000. Mining the human genome using microarrays of open reading frames. *Nat. Genetics* 26: 315–318.

Petrov, D., T. Sangster, J. Johnston, D. Hartl and K. Shaw. 2000. Evidence for DNA loss as a determinant of genome size. *Science* 287: 1060–1062.

Rajewsky, N., M. Vergassola, U. Gaul and E. D. Siggia. 2002. Computational detection of genomic *cis*-regulatory modules applied to body patterning in the early *Drosophila* embryo. *BMC Bioinformatics* 3: 30.

Reboul, J. et al. 2001. Open-reading-frame sequence tags (OSTs) support the existence of at least 17,300 genes in *C. elegans. Nat. Genetics* 27: 332–336.

Rogic, S., A. Mackworth and F. Ouellette. 2001. Evaluation of gene finding programs on mammalian sequences. *Genome Res* 11: 817–832.

Sanger, F., J. Donelson, A. Coulson, H. Kossel and D. Fischer. 1974. Determination of a nucleotide sequence in bacteriophage φ 1 DNA by primed synthesis with DNA polymerase. *J. Mol. Biol.* 90: 315–333.

Smith, T. and M. Waterman. 1981. Identification of common molecular subsequences. *J. Mol. Biol.* 147: 195–197.

Shoemaker, D. D. et al. 2001. Experimental annotation of the human genome using microarray technology. *Nature* 409: 922–927.

Sonnhammer, E., S. Eddy, E. Birney, A. Bateman and R. Durbin. 1998. Pfam: Multiple sequence alignments and HMM profiles of protein domains. *Nucl. Acids Res.* 26: 320–322.

Swofford, D. L. 2002. *PAUP*: Phylogenetic Analysis Using Parsimony (and Other Methods) 4.0 Beta*. Sinauer Associates, Sunderland, MA.

Swofford, D. L., G. J. Olson, P. J. Waddell and D. M. Hillis. 1996. Phylogenetic inference. *In* D. M. Hillis, C. Moritz and B. K. Mable (eds.), *Molecular Systematics*, 2nd Ed., pp. 407–514. Sinauer Associates, Sunderland, MA.

Tatusov, R. et al. 2001. The COG database: new developments in phylogenetic classification of proteins from complete genomes. *Nucl. Acids Res.* 29: 22–28.

Tatusov, R. et al. 2003. The COG database: an updated version includes eukaryotes. *BMC Bioinformatics.* 4: 41.

Thomas, J. W. et al. 2003. Comparative analyses of multi-species sequences from targeted genomic regions. *Nature* 424: 788–793.

Venter, J. C., M. Adams, G. Sutton, A. Kerlavage, H. Smith and M. Hunkapiller 1998. Shotgun sequencing of the human genome. *Science* 280: 1540–1542.

Venter, J. C. et al. 2001. The sequence of the human genome. *Science* 291: 1304–1351.

Waterman, M. S. 1995. *Introduction to Computational Biology*. Chapman and Hall, London.

Weber, J. and H. Myers. 1997. Human whole-genome shotgun sequencing. *Genome Res.* 7: 401–409.

Zhang, Z., A. Schaffer, W. Miller, T. Madden, D. Lipman, E. Koonin and S. Altschul. 1998. Protein sequence similarity searches using patterns as seeds. *Nucl. Acids Res.* 26: 3986–3990.

Zhu, J., J. Shendure, R. D. Mitra and G. M. Church. 2003. Single molecule profiling of alternative pre-mRNA splicing. *Science* 301: 836–838.

3 *SNPs and Variation*

Having sequenced the genome of one individual of a species, one of the next tasks in genomics is to describe the nature and distribution of variation among individuals. Most variation at the DNA level comes in the form of small insertions and deletions and **single nucleotide polymorphisms (SNPs).** The acronym SNP (pronounced "snip") refers to any site at which two or more different nucleotides are segregating in a population. A cluster of linked SNPs on one chromosome makes up a **haplotype.** SNPs and haplotypes are an increasingly important component of a wide range of biological investigation, ranging from biomedical research to ecology and evolutionary biology. They have many applications in the characterization of population structure and the history of genes and individuals and are indispensable for recombination mapping purposes. Much attention is also focused on the use of SNPs to identify loci responsible for the genetic component of multifactorial phenotypic variation, including disease susceptibility.

In this chapter, we start by considering the nature and distribution of SNPs and the theory of SNP applications, then move on to a description of current methods for SNP discovery and, finally, SNP genotyping. Thorough treatment of basic theoretical population and quantitative genetics can be found in textbooks by Hartl and Clark (1998), Falconer and Mackay (1996), and Lynch and Walsh (1998), while the primary literature is the best reference for emerging methods of SNP detection.

The Nature of Single Nucleotide Polymorphisms

Classification of SNPs

Single nucleotide polymorphisms are naturally occurring variants that affect a single nucleotide. They are most commonly changes from one base to another—transitions and transversions—but single-base insertions and deletions ("indels") are also SNPs. Some authors also regard two-nucleotide changes and small indels up to a few nucleotides as SNPs, in which case the term **simple nucleotide polymorphism** may be preferred. Microsatellites, longer simple sequence repeats, and all other classes of molecular polymorphism (including transposable element insertions, deletions from several bases to megabases in length, chromosome inversions and translocations, and aneuploidy), are not SNPs. Even so, the vast majority of sequence polymorphisms in the euchromatic portion of most genomes are SNPs.

SNPs are classified according to the nature of the nucleotide that is affected. **Noncoding SNPs** may be located in a 5′ or 3′ nontranscribed region (NTR), in a 5′ or 3′ untranslated region (UTR), in an intron, or they may be intergenic. **Coding SNPs** may be **replacement polymorphisms** (they change the amino acid that is encoded) or **synonymous polymorphisms** (they change the codon but not the amino acid). **Nonreplacement polymorphisms** include synonymous and noncoding polymorphisms, but many of these may affect gene function through an effect on transcriptional or translational regulation, splicing, or RNA stability. The importance of nonreplacement polymorphisms as a source of genetic variation is shown in Figure 3.1, which is compiled from a collection of over 140 human promoter SNPs that have been associated with an effect on gene expression or transcription factor binding, and in many cases, a clinical outcome (Rockman and Wray 2002).

Alternatively, SNPs can be classified as transitions or transversions. **Transitions** change a purine to a purine or a pyrimidine to a pyrimidine—A to G or C to T, and vice versa. **Transversions** change a purine to a pyrimidine and vice versa—A or G to C or T; C or T to A or G. Even though there are twice as many possible transversions as transitions, transitions tend to be at least equal to transversions in frequency, and often are more prevalent. This is broadly true of both coding and noncoding SNPs, and arises in part as a result of differences in the *ab initio* mechanisms by which particular types of mutations arise and are repaired. In addition, due to the nature of the genetic code, transitions are less likely to affect amino acids than are transversions, so transitions are thought to have a higher probability of retention in coding regions. As a result, the transition/transversion ratio is affected by constraints on sequence evolution, and must be accounted for in the computation of genetic distances based on the amount of sequence divergence (Graur and Li 2000).

In classical population genetic theory, genetic loci are only regarded as polymorphic if the frequency of the most common allele is less than some upper threshold, often set arbitrarily as 95%. By this criterion, **singletons,** which are variants detected in just one individual in a sample, are not polymorphisms. However, on the genome scale, most SNPs are first detected in a sample of fewer than 10 individuals, so the frequency criterion is not applied; all single nucleotide changes are described initially as candidate SNPs. A useful database of SNPs can be found at the NCBI's dbSNP site, http://www.ncbi.nlm.nih.gov/SNP/index.html. For human geneticists a

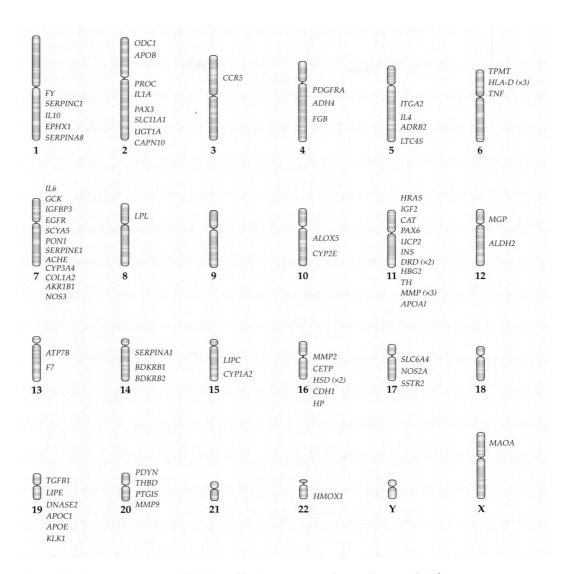

Figure 3.1 Human promoter SNPs that affect gene expression.   These are loci for which a SNP has been implicated in modulation of transcript levels, either by statistical association or using a biochemical assay in cell lines that are dispersed throughout the human genome. (Data from a literature survey by Rockman and Wray 2002.)

curated database of SNPs in candidate disease loci can be found at the Seattle SNP site, http://pga.mbt.washington.edu.

Confusion also arises over the distinction between *polymorphisms* and *mutations*, largely due to the dual usage of the term "mutation." All SNPs arise as mutations, in the sense that the conversion of one nucleotide into another is a mutational event. But by the time a sequence variant is observed in a population, the event that created it is usually long past, so the observed SNP is no longer a mutation—it is just a rare sequence variant or a polymorphism. However, "mutation" is also often used to describe an allele that deviates from the majority type, particularly where the aberrant allele is thought to affect some phenotype or disease status. Since this distinction only applies to a small fraction of all SNPs, the term "polymorphism" is more general.

Furthermore, most loci in natural populations harbor multiple haplotypes (a haplotype is a distinct combination of polymorphisms on a single chromosome at a locus; in highly polymorphic species, there are generally many different haplotypes per locus), so the "wild-type" is actually a set of different haplotypes. Even where one or more haplotypes are associated with a quantitative phenotype, it is not trivial to make the distinction between wild-type and mutant, and in fact the more rare form may be ancestral, so the term "polymorphism" is again preferred.

Distribution of SNPs

The study of the distribution of genetic variants, including SNPs, lies within the domain of population genetics, and the study of the relationship between SNPs and phenotypic variation lies in the domain of quantitative genetics. Genomic methods have led to a renaissance of interest in both fields of enquiry. While the earliest applications of genomics were in the mapping of Mendelian loci, the emerging importance of SNPs lies in the mapping and identification of **quantitative trait loci (QTL),** which are loci that contribute to polygenic phenotypic variation. On the one hand, SNPs provide the wherewithal to scan genomes for linkage to QTL; on the other hand, the vast majority of QTL effects are almost certainly due to as yet unidentified SNPs. For both reasons, characterization of the distribution of SNP variation is a major goal of most genome projects.

According to the **neutral theory** of molecular evolution, most SNPs are maintained in natural populations as a result of a balance between mutation and genetic drift. That is to say, the rate at which mutations are introduced into a population is in equilibrium with the rate at which polymorphisms are lost as a result of random sampling effects. Most mutations, whether deleterious, advantageous, or neutral in effect, are lost within a few generations, but some attain appreciable frequencies and even move to fixation, by chance. To this theoretical formulation must be added the effect of selection, which most commonly acts to reduce the frequency of slightly deleterious alleles, but on occasion tends to favor a new allele (**positive selection**) or to maintain two or more polymorphisms (**balancing selection**) at some loci.

Population structure is also a crucial determinant of allele frequencies within and among populations, as migration is a potent source of diversity, isolation affects the rate at which variation is lost due to drift, and rapid population growth alters the spectrum of haplotype frequencies. While the dynamics of genotypic variation are described by the balance of mutation, drift, migration, and selection, neutral theory supplies the null hypothesis by which observed patterns of molecular variation are assessed, and for the most part is regarded as a successful explanation for the distribution of SNP variation in genomes. However, the concepts of population genetics are often counterintuitive and seemingly self-contradictory, and we urge the reader to refer to textbooks such as Hartl and Clark (1998) for a full treatment.

Three key concepts are important in characterizing SNP variation:

- Allele frequency distribution
- Linkage disequilibrium
- Population stratification

While a common body of theory describes each of these attributes, the proximate reasons for differences in each quantity among species and within genomes are not yet well characterized. Among the prominent causes of variation in SNP diversity between species are population size, including bottlenecks during speciation; mating system; population structure; and migration, including admixture. Within a genome, SNP diversity is also heavily impacted by genetic factors impinging on the neutral mutation rate, such as variation in recombination rate along and between chromosomes, and structural and functional constraints on genes.

As a result of the balance between mutation and drift, the vast majority of SNPs are rare, and in fact the frequency distribution of SNPs is expected to be L-shaped (Figure 3.2A). In a sample of several hundred alleles, the most common class of SNPs are typically singletons (which appear only once in the sample), followed by doubletons, tripletons, and so on up to two equally common variants. Only between one-third and one-half of all SNPs are "common" in the sense that the more rare allele is present in more than 5% of the individuals. This in turn means that a SNP that is useful for mapping in one pedigree or cross is not necessarily going to be useful in another set of comparisons. Similarly, population structure affects the relative frequencies of alleles among populations and, while "private alleles" that are unique to a single population are useful for many applications, differences in allele frequency greatly complicate quantitative genetic inference.

The second relevant aspect of the frequency distribution is **nucleotide diversity**: the average fraction of nucleotides that differ between a pair of alleles chosen at random from a population. Nucleotide diversity is a function of the number of segregating sites and the frequency of each allele at these sites. Humans have relatively low nucleotide diversity, with an average of one SNP every kilobase between the two chromosomes of any individual. *Drosophila* and maize each have more than an order of magnitude greater polymorphism, with one SNP every 50–100 bases on average. These numbers vary over a wide range within genomes: nucleotide diversity exceeds 10% in some parts of the human major histocompatibility complex, but several kilobases can be monomorphic in other parts of the genome.

Figure 3.2 Nucleotide diversity in natural populations. (A) Observed and expected distributions of SNP frequencies for 874 SNPs from 75 candidate human hypertension loci. Rare alleles are the most frequent, and the number of SNPs in each frequency class declines as the more rare allele becomes more common. (B) Linkage disequilibrium (D′) decays with time (number of generations) in proportion to the recombination rate r. (C) As a result, the level of nucleotide diversity is a function of recombination rate, and hence chromosomal position, as in this example for *Drosophila melanogaster*. (A after Halushka et al. 1999; B after Hartl and Clark 1988, C after Begun and Aquadro 1992.)

(A)

(B)

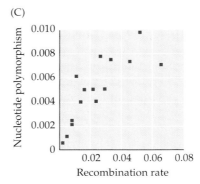

(C)

Wide variation in nucleotide diversity is also seen within loci, often between exons and introns. Some regulatory sequences in introns and 5′ NTRs are among the least polymorphic of all sequences, despite the general trend for coding sequences to be more highly constrained. Nucleotide diversity is strongly affected by recombination rate, as a consequence of its effect on the duration over which positive or negative selection affects linked sites: the faster recombination separates two SNPs, the less effect selection on one will have on the frequency of the other. As a result, diversity tends to be lower around centromeres and at the telomeres, and more generally to vary according to chromosomal location, as shown in Figure 3.2C (Begun and Aquadro 1992).

Linkage Disequilibrium and Haplotype Maps

Linkage disequilibrium (LD) refers to the nonrandom association of alleles. If two SNPs assort at random, then the expected fraction of each of the four possible two-locus genotypes can be calculated simply by multiplying the respective SNP allele frequencies. If the proportion of, for example, double homozygotes is greater than predicted, then there is linkage disequilibrium between the SNPs. As described in Box 3.1, statistical tests have been devised to detect LD. As will be described later in this chapter, LD is important in that it allows mapping of disease loci in large populations; however, it may also obscure correct inference of the meaning of association between a SNP genotype and a phenotype. Consequently, the characterization of the extent of LD is crucial for most quantitative applications of genomics.

The primary cause of linkage disequilibrium is thought to be historical contingency. When a mutation arises, the SNP it creates is in complete LD with every other site on the chromosome. As the new SNP rises in frequency in a population, recombination breaks up the LD in proportion to the genetic—and hence the physical—distance between sites. For this reason, LD is generally expected to decay monotonically on either side of each SNP (Figure 3.2B). However, this expectation is only observed approximately. It is not uncommon for two sites several kilobases apart to be in statistically significant LD, while numerous SNPs between them are in linkage equilibrium with one another and with the sites that are in LD (Stephens et al. 2001). This situation dramatically complicates the interpretation of genotype-phenotype association studies, but can only be detected after thorough sampling across the full extent of a locus. Such thorough sampling has only been performed in a few cases—for example, the human LPL locus, as demonstrated in Figure 3.3 (Clark et al. 1998).

Figure 3.3 Distribution of linkage disequilibrium across the human *lipoprotein lipase* (*LPL*) gene. LD plots show the distribution of significant scores between pairs of sites, in this example 66 sites spread over almost 10 kb of a sample of 142 chromosomes. Blue boxes indicate significance of Fisher's exact test ($p < 0.001$), yellow boxes indicate nonsignificance, and white boxes are cases where there was insufficient power to test for LD at this level. Note that the extent of LD varies across the locus, and is not restricted to exon sequences. (After Clark et al. 1998.)

BOX 3.1 Disequilibrium between Alleles at Two Loci

For a variety of reasons, genome scientists are interested in studying the independence (or *lack* of independence) of allelic combinations. A deviation from independence among alleles at a single locus is called **Hardy-Weinberg disequilibrium.** When alleles at two or more loci do not segregate independently, the phenomenon is called **gametic phase disequilibrium** or, more commonly, **linkage disequilibrium.**

Linkage Disequilibrium

As described in the text, linkage disequilibrium (LD) may arise from a variety of sources including nonrandom mating, population admixture, and epistatic selection. The last is of particular relevance for genome scientists, since linkage disequilibrium may indicate a functional interaction between loci associated with a phenotype of interest. For this reason, we now discuss LD and statistical tests for its presence.

Consider two loci, A and B, and for simplicity suppose they each segregate two alleles, A_1 and A_2 at the A locus, B_1 and B_2 at the B locus. Let the allele frequencies at A be p_1 and p_2, while q_1 and q_2 denote the frequencies at locus B (implying $p_1 + p_2 = 1$, $q_1 + q_2 = 1$). If the alleles at the two loci segregate randomly to form gametes, then gametes of the form A_iB_j occur with frequency p_iq_j. If this is the case, loci A and B are said to be in linkage equilibrium. If, on the other hand, the loci are in linkage disequilibrium, the frequency of A_iB_j gametes can be described with the introduction of a linkage disequilibrium coefficient, D (Lewontin and Kojima 1960), as shown in Table A. Those with a mathematical inclination should verify that the marginal totals are correct.

Table A

	B_1	B_2	Total
A_1	$p_{11} = p_1q_1 + D$	$p_{12} = p_1q_2 - D$	p_1
A_2	$p_{21} = p_2q_1 - D$	$p_{22} = p_2q_2 + D$	p_2
Total	q_1	q_2	1

Other measures of linkage disequilibrium have been suggested (see Hedrick 1985 for a review). Recognizing that the range of D depends on the allele frequencies at A and B, Lewontin (1964) proposed the measure:

$$D' = \frac{D}{D_{\max}}$$

where $D_{\max}$ is the maximum possible value of D given the allele frequencies. To illustrate this concept, consider the following 3 sets of 10 haplotypes in which $p_1 = 0.6$, $p_2 = 0.4$, $q_1 = 0.6$, and $q_1 = 0.4$:

Equilibrium	Coupling	Repulsion
A_1B_1 A_1B_2	A_1B_1 A_1B_1	A_1B_2 A_1B_2
A_1B_1 A_1B_2	A_1B_1 A_1B_1	A_1B_2 A_1B_2
A_1B_1 A_1B_1	A_1B_1 A_1B_1	A_1B_1 A_1B_1
A_2B_1 A_2B_1	A_2B_2 A_2B_2	A_2B_1 A_2B_1
A_2B_2 A_2B_2	A_2B_2 A_2B_2	A_2B_1 A_2B_1

The left-hand group has what looks to be a random assortment of alleles such as would be expected under Hardy-Weinberg equilibrium; the middle group has *all* A_1B_1 and A_2B_2 haplotypes; and the right-hand group has *no* A_2B_2. You should be able to calculate that D for these three groups is 0.04, 0.24, and –0.16, respectively, and to see that in fact there is no way that the absolute value of D could be any greater for the latter two populations. We say that the middle group, where the two common alleles are always on the same chromosome, exhibits complete linkage disequilibrium in coupling phase, whereas the right-hand group, in which the rare allele of one locus is always with the common allele of the other, exhibits complete LD in repulsion phase. The quantity D' takes on values of 0.17, 1.0, and –1.0 for these three groups, which is useful because, ranging from –1 to 1, it allows us to compare the relative amounts of LD in either phase. Furthermore, unlike D, D' is independent of allele frequencies, so it allows us to compare LD at different combinations of loci. For a similar reason, Hill and Robertson (1968) suggested the use of the squared correlation coefficient,

$$r^2 = \frac{D^2}{p_1 p_2 q_1 q_2}$$

For discussions of the relative merits of these and other measures, see Clegg et al. (1976) or Devlin and Risch (1995). Weir (1996) provides a theoretical treatment of statistical estimation of linkage disequilibrium coefficients.

Statistical Tests for Linkage Disequilibrium

We now focus on statistical tests for detecting the presence of linkage disequilibrium. The basic objective is a two-way test of independence, leading naturally to the use of traditional contingency table procedures. Consider a random sample consisting of $2n$ gametes. Let n_{ij} denote the number of $A_i B_j$ gametes in the sample, with $n_{i\cdot}$ and $n_{\cdot j}$ being the observed number of A_i and B_j alleles, respectively:

	B_1	B_2	Total
A_1	n_{11}	n_{12}	$n_{1\cdot}$
A_2	n_{21}	n_{22}	$n_{2\cdot}$
Total	$n_{\cdot 1}$	$n_{\cdot 2}$	$2n$

If alleles at the two loci segregated independently, the expected frequency of $A_i B_j$ gametes is $p_i q_j$, and the expected number of $A_i B_j$ gametes in a sample of size $2n$ is $e_{ij} = 2n p_i q_j$. If we estimate the unknown allele frequencies with the observed frequencies (e.g., $p_1 = n_{1\cdot}/2n$), we can then construct a chi-squared test in the typical fashion:

$$\chi^2 = \sum_{i,j} \frac{\left(n_{ij} - e_{ij}\right)^2}{e_{ij}}$$

The χ^2 statistic (the summation of the squared differences between observed and expected divided by the expected values) is compared to a chi-squared distribution with 1 degree of freedom to determine significance.

Suppose we have an observed data set with the following counts:

	B_1	B_2	Total
A_1	9	1	10
A_2	2	4	6
Total	11	5	16

The estimated value for p_1 is

$$\hat{p}_1 = {}^{n_{1\cdot}}\!/_{2n} = {}^{10}\!/_{16} = 0.625$$

Similarly, the estimate for q_1 is

$$\hat{q}_1 = {}^{n_{\cdot 1}}\!/_{2n} = {}^{11}\!/_{16} = 0.6875$$

The disequilibrium coefficient is $D = 0.5625 - 0.625 \times 0.6875 = 0.1328$. To carry out the chi-squared test we must first compute the expected counts. As an example, $e_{11} = 16 \times 0.625 \times 0.6875 = 6.875$, and the corresponding term in the χ^2 formula is $(9 - 6.875)^2 / 6.875 = 0.657$. The complete value is $\chi^2 = 0.657 + 1.445 + 1.095 + 2.408 = 5.605$, which is greater than the 3.84 critical value. Thus, we conclude that loci A and B are in linkage disequilibrium. The data contain an excess of $A_1 B_1$ and $A_2 B_2$ gametes (i.e., a paucity of heterozygotes).

An alternate procedure, particularly useful for small samples, is to perform a Fisher exact test. The idea of the Fisher exact test is to enumerate all possible samples of size $2n$ having the same allele counts as the observed sample. For each of these possible samples, the value of D is computed along with the probability of observing that sample if alleles segregated independently. Finally, a p-value is calculated by ranking the possible samples according to the value of D and, then tabulating the cumulative probability of samples with D greater than the value actually observed in the data. Weir (1996) shows that the conditional probability of a sample with counts n_{11}, n_{12}, n_{21}, and n_{22} given marginal alleles totals $n_{1\cdot}$ and $n_{\cdot 1}$ is

$$\frac{n_{1\cdot}! n_{2\cdot}! n_{\cdot 1}! n_{\cdot 2}!}{n_{11}! n_{12}! n_{21}! n_{22}! (2n)!}$$

(Continued on next page)

BOX 3.1 *(continued)*

Using that formula, the Fisher exact test for linkage disequilibrium can be computed as shown in Table B. The observed data (shaded) produce an exact test with a *p*-value of 0.035, suggesting that the two loci are in linkage disequilibrium.

The discussion above treats the case of two loci with two alleles, and assumes that data on gametes are available. Treatments accounting for multiple alleles and for un-

known linkage phase of double heterozygotes, along with a comprehensive discussion of tests for linkage disequilibrium, may be found in Weir (1996). Lewontin (1995) discusses a different approach to detection of linkage disequilibrium across a large sample of rare alleles such as are typically encountered in the pooled genome sequences of many individuals.

Table B

n_{11}	n_{12}	n_{21}	n_{22}	D	Probability	Cumulative probability
10	0	1	5	0.195	0.001	0.001
9	1	2	4	0.133	0.034	0.035
8	2	3	3	0.070	0.206	0.241
7	3	4	2	0.008	0.412	0.653
6	4	5	1	−0.055	0.288	0.941
5	5	6	0	−0.117	0.058	1.000

The extent of LD varies within and among species. In mammals, LD is commonly observed for several tens, and in many cases hundreds, of kilobases on either side of a SNP. Figure 3.4 shows that, on average, detectable LD extends out to 160 kb between randomly sampled sites in the human genome, but there is wide variation in how quickly it decays. In flies, by contrast, LD is rarely observed more than one or two kilobases away from a site, and often decays within a few hundred bases. Furthermore, the haplotypes that exist in the human genome as a result of LD have a somewhat blocklike, or clustered, distribution. To a broad approximation, our genome consists of tens of thousands of blocks, each up to a hundred kilobases long, containing three to five common haplotypes formed by tens or even hundreds of SNPs in LD. The extent of linkage disequilibrium is also a function of local recombination rates. LD is greater in nonrecombining regions of the genome, including the Y chromosome and parts of the X chromosome, as well as centromere-proximal regions of autosomes. An absence of recombination allows diversification of haplotypes, which in turn makes these portions of the genome suitable for inferring phylogenetic relationships over large timescales. Even over just a few centiMorgans, recombination rates are now known to vary widely, and hotspots just kilobases in length along the chromosome probably affect the determination of the limits of haplotype blocks. The boundaries between haplotype blocks may coincide with

(A)

(B)

Figure 3.4 Distribution of LD in the human genome.   (A) Plot of average level of LD against distance in kilobases for 19 randomly sampled regions of the human genome. Linkage disequilibrium typically decays to background levels after more than 100 kb, although individual loci can show very different patterns, (B) as exemplified by the loci ACVR2B and PAI2. The dotted line in (B) is the same overall plot as shown in green in (A). (After Reich et al. 2001.)

such hotspots, though several other factors also influence haplotype structure and much theoretical and empirical analysis of haplotype structure remains to be performed.

EXERCISE 3.1 *Quantifying heterozygosity and LD*

Using the following ten sequences:

```
 1 gctgcatcag aagaggccat caagcgcatc actgtacttc tgccatggcc
 2 gctgtatcag aacaggccat caagcgcatc actgtacttc tgccatggcc
 3 gctgtatcag aacaggccat caagcacctc actgtacttc tgccatggac
 4 gctgcatcag aagaggccat caagcacatc actgtccttc tgccatggcc
 5 gctgcatcag aagaggccat caagcacatc actgtccttc tgccatggcc
 6 gctgtatcag aacaggccat caagcgcatc actgtccttc tgccatggcc
 7 gcggcatcag aagaggcgat caagcacatc actgtccttc tgccatggac
 8 gctgcatcag aagaggccat caagcacatc actctacttc tgccatggcc
 9 gctgtatcag aacaggccat caagcgcatc actgtccttc tgccatggcc
10 gctgcatcag aagaggccat caagcgcatc actctccttc tgccatggcc
```

a. count the number of segregating sites;
b. estimate the nucleotide diversity (what assumptions must you make?)
c. measure the level of linkage disequilibrium between the common polymorphisms.

(continued on next page)

Exercise 3.1 (continued)

ANSWER: (a) There are 9 segregating sites, at positions 3, 5, 13, 18, 26, 28, 34, 36, and 49. (b) The minor allele frequencies at those 9 sites are, respectively, 0.1, 0.4, 0.4, 0.1, 0.5, 0.1, 0.2, 0.4, and 0.2. If a locus is in Hardy-Weinberg equilibrium, the predicted level of heterozygosity in the population is $2pq$, where p and q are the frequencies of the two alleles at that site. The nucleotide diversity can be estimated as the average over the 50 sites:

$$H = 1/50[(41 \times 0) + (3 \times 0.18) + (3 \times 0.48) + (2 \times 0.32) + (1 \times 0.5)] = 0.0624$$

Thus, if we pick two of these 50-nucleotide sequences at random from a population, we expect to see just over three site differences.

(c) There are four common polymorphisms, at positions 5, 13, 26, and 36.

cgga	If we extract these sites, it is easier to see how they are related.
tcga	Following the procedure in Box 3.1, draw a table of haplotype
tcaa	frequencies for each pairwise combination. For example,
cgac	for sites 5 and 13:

cgac
tcgc
cgaa
cgaa
tcgc
cggc

Site 5	**Site 13**	
	c $(q_1 = 0.4)$	**g** $(q_2 = 0.6)$
t $(p_1 = 0.4)$	$p_{11} = 0.4$	$p_{12} = 0.0$
c $(p_2 = 0.6)$	$p_{21} = 0.0$	$p_{22} = 0.6$

Since $D = p_{11} - p_1 q_1$, for this pair $D_{5,13} = 0.4 - (0.4 \times 0.4) = 0.24$.

This is also the maximal value, D_{max}, since all of the less common alleles at both sites always segregate together. Consequently, D' is equal to 1. You should be able to calculate the following table for the other linkage disequilibrium estimates. D_{max} is just the maximum value p_{11} could take, minus $p_1 q_1$.

Site pair	p_1	q_1	p_{11}	D	D_{max}	D'
5, 13	0.4	0.4	0.4	0.24	0.24	1.00
5, 26	0.4	0.5	0.3	0.10	0.20	0.50
5, 36	0.4	0.4	0.2	0.04	0.24	0.17
13, 26	0.4	0.5	0.3	0.10	0.20	0.50
13, 36	0.4	0.4	0.2	0.04	0.24	0.17
26, 36	0.5	0.4	0.2	0.00	0.20	0.00

This means that there is complete LD between the first two sites, but linkage equilibrium between the last two sites, with partial LD for all other comparisons.

Sequences

```
. . . T C A A G T C A A G C G A T C A T G . . .
. . . T C A A G T C A A G C G A T C A G G . . .
. . . T C A G G T C A A G T G A T C A T G . . .
. . . T C A G G T C A A G T G A T C A T G . . .
. . . T C A A G T C A A G C G A T C A G G . . .
. . . T C A A G T C A A G C G A A C A G G . . .
```

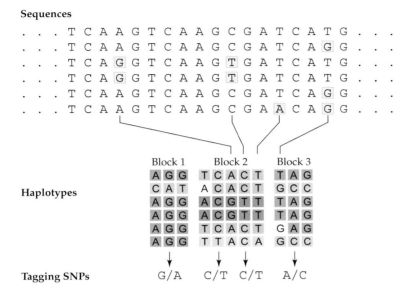

Figure 3.5 Tagging SNPs and the haplotype clusters.   Extraction of the polymorphisms from a set of sequences typically reveals a block-like pattern of haplotypes. In this hypothetical example, Block 1 has two classes of haplotypes, one rare and one common; Block 2 has three classes of haplotypes; and Block 3 has two classes of haplotypes. Note that the boundaries between blocks are relatively sharp. The tagging SNPs can be used to define all of the variation in the sample.

The International HapMap Project is an effort to map all of the common haplotypes in the human genome for the purposes of facilitating discovery of disease-promoting alleles (IHMC 2003). As of August 2004, over 60 million genotyping reactions involving 660,000 SNPs in hundreds of individuals of diverse ethnicity had been performed, and the data are available to the public in a browsable display of haplotypes at http://www.hapmap.org. Figure 3.5 illustrates the concept of **tagging SNPs,** which are SNPs that capture most of the nucleotide variation included in haplotypes. It is thought that between 50,000 and 100,000 SNPs should be sufficient to represent enough genetic diversity to support "genome scans" for linkage disequilibrium between haplotypes and disease susceptibility. Consequently, several massive public and private genome initiatives are directed at typing these markers in tens of thousands of individuals.

An important parameter affecting the utility of such data is the degree of **population stratification,** which is the partitioning of genetic variation among populations within species. Statistics for the identification and quantification of population structure have been in use by population geneticists for well over a half a century, and there is a wealth of data derived from classical markers such as allozymes. Outbreeding species show wide variation in the degree of population structure, reflecting constraints on migration due to geographic, ecological, and behavioral factors, as well as historical

Individuals within populations

Mississippi   Finland   Minnesota

Figure 3.6 Haplotype structure in the human *lipoprotein lipase* (*LPL*) gene.
Plot of 9.7 kb from each of 71 individuals from three populations (Jackson, MS; North
Karelia, Finland; Rochester, MN) were sequenced. The figure shows the distribution of
genotypes at each of 88 SNPs, with homozygotes for the more common allele in
blue, homozygotes for the more rare allele in yellow, heterozygotes red, and missing
data black. Two major classes of haplotype are visible in the data; these are found in
each of the populations, possibly at different frequencies and against a background
of diversity. (After Nickerson et al. 1998.)

contingency and local adaptation. Self-fertilizing and asexual species are
likely to exhibit very high levels of population structure, which can have a
marked effect on plant and parasite diversity in particular. Other species
(including humans) have more limited geographic variation.

The majority of all human SNPs are shared among the major racial
groups. The few detailed studies of human loci, however, are starting to
reveal significant population structure at the level of haplotype frequencies,
as shown for the human *LPL* locus in Figure 3.6. A survey of sequence diversity in at least 2 kb of 313 loci in 82 unrelated humans found that only 8%
of all haplotypes were unique to one of the four ethnic groups surveyed,
and that 71% were common to all groups, even in this moderate sample

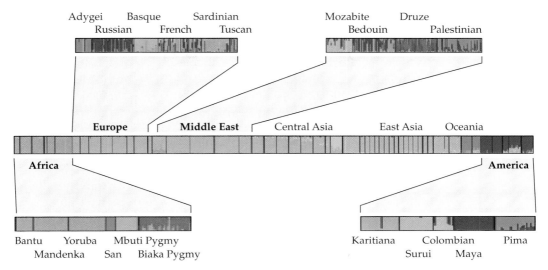

Figure 3.7 Human diversity and population structure.   Assuming six different human population groups, the Structure algorithm distinguishes Africans, Europeans, Middle Easterners, Central and East Asians, Oceanics, and Americans. In the central bar, each individual within a population is represented by a thin vertical line, and colors identify what proportion of the microsatellite alleles is attributable to each population group. More detailed analysis of each population clearly differentiates populations within groups in Africa and America, but there appears to be more mixture in Europeans and Middle Easterners, which adds ambiguity to those classifications. (After Rosenberg et al. 2002.)

(Stephens et al. 2001). A more extensive survey of 377 microsatellites in 1,056 individuals showed that differences in allele frequency among six major population groups account for, at most, 5% of human variation, yet with this number of markers it is possible to assign most humans to a genetic cluster that broadly reflects geographic distribution (Rosenberg et al. 2002; Figure 3.7). Similarly, typing of hundreds of microsatellites allowed individual dogs to be assigned with high probability to the appropriate breed, also resolving the genetic relatedness among breeds (Parker et al. 2004).

Applications of SNP Technology

Population Genetics

High-throughput DNA sequencing has numerous applications in classical population genetic research. First among these is the study of population structure at greater resolution than can be achieved with established methods such as allozymes and microsatellites. As the cost of DNA sequencing has come down and the throughput increased, sequence comparison using markers dispersed throughout a genome becomes a practical option for characterizing the degree to which genetic material is stratified among populations. The only limitation lies in the identification of polymorphic markers. Techniques now exist for rapid identification of fragments with high nucleotide diversity prior to sequencing; once these have been identified, it is possible to generate rapid and unambiguous genotype data for hundreds of organisms in a cost-effective manner. The coalescent framework that underlies much modern population genetic theory and inference from polymorphic sequence data is outlined in Box 3.2.

Applications of sequence comparison among populations include inference of demographic history; studies of mating systems; conservation genetics; paternity and maternity testing; and analysis of the evolutionary factors experienced by individual loci. Advantages of sequence comparison over allozyme analysis (**allozymes** are protein isoforms due to different alleles, and are usually detected by starch gel electrophoresis) are the ability to distinguish among alleles with similar allozyme profiles; to discriminate the complete range of SNP types at a locus and resolve haplotypes; to study noncoding sequences; and to extend population genetic surveys to all classes of gene, not just proteins for which biochemical assays or antibodies exist. Advantages of sequence data over microsatellites include the relatively low level of homoplasy (recurrent and back mutation) that complicates interpretation of microsatellite distributions, and the fact that sequence data allows comparisons among species and higher taxonomic levels that can supplement intraspecific data.

The utility of DNA sequence comparisons for inferring demographic history is well illustrated by findings of molecular anthropology. Studies of mitochondrial DNA have established that human genetic diversity is greatest in Africa, and suggest that *Homo sapiens* emerged there between 150,000 and 200,000 years ago (Cann et al. 1987), in contrast with the alternate view that humans evolved multiregionally across the globe from a species that emerged closer to a million years ago. Multiple Neandertal specimens have been found to be significantly more similar to one another than to modern human specimens, suggesting that Neandertals did not contribute to the contemporary human gene pool (Krings et al. 2000). Studies of other ancient fossil specimens, as well as from Y-chromosomal markers and numerous autosomal loci and microsatellites, continue to lead to refinement of the "Out-of-Africa" hypothesis, and to fill the gaps in our understanding of human migration patterns into Eurasia, the Americas, and Australasia over the past 50,000 years.

High-volume sequencing and genotyping is expanding the range and volume of traditional population genetic applications. One such area is the ascertainment of the breeding structure and degree of dispersal of soil microbes, pathogenic bacteria, parasitic nematodes, and numerous types of insect pests. Sequence comparisons can reveal the degree of haplotype divergence in microbial genomes, providing clues as to whether the species spread clonally or sexually, by epidemics or pandemics. Whole genome sequences provide hints as to how microbes have adapted to particular niches and may highlight genes that are likely to regulate parasitism, virulence, and toxicity. Comparing host and parasite phylogenies can indicate whether associations are ancient or recent, restricted or labile. Viral evolution can be studied at the whole genome level; in the case of RNA viruses such as HIV, it may even provide clues as to how the virus responds over a period of months to antiviral therapy. Numerous applications in conservation genetics will become increasingly feasible as costs decline and standard technologies emerge.

The second broad class of application of SNP diversity studies is the inference of the evolutionary factors experienced by individual loci. Numerous "tests of neutrality" have been developed that compare observed patterns of sequence variation within and among genes, with expectations derived from neutral theory (reviewed in Kreitman 2000). The simplest test, developed by Tajima, is based on the expectation that the number of segregating sites in a sample of alleles from a population should be proportional to the average pairwise distance between alleles. This test has low power and is often supplemented with more sophisticated analyses that are based on the notion diagrammed in Figure 3.8. The **HKA test,** for example, contrasts the ratio of polymorphic sites within a species to the number of fixed differences between species, between two or more loci, using an evolutionary model to predict the most likely relationship between these parameters. An excess of intraspecific polymorphism may indicate that balancing selection is acting on a site or sites and retaining linked neutral SNPs at the locus, whereas a deficit of intraspecific polymorphisms may under some circumstances indicate that a selective sweep recently removed the expected vari-

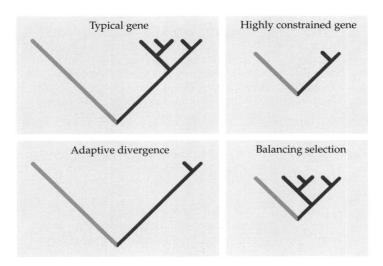

Figure 3.8 Basis for tests of neutrality. Four possible scenarios for comparing the amount of intraspecific polymorphism (blue; right-hand arm of each panel) with the extent of divergence to a sibling species (orange; left arm). Under neutrality, there is an expectation that highly constrained genes are less polymorphic and diverge less between species. After adaptive divergence, diversity is reduced within a species, whereas balancing selection is one possible explanation for the maintenance of polymorphism despite low divergence.

BOX 3.2 The Coalescent

The ability to collect the DNA sequence of a specific gene or genomic region from multiple individuals in a species cheaply and quickly has led to a rebirth of the field of population genetics. The data offer the potential to study evolution at the finest level of detail, and the promise of identifying individual nucleotide mutations that influence variation at the phenotypic level. In order to make use of DNA sequence surveys, it was necessary to develop a new body of population genetic theory and a collection of data analysis tools based on that theory. The driving force behind this effort has been the emergence of *coalescent theory*.

The most fundamental notion behind the coalescent is the idea that if one takes a random sample of DNA sequences from a population, then the number of generations one must trace back to find the **most recent common ancestor (MRCA)** of all the sampled sequences is a random variable. Figure A illustrates the case of two samples of $n = 3$ taken from a very small population of size $N = 6$ diploid individuals, which have a total of $2N = 12$ alleles. For a simple random mating population with constant population size, it is of no concern which individual carries an allele, and thus, the study of a diploid population of N individuals is the same as that of a haploid population with $2N$ individuals. Tracing the history of the three sampled blue sequences, we see that two of them shared a common ancestor two generations ago and that the MRCA of the sample of all three was six generations in the past. By contrast, when a second sample of size three is taken from the same population, such as the red sample, the time back to the MRCA is likely to be different than six (in this case, the three different red individuals coalesce to a common ancestor four generations in the past, and any other sample will also have a slightly different history). In other words, the number of generations back to the MRCA of the n sampled sequences, t_{MRCA},

Figure A

is a random variable with a probability distribution.

A large body of literature has been devoted to understanding the probabilistic behavior of t_{MRCA}. Upon reflection, one sees that t_{MRCA} has a significant impact on the amount of variability observed in the sample data: When t_{MRCA} is large, there are more opportunities for mutations, which in turn result in more sample variation. It is this link between variation and t_{MRCA} that make understanding its properties so important. When dealing with simple random mating populations it is quite simple to show that t_{MRCA} is described well by properties of exponential distributions.

Returning to the example from the figure, consider the probability that none of the sampled sequences share a common ancestor in the previous generation. This event occurs only if all of the $n = 3$ sampled alleles have unique ancestors in the previous generation, and has probability

$$\Pr(\text{3 unique ancestors in previous generation}) =$$

$$1 \times \frac{2N-1}{2N} \times \frac{2N-2}{2N}$$

The form of the expression reflects the fact that the first individual can trace back to any of the $2N$ sequences in the previous generation. In order to have unique ancestors, the second sampled sequence must be a descendant of one of the $2N - 1$ remaining sequences. Likewise, the third sampled sequence must trace to one of the remaining $2N - 2$. In general, if n sequences are sampled from a population of $2N$, the probability that none of the sequences share a common ancestor in the previous generation is:

$$p_n = \text{Pr}\,(n \text{ unique ancestors in previous generation})$$

$$= \prod_{i=1}^{n} \frac{2N-(i-1)}{2N}$$

This expression can be used to find the probability that the first ancestor of any pair of individuals in the sample occurred t generations ago:

$$\text{Pr}\,(\text{first ancestor of any pair was } t \text{ generations ago}) = p_n^{\,t-1}\,(1-p_n)$$

This formula arises from the requirement of $t-1$ consecutive generations with no common ancestors, followed by a single generation with two sequences tracing to a common ancestor, and it can be used to show that the times between coalescent events follow exponential distributions.

Once the first coalescent event has occurred it is as if the entire process restarts with a sample size of $n-1$, because two original lineages have been combined into one. Indeed, this recursive nature continues after each coalescent event until all lineages converge into a single one at the MRCA. Figure A shows the lengths of the periods of time between each event for the blue sample, $t_n, t_{n-1}, \ldots, t_2$, with time t_k denoting the length of time during which the sample is represented by k lineages. Using properties of the exponential distribution it is simple to verify that the expected value of t_k is:

$$E(t_k) = \frac{4N}{k(k-1)}$$

and that the expected value of t_{MRCA} is:

$$E(t_{MRCA}) = E(t_n) + E(t_{n-1}) + \ldots + E(t_2)$$

$$= 4N\left(1 - \frac{1}{n}\right)$$

Consider next the total length of the genealogy. Noting that during the kth interval there are k lineages of length t_k, the total length is:

$$T = \sum_{k=2}^{n} k t_k$$

The value of T is intimately related to the amount of variability in a sample. Because it measures the total amount of time "captured" in a sample of sequences, the genealogy length provides the number of opportunities for mutations to occur. Note, however, that T is also a random quantity, its distribution being described by sums of exponential random variables as described above. If mutations arise at constant rate μ, then the expected number of mutations on a genealogy of length T is simply μT. If one assumes the infinite sites population genetic model, then careful thought reveals that the number of mutations on the genealogy is exactly the same as the number of variable, or segregating sites in the sample of DNA sequences, S. Some algebra shows that the expected number of segregating sites in a sample of size n taken from a population of N diploid individuals with per-site mutation rate μ is:

$$E(S) = \mu E(T) = \theta \sum_{k=1}^{n-1} \frac{1}{k}$$

where $\theta = 4N\mu$.

This formula is a cornerstone of coalescent theory, because it provides a direct relationship between a quantity observable in sampled data (S) and a population parameter that governs the accumulation of variation (θ). The relationship also offers a statistical estimator for θ,

$$\hat{\theta} = \frac{S}{\displaystyle\sum_{k=1}^{n-1} \frac{1}{k}}$$

Only the product of mutation rate and population size is estimable. However, if additional information is available, the two components can be separated. If, for example, an empirical estimate of the mutation rate is available, the estimation formula can also be used to estimate the population size.

Consider the following three sequences that differ at five segregating sites:

```
CGCTACAAAGTCATACCTGGTCAGATAGA
CGTTACAAAGTCACACCTGGTCTGATAAA
CGTTACAAGGTCACACCTGGTCAGATAAA
  *       *     *        *    *
```

(Continued on next page)

BOX 3.2 (continued)

The estimate of $\theta = 4N\mu$ is simply:

$$\hat{\theta} = \frac{S}{\sum\limits_{k=1}^{n-1}\frac{1}{k}} = \frac{5}{\frac{1}{1}+\frac{1}{2}} = \frac{5}{\frac{3}{2}} = 3.33$$

If available experimental data provided an estimate of the mutation rate to be 0.0005, then an estimate of the effective size of the population from which the sample is drawn is:

$$\hat{N} = \frac{\hat{\theta}}{4\hat{\mu}} = \frac{3.33}{(4\times 0.0005)} = 1{,}666$$

The process and concepts described above lay the foundation for a large body of theory relating sequence variation to population and mutation processes. Coalescent theory in recent years has evolved to offer estimators of recombination rates, selection coefficients, migration rates, and rates of population growth. It has been the basis of methods for identifying genes undergoing adaptive evolution. In conjunction with rapid improvements in desktop computing power, the coalescent has become the most important theoretical tool in modern molecular population genetics.

---

ation at the locus. Similarly, the **McDonald-Kreitman statistic** tests whether the ratio of polymorphism to divergence within a single locus is the same for synonymous and replacement sites, as it should be under neutrality.

Such tests are of theoretical importance in further delineation of the balance of mutation, drift, and selection across the genome. They may also have practical applications in the identification of target loci for advanced plant and animal breeding, as well as control of pathogens. Domestication of plants, notably maize (see Figure 1.18), has been shown to leave a "footprint" of reduced genetic diversity within a few kilobases of loci that experienced strong artificial selection (Doebley et al. 1997). Similarly, scans of sequence diversity in microbial parasite genomes are likely to contribute to the identification of strongly selected genes that are required for infectivity, immune system evasion, and pathogenicity.

Sequence analysis has allowed inference of selection on genes involved in human divergence from other primates. As many as 500 proteins have more amino acids substitutions than expected given observed numbers of synonymous substitutions between humans and chimps. These include many proteins involved in olfaction and hearing, amino acid catabolism, and early development (Clark et al. 2003), as well as an excess of genes represented in the OMIM disease database. By contrast, the high nucleotide diversity of the human immune complex loci at the MHC and across the immunoglobulin superfamily is likely, at least in part, to reflect maintenance of variation due to a combination of heterozygote advantage and frequency-dependent selection during disease epidemics (Vogel et al. 1999). Yet another application of population genetics lies in inference of the history of disease-promoting polymorphisms. For example, a common regulatory SNP that affects expression of the IL-4 cytokine has opposing effects on susceptibility to HIV infection and asthma, among other immune disorders, and occurs at such different frequencies in different populations that it seems that different selective forces act on the gene in different parts of the globe (Rockman et al. 2003).

Recombination Mapping

The major application of SNPs in human genetics is in the mapping and identification of disease loci. **Recombination mapping,** also known as linkage mapping, has been used as the basis for positional cloning of Mendelian disease genes, namely single-gene disorders such as cystic fibrosis, Huntington's disease, and heritable long-QT syndrome. **Positional cloning** refers to the isolation of a gene that is responsible for a phenotype on the basis of its map position (as opposed to a biochemical assay, for example). It relies on complete association between the mutation and the disease. Given a sufficiently high density of SNPs and a large enough set of pedigrees, identification of the disease-causing mutation is (in theory) simply a sampling and genotyping problem. For this reason, the completion of the human genome sequence has been hailed as the first step toward mapping every one of the thousands of known inborn errors of metabolism and other single-gene disorders, as it provides the framework for comprehensive SNP identification.

In practice, most of these disease genes are actually mapped in two steps: localization of the locus to several hundred kilobases, followed by sequencing of candidate genes in the interval to identify candidate disease-causing SNPs (such as premature stop codons). In humans, the mapping is performed in pedigrees, an example of which is shown in Figure 3.9. Each meiosis has

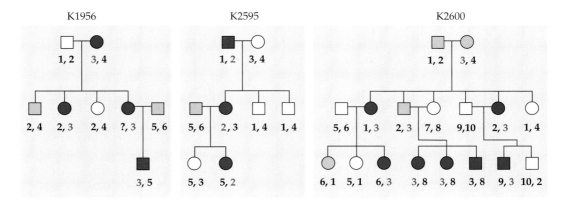

Figure 3.9 Pedigree mapping. Portions of three of the nonrelated pedigrees that contributed to the mapping of the gene for a dominant form of long-QT syndrome are shown. In each case, one haplotype, highlighted in red, co-segregates with the disease. (Note that the numbers refer to different haplotypes in each family: it is only identity by descent within each pedigree that is relevant.) Squares represent males, circles females; dark color indicates symptomatic individuals, shading equivocal or unknown disease status. Each haplotype was inferred from five linked markers. (After Curran et al. 1995.)

the potential to result in recombination between the disease locus and any of the anonymous markers being used in the mapping. Failure of a marker to segregate with the disease in affected offspring indicates that a recombination event has occurred, and allows the investigator to deduce whether the disease locus lies proximal or distal to the marker on the chromosome. Similarly, segregation of the marker in a nonaffected sibling provides mapping information as does cosegregation of markers in affected sib-pairs (Asps). Clearly, the greater the number of affected individuals that can be screened, the more refined the mapping. Affected individuals can be identified in extended pedigrees covering multiple generations where DNA and medical records are available, or they can be contrasted across multiple distinct pedigrees. In the latter case, different markers might be used for each pedigree, but so long as the markers are mapped relative to one another, relative positional information can be extracted.

EXERCISE 3.2 *Inferring haplotype structure*

Suppose that a couple have three children. The genotypes at four sites in a gene of interest for the parents and children are given below. Assuming there was no recombination between the sites during transmission from parent to child, write down the 4 possible haplotypes that each parent could have, and then infer the haplotypes that the three children have, using the parents' genotypes as a guide. Is there anything about this data that might lead a forensic scientist to suspect that the father is not the biological father of one of the children? If the mother's genotype had been CT at the fourth site, would you be able to infer the haplotype phases?

Mother:	GG	AT	CA	TT
Father:	CC	AA	AC	CT
First child:	GC	AA	CC	CT
Second child:	GC	AT	AA	TT
Third child:	GC	AA	AC	CT

ANSWER: The four possible haplotypes that each parent could have are as follows:

Mother:	G	A	C	T	(M1)
	G	T	A	T	(M2)
	G	T	C	T	(M3)
	G	A	A	T	(M4)
Father:	C	A	A	C	(F1)
	C	A	C	T	(F2)
	C	A	A	T	(F3)
	C	A	C	C	(F4)

The first child must have received haplotypes M1 and F4, the second child M2 and F3, and the third child either M1 and F1 or M4 and F4. Since each parent can only transmit two haplotypes, but three haplotypes from each parent are required to explain the three children's genotypes, something is amiss. There is a further constraint that the mother must be M1 and M2 or M3 and M4, and similarly the father F1 and F2 or F3 and F4. From this we can infer that the third child must have received either M4 from another mother, or F1 from another father (or must have been adopted).

If the mother were heterozygous at the fourth site, she would have 8 possible haplotypes, and inferring the linkage phases in the parents would be ambiguous.

Numerous strategies now exist for jumping straight to candidate genes within mapped intervals. Several hundred kilobases may include tens of different genes. With the complete and annotated genome sequence, the sequences of predicted ORFs may provide an immediate clue as to which gene has a disease-promoting mutation. This proved to be the case with long-QT syndrome, a fatal heart arrhythmia condition for which sodium and potassium channels were identified as potential modulators of cardiac rhythmicity (Curran et al. 1995).

Synteny with other mammalian genomes may be used to identify known candidate genes with related disease phenotypes, as with a number of congenital diseases in dogs. Alternatively, each of the genes in the region of synteny of the mouse may be tested by Northern blots or in situ hybridization to assess whether the gene is expressed in the affected tissue, such as muscle, bone marrow, or liver. More divergent model systems such as zebrafish, or even invertebrates, might also suggest functional associations that hint at a disease association, as described in Chapter 5. Or, microarray data may suggest that the one candidate gene is co-regulated with other genes that are implicated in the etiology of the disease. Failing such circumstantial evidence, it is now feasible simply to sequence each of the exons of the genes in the candidate region to search for candidate causative SNPs, assuming that Mendelian mutations are likely to completely knock out or otherwise dramatically alter the protein product.

Recombination mapping has also been used to identify major susceptibility factors for a small number of more complex disorders. Examples include the *LDL* receptor gene for heart disease, *APC* for colon cancer, and the *BRCA1* and *BRCA2* genes for heritable forms of breast cancer. The first two cases were aided by clear functional roles of the mutated candidate genes—in one case a receptor for the low-density lipoproteins that are a major risk factor for heart disease, and in the other a DNA repair enzyme. Neither of the *BRCA* genes, by contrast, were obvious candidates, and in fact evidence that the identified genes are the true causative loci was debated until multiple independent mutations were associated with breast cancer in different pedigrees. Formal genetic proof that a mutation causes a particular phenotype consists either of reversion of the mutation, or rescue of the defect with a wild-type transgene, neither of which are practicable in humans.

Several countries have taken steps to coordinate public health databases with collection of blood samples from tens of thousands of individuals, with the aim of accelerating disease gene discovery. The first example of this was an agreement between the government of Iceland and the deCODE company, which received access to generations of health records in return for their investment in high volume genotyping of samples that had been made anonymous to the scientists. Successes in mapping genes for psoriasis and schizophrenia have led to similar commercial/public collaborative projects in populations with particularly interesting genetic or epidemiological attributes.

QTL Mapping

Techniques for recombination mapping in plants and nonhuman animals are more flexible and in many cases offer increased power to map quantitative trait loci (QTL)—polygenes that affect complex traits. The key features of successful **QTL mapping** designs are the ability to control the starting genetic variance; to reduce the environmental variance (in order to *increase* the proportion of overall phenotypic variance that is due to each QTL); and to increase the number of meioses as desired. Most designs start with two inbred parents that differ with respect to the trait of interest, either by chance or as a result of divergent artificial selection. Recombination is then allowed to break up linkage disequilibrium between markers over succeeding generations. In F_2 or backcross designs, two F_1 offspring of a cross between the parents are themselves crossed (or one or more F_1 individuals are backcrossed to a parent), and several hundred F_2 grandchildren are scored for both the phenotype and genotype at SNPs distributed throughout the genome (Figure 3.10).

Alternatively, F_2 offspring can be further sib-mated (that is, crosses set up between siblings) for 10 or more generations to create a set of **recombinant inbred lines (RIL),** each of which contains a random but nearly homozygous set of chromosome segments derived from each parent. The advantages of RIL are that multiple genetically identical individuals can be scored for each line, decreasing the environmental contribution to the trait measure; and that the same lines can be used by different investigators to characterize numerous traits.

Breakpoints between segments can be mapped using SNPs or other molecular markers. A further variant on this strategy is to **introgress** one parental haplotype into the genetic background of the other parent either by **marker-assisted selection** in each generation, or by artificial selection on the trait of interest. In the former case, the investigator can then ask whether the introgressed portion of the genome affects the trait; in the latter, marker genotyping will delimit the portion of the chromosome that is responsible for the quantitative effect. A prominent example is the cloning of a gene for small

(A) F$_2$ design   (B) Backcross design   (C) RIL

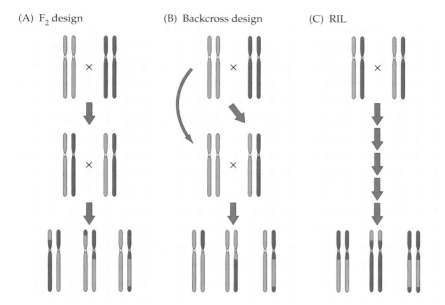

Figure 3.10 F$_2$, BC, and RIL experimental designs.   Controlled mapping crosses are designed to segregate variation that is fixed in two divergent parental lines. (A) In an F$_2$ experiment, combinations of all three possible single-locus genotypes are generated by sib-mating of F$_1$ progeny. (B) In a backcross, F$_1$ individuals are crossed back to one parent, generating heterozygotes and homozygotes of a single class. (C) Recombinant inbred lines are produced by repeated sib-mating for at least 15 generations, which results in homozygous lines that allow testing of multiple genetically similar individuals.

fruit size in tomato, which was performed by introgression mapping to several hundred kilobases, followed by regeneration of the phenotype in transgenic plants with a cosmid that carried the QTL (Frary et al. 2000).

Relatively high-resolution QTL mapping was enabled in the 1990s by the development of new techniques for marker genotyping (including SNP detection) and by novel statistical methods. The core idea of QTL mapping is no different from that of Mendelian recombination mapping, except that multiple loci, none of which is solely responsible for the trait, are mapped simultaneously. Instead of looking for perfect association between a marker and the phenotype, the genome is scanned for statistically significant associations between markers and the phenotype. Initially, significance was judged by *t*-tests for a difference between the mean phenotypes of the two allele classes at each marker. This procedure underestimates the magnitude of QTL effects, is unable to separate closely linked QTL, and has low precision. The **interval mapping** procedure (Figure 3.11; Lander and Botstein 1989) improved resolution and power by estimating marker genotypes and association at each position in the interval between adjacent pairs of markers. Subsequent modifications include conditioning of marker effects on

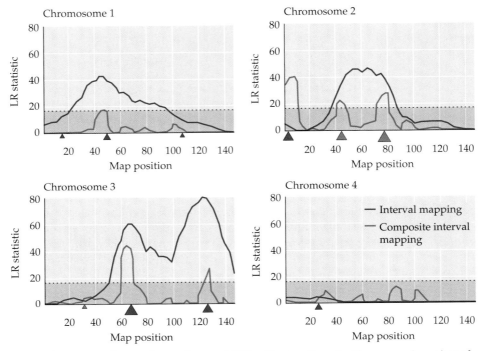

Figure 3.11 Interval mapping and CIM. The location, significance, and number of peaks detected in QTL analysis is a function of the analytical method used. For this simulation of 10 QTL of varying magnitude of the effect of the QTL (size of triangle) and sign (red triangles positive, green negative) distributed on four equal-length chromosomes with markers every 10 cM, composite interval mapping (CIM) resolved the six strongest QTL with more precise positions (red peaks that exceed the threshold dotted horizontal line for each chromosome) than did ordinary interval mapping (blue curves). Significance thresholds were determined by permutation testing. (After Zeng 1994.)

other significant markers in the genome (composite interval mapping; Zeng 1994) and simultaneous fitting of multiple QTL effects.

In these cases, statistical support is measured using a likelihood ratio, or by a logarithm of the odds (LOD) score, which is proportional to the logarithm of the likelihood ratio. For many purposes in human genetics, a LOD of 3 is taken as significant; however, given the number of tests performed, it is not always clear that this is appropriate, and permutation tests are now commonly used to define significance thresholds.

QTL are said to lie close to the significant peaks in a LOD profile, but to have a confidence interval of ±2 LOD units—which even in a study involving more than a hundred individuals can be anywhere from several to tens of centiMorgans. While useful for estimating the distribution and magnitude of allelic effects, such mapping is not generally sufficient to identify candidate genes since the interval will span several megabases and hundreds of genes. For first-pass mapping, markers spaced between 10 and 20 cM are appropriate. Finer resolution can be obtained using a higher density of markers in the vicinity of each QTL and thousands of meioses, but even so, resolution of genes corresponding to QTL that individually account for less than 10% of the phenotypic variation has not yet been achieved in more than a couple of cases.

It should be emphasized that QTL mapping can be extremely useful even if it does not lead to cloning of the gene that is responsible for the effect (Lynch and Walsh 1998). The localization of a QTL to within 10 cM is sufficient to initiate marker-assisted breeding by introgression of the QTL interval flanked by markers into desired genetic backgrounds. For example, introducing tropical genetic material into a wide range of crop plants for the genetic improvement of traits such as yield, disease resistance, and fruit quality will depend on QTL mapping technology. With mapped QTL, it is possible to estimate the degree of dominance of segregating alleles. Mapped QTL also make it possible to measure **epistasis,** which is the effect on a trait of interactions between genotypes at two or more loci. Such data are fundamental to modeling evolution at the species level, and can also be utilized for agricultural breeding purposes.

Comparison of QTL positions affecting similar traits across species may be useful in further characterization of the loci, and for testing the role of homologous genes in domestication (for example, in relation to flowering time in grasses). Alternative methods for identification of QTL, such as transposable element insertional mutagenesis and mapping the haploinsufficient effects of small deletions (also known as **deficiency complementation mapping;** Long et al. 1996), are likely to merge with interval mapping methods and functional genomics in the coming years to resolve the molecular basis for quantitative traits to levels that could barely be imagined in the pregenomic era.

Linkage Disequilibrium Mapping

Genic resolution of polygenic complex disease loci is now being pursued using a series of methods that are loosely categorized as **linkage disequilibrium mapping.** The idea behind each of these procedures is to ask whether a particular class of SNP is more commonly seen in unrelated affected individuals than you would expect by chance. Aside from sampling artifacts, significant results can arise either because the SNP contributes to the trait, or (more commonly) because the sampled SNP is in linkage disequilibrium with a site that contributes to the trait.

Linkage disequilibrium is the nonrandom association of alleles, as described in Box 3.1. If there were no linkage disequilibrium, it would be necessary to test every SNP in the genome for association. On the other hand, if every SNP were in LD with every other SNP, there would be no way to resolve the location of a significant effect, since all of the SNPs would show the association. If, however, there is local LD over tens of kilobases (as is often the case), in theory it is only necessary to sample one of a cluster of sites that tend to segregate together. Thus it should be sufficient to sample

SNPs at a density similar to the level of local LD, or 50–100 kb in humans. Genome scans are predicated on this principle: that if there is LD, then it is only necessary to sample well-spaced tagging SNPs for association tests. As pointed out above, the caveat is that there is wide variation in levels of LD, so there is no guarantee that one SNP per 50-kb interval is sufficient to represent any particular haplotype block.

Once a significant site is identified, more detailed sampling is employed to identify the gene and eventually QTN (quantitative trait nucleotides), namely the polymorphic nucleotides that are responsible for a QTL effect. Two general classes of procedure have been adopted: **case-control population sampling,** which is essentially looking for associations between SNPs and the disease in a large population; and pedigree-based **transmission disequilibrium testing,** which is essentially looking for unequal transmission of SNP alleles to affected and nonaffected siblings.

Pitfalls of association studies. Before giving a very general overview of these methods, it is worth reviewing some of the hurdles that must be overcome. An important distinction must be made between linkage and association. *Linkage* means that the marker SNP is inferred to be within 50 cM of the disease allele by virtue of physical placement on the same chromosome, whereas *association* means that the SNP is significantly associated with the disease or trait. It is possible to have association without linkage—a chance false positive, for example, or less trivially as a result of population stratification or association between the SNP and an environment that contributes to the disease. It is also possible to have linkage without association; for example, where there is insufficient evidence to detect the association between the disease allele and the disease (a false negative), or because the disease association is weak and limited recombination between the marker and the disease allele has been sufficient to break up the association. In general, significant results of association tests between genotype and phenotype may be due to both linkage and association, and disentangling these effects is nontrivial.

As reviewed in Zondervan and Cardon (2004), numerous factors will affect the success of a linkage disequilibrium mapping experiment. Among these factors are:

- *The heritability of the disease or trait.* If the genetic contribution to the disease is too small, no sample size or design will be sufficient to map the individual loci. A closely related concept is that of **relative risk,** which describes the increased likelihood that one group of individuals will get a disease relative to the risk factor for another group.

- *The number of genes affecting the disease or trait.* A ballpark estimate of the number of genes affecting quantitative traits can be obtained by comparing the variance among generations, but this is extremely difficult for threshold-dependent traits (probably many diseases) and hard to obtain from pedigree data. The larger the number of genes, the smaller will be their average contributions, and the harder they will be to detect.

- *The penetrance and expressivity of the QTL effects.* **Penetrance** refers to the fraction of individuals with the QTL who show the trait, and **expressivity** refers to the severity of the effect. The lower the penetrance or expressivity, the harder it will be to map the gene. For human diseases, penetrance can be strongly age-dependent, leading to misascertainment; and expressivity can be modulated by the environment, which for psychological diseases in particular can lead to misdiagnosis or misclassification of severity.

- *The extent and uniformity of linkage disequilibrium between SNP markers, and between markers and the disease locus.* Variable LD can result in large sections of the genome remaining unscored. The major factors affecting LD are the age of the disease allele; the physical and hence genetic distance between the marker and disease allele; and the effective population size and stability of the population being typed.

- *Genetic heterogeneity.* **Genetic heterogeneity** describes the frequent observation that the same disease is affected by different loci in different pedigrees; for example, more than a dozen loci can mutate to cause retinitis pigmentosum. A different type of heterogeneity that causes different problems for association and linkage mapping is **allelic heterogeneity** within a locus (for example, nearly 100 different mutations in the *CFTR* gene cause cystic fibrosis).

- *Population stratification.* Differences in allele and haplotype frequencies among populations can lead to false positive associations (Figure 3.12), or in cases of cultural or environmental transmission, to false attribution of

Figure 3.12 Population stratification. Suppose two populations differ in both the frequency of a particular allele and the prevalence of a phenotype (such as diabetes). If the two populations are pooled (in equal ratios in this example), the measured association between the two genotypes and the disease prevalence may be significantly different, but not necessarily as a result of linkage to the marker. Since 73% [0.8/(0.8 + 0.3) = 0.73] of the A alleles are in population 1, which has a higher disease prevalence, the A allele would seem to be associated with disease, but it need not be linked to a true causative site. Environmental, cultural, or genetic factors may contribute to the apparent association, even when the population structure is not apparent to the observer.

an association to likely causation. A hypothetical example of the latter would be the improper inference that a Y-chromosomal haplotype found uniquely in a particular Jewish lineage actually contributes to the holding of specific religious beliefs.

- *Admixture.* Recent contact between previously isolated populations can result in transient linkage disequilibrium and novel population structures that often go undetected. The effects of such **admixture** will change over a succession of generations as heterozygosity returns to normal and allele frequencies attain a new equilibrium.

- *Hidden environmental structure.* This is particularly relevant in relation to assessment of risk for physiological diseases such as obesity, diabetes, and heart disease. Undetected pathogens or incidence of exposure to pathogens and toxins may also have an enormous impact on human health and psychology.

- *Genotype by environment and genotype by genotype interactions.* These interactions are of essentially unknown magnitude, are difficult to detect statistically, and may affect the ability to detect main effects of susceptibility loci. Phenotypes are a function not just of genotype, but of the variable expression of genotypes in different environments (a plot of which is sometimes called a norm of reaction) and different genetic backgrounds (epistasis). These effects may be particularly prevalent in relation to categorical and threshold-dependent traits, as well as in certain types of cancer.

- *Incomplete genotyping.* This includes both gaps in information due to missing and deceased individuals, and genotyping errors due to factors such as undetected mispaternity and mismaternity (the latter associated with incomplete documentation of adoptions).

Despite all of these potential obstacles, association studies offer the best prospects for mapping of complex disease loci to the level of single genes. Dozens of different tests have been proposed and applied with varying degrees of success. All of these generally require sample sizes in excess of 500 individuals to achieve over 80% power to detect a locus that accounts for at least 5% of the disease susceptibility (Long and Langley 1999). Studies of this magnitude are now routine, and in fact several journals now require independent replication on a second study population. Moreover, interesting associations are typically pursued by multiple investigators, and since there is a reasonable likelihood that any one study will fail to see an effect, proof that a locus affects disease susceptibility, and estimation of the magnitude of the effect, generally requires multiple studies (Ioannidis et al. 2001). A useful online searchable database of human disease association studies can be accessed at http://geneticassociationdb.nih.gov.

Population-based case-control design.   The most direct and efficient form of association mapping is population-based case-control design. Marker frequencies are determined in a sample of affected individuals (the "cases") and compared with marker frequencies in an age, sex, and population-matched sample of unaffected "controls." In nonparametric approaches, the test statistic simply contrasts the observed frequencies with expected frequencies based on the proportions of the alleles at the marker.

Suppose a disease is thought to be affected by the status of a biallelic locus with alleles *A* and *a*. Assuming Hardy-Weinberg equilibrium, observed and expected frequencies can be contrasted most simply by a contingency chi-square analysis:

	Observed		**Expected**	
	Allele *A*	Allele *a*	Allele *A*	Allele *a*
Affected	24	278	49	253
Unaffected	86	296	61	321

$\chi^2 = 27.5$, $p < 0.001$

In reality, continuity adjustments must be made for the fact that the alleles are not drawn from the same sample, so the distribution of expected alleles is not so straightforward. Nevertheless, in this hypothetical example, the *A* allele appears to be significantly less frequent in affected individuals than you would expect given its frequency in the population, and one possible explanation is that this allele may be protective against the disease. Variants of this approach discussed in Box 3.3 include comparison of genotype (*AA*, *Aa*, *aa*) frequencies, or estimation of multimarker haplotypes and testing for differences in disease prevalence by analysis of variance.

More sophisticated case-control designs incorporate parametric model testing. Parameters describing the probable magnitude and dominance of the effect, as well as the recombination distance from the disease locus, can be estimated simultaneously using maximum likelihood methods. Prior knowledge of the inheritance patterns may also be incorporated, such as the degree of dominance of the trait. Significance is tested by assessing the likelihood ratio of the probability of obtaining the observed frequencies given the data and best-fit parameters, to the probability given a null hypothesis of no association.

Three major problems in establishing statistical significance are (1) dealing with the fact that multiple comparisons are performed, (2) controlling for the nonindependence of tests, and (3) dealing with the fact that power is a function of marker and disease allele frequencies (Freimer and Sabatti 2004). The traditional Bonferroni correction for multiple comparisons is to divide the nominal testwise significance threshold by the number of contrasts to obtain an experiment-wide acceptable false positive rate. This is impractical for a genome scan, where perhaps a hundred thousand markers will be tested—in turn requiring that a single marker *p*-value be less than 0.0000001 to avoid a 5% false-positive acceptance rate. Even for a localized scan of 50 markers, an adjusted threshold of 0.001 is quite stringent, given the small magnitude of risk factors that are likely to accompany many disease loci.

One proposed solution is to accept a high false-positive result in initial scans, and then attempt to confirm the significant association in a replicate study involving independent samples (although a negative result does not

BOX 3.3 Case-Control Association Studies

In a sizeable population of individuals who tend not to mate with relatives, and who otherwise choose their partners at random, only very closely linked polymorphic sites tend to be in linkage disequilibrium. If we ask the question "Is a particular marker more or less frequent in individuals affected with a disease than in unaffected people?" the answer will almost always be independent of the answer obtained for another marker. That is to say, even if there is a significant difference for one marker, chances are that any marker more than about 100 kb away will not show a difference. Or, to put it the other way around, if you wish to demonstrate a statistically robust marker-disease association, you had better sample a site within 100 kb of any SNP that actually affects susceptibility to the disease.

For any disease, there are between a couple and about two dozen polymorphic loci that confer detectably elevated risk of disease occurrence. Detecting these sites among the millions of SNPs in the human genome is quite a challenge. Assuming that a very dense SNP map is available, then more than 25,000 tests must be performed to ensure that at least one marker in every 100-kb interval is sampled. This creates what is known as a multiple comparison problem, as follows.

We demand that there be less than a 1 in 20 chance that any one of the tests is positive before we accept the result as evidence for an association. But, since we would actually *expect* to see a significant result at the 0.05 level once in every 20 tests, we must further divide the testwise significance level of 0.05 by the number of tests performed (25,000), and hence seek a p-value of less than 1 in 500,000. The problem is that this is extremely stringent, so a risk factor would have to be extremely large to be detectable in a sample of even several hundred individuals. Alternatively, several thousand individuals must be sampled. For multifactorial diseases in which the environment interacts strongly with genotype, the task is even more daunting. The statistical issues associated with controlling for the number of tests performed are discussed in detail in Devlin and Roeder (1999).

So, while the technology for genome scans of thousands of individuals is on the horizon, it should be clear that statistical associations are just a first step toward identifying candidate genes. With multibillion-dollar markets as an incentive, the investment of hundreds of millions of dollars can be justified even though there is no guarantee of success, and even though some positive associations will turn out to be statistical artifacts. In practice, researchers may address the latter possibility by accepting an elevated false-positive rate in a first genome scan, in the hope that replication of the study focusing just on the suggestive regions will provide more robust evidence. The number of tests can also be reduced by initially focusing the study on a candidate gene that has been identified by linkage mapping in pedigrees or by

necessarily negate the biological meaning of the initial result). Another approach is to use permutation testing against simulated data sets to establish the expected false-positive distribution and adjust the threshold accordingly. This approach also accounts for nonindependence among markers due to LD, as well as variability of marker frequencies, but the procedures are in the early stages of development.

Steps can be taken to increase power by appropriate experimental design. Two goals of a good design are (1) to decrease the amount of variance of the trait that is due to environmental differences and genetic heterogeneity, and biochemical or comparative genomic approaches. Positive results from such tests are likely to suggest molecular biological approaches to understanding why variation at the locus may affect the disease.

Association tests can be performed in several ways (Sasieni 1997). In all cases, the strategy is to collect genotype data from several hundred affected individuals and from several hundred age- and sex-matched unaffecteds from the same or a similar population. A table can then be drawn documenting the number of times in the case and control groups that each genotype class occurs (homozygotes with 0 or 2 copies of the allele of interest, or heterozygotes with 1):

Number of individuals in each genotype class

	0	1	2	Total
Case	r_0	r_1	r_2	R
Control	s_0	s_1	s_2	S
Total	n_0	n_1	n_2	N

Several χ^2-distributed test statistics with 1 degree of freedom have been developed. We describe one, known as a trend test, here. A trend test models the effect of an additive allele—that is, a similar increment in the likelihood of having the disease is observed when we compare heterozygotes to homozygotes *without* the allele as when we compare heterozygotes to homozygotes *with* the allele. The statistic is:

$$\chi^2 = \frac{N[N(r_1 + 2r_2) - R(n_1 + 2n_2)]^2}{RS[N(n_1 + 4n_2) - (n_1 + 2n_2)^2]}$$

For the example given in the text, if none of the 151 affected individuals were homozygous for the A allele ($r_0 = 0$, $r_1 = 24$, $r_2 = 127$), while 9 of the 191 unaffecteds were homozygous ($s_0 = 9$, $s_1 = 68$, $s_2 = 114$), then $\chi^2 = 26.4$. This is essentially the same result we obtain with allele counts alone when Hardy-Weinberg conditions apply.

If nonadditive effects are suspected, a genotype-case control statistic takes the form:

$$\chi^2 = \sum_i \left[\frac{(Nr_i - Rn_i)^2}{NRn_i} + \frac{(Ns_i - Sn_i)^2}{NSn_i} \right]$$

where the summation is over each of the three genotype classes ($i = 0$, 1, or 2). This example yields $\chi^2 = 27.8$ with 2 degrees of freedom.

Adjustments can also be made for multi-allelic markers (for example, microsatellites as opposed to biallelic SNPs; Nielsen and Weir 1999) and for subpopulation structure if it is detected (Pritchard et al. 2000). An emerging approach is to perform haplotype association studies, in which multisite genotypes are reduced to haplotypes and treated as if they define a single multiallelic locus.

Much statistical theory is still being developed in relation to association studies, and human geneticists are just beginning to get a feel for the structure of large data samples. Sampling, variable patterns of linkage disequilibrium, population divergence, and cultural factors all complicate the analyses, but there is much excitement about the potential for dissecting the genetic components of complex diseases.

(2) to increase the level of linkage disequilibrium in the sample. One strategy for achieving the first is to sample extreme individuals, such as comparing individuals who exhibit the disease at a young age or have a positive family history of disease with hypernormal individuals who are disease-free despite unfavorable indicators or who show extreme physiological covariates. An approach to the second goal is to choose young, isolated populations for study, such as Finns or Pacific islanders. Population expansion from a restricted base allows a disease allele to increase in frequency and carry with it adjacent markers for a period of several hundred generations before

(A)

Linkage mapping	7 cM at Chr 2q37
Construct physical contig	~1.7 Mb, 22 genes
Low resolution case-control association scan	~100 kb, 3 genes
Sequence 10 individuals over 66 kb	179 variants, including 161 SNPs
Strong association with SNP 43	Calpain 10, Intron 3
Haplotype tests → complexity	3 SNP heterozygotes
Population attributable risk	4% Northern Europeans 14% Mexican Americans

SNP affects binding to nuclear factor in human pancreatic extracts and transcriptional activation of a reporter gene in a cell line

Hypothesis: Calpain 10 protease involvement in NIDDM

Figure 3.13 Positional cloning of a candidate complex disease gene. Though falling short of definitive proof, the study schematized here for the NIDDM1 type 2 diabetes susceptibility locus demonstrates one strategy for identifying candidate complex disease genes. After standard linkage analysis (A), low-resolution scanning of the 2-Mb region for SNPs focused the search on 100 kb (B) that included 3 genes and a handful of SNPs with replicated disease association across a number of case-control study populations. The strongest association was observed for heterozygotes for a particular haplotype involving three linked SNPs (red arrows). Follow-up biochemical studies are being used to test the hypothesis that variable transcription of the calpain10 protease contributes up to 15% of the population relative risk for the disease. (After Horikawa et al. 2000.)

drift and recombination begin to obscure associations. This strategy has already been very successful on a handful of occasions, leading for example to the cloning of the diastrophic dysplasia gene (Hästbacka et al. 1994) and a susceptibility locus for type II diabetes in Mexican American and Northern European populations (Figure 3.13; Horikawa et al. 2000).

The major drawback of population-based association studies is their susceptibility to the effects of population stratification. In many cases this may go undetected, but in others—for example, studies of diabetes in Native American populations that have experienced extensive admixture with Europeans—it introduces biases that are difficult to control. Genomic control methods attempt to account for population artifacts by incorporating a parameter that estimates the effect of admixture into the assessment of significance from the genetic model (Devlin and Roeder 1999). Another proposal is to condition marker genotypes from each individual on neutral marker genotypes that are known to differentiate aspects of population structure. These procedures are also highly applicable to mapping of yield and other desirable QTL in crop plants for which accessions with well-defined histories are known (Thornsberry et al. 2001).

Pedigree-based analysis.   The alternative to population-based analysis is to adopt family-based linkage disequilibrium methods such as transmission disequilibrium tests (TDTs), which are independent of population structure but have high resolution because they do not depend on meiosis in short pedigrees. The idea is to test whether heterozygous parents transmit either of their two alleles to a single affected child with equal probability, as shown for the following data from a study of the association between a SNP in the gene that encodes insulin, and IDDM diabetes (Spielman et al. 1993).

	Transmitted allele	
	A	*a*
Observed	78	46
Expected	62	62

$\chi^2 = 8.2, p < 0.005$

The test is inefficient in the sense that only one child is tested per family, and only heterozygous parents are informative. If the parent is homozygous *AA*, all offspring receive the *A* allele and there is no basis for comparison. By contrast, if the parent is heterozygous, under normal circumstances both alleles should be transmitted with equal probability to affected and to unaffected children. In some cases, there may be a bias for one allele to be transmitted more frequently than the other to all offspring, but adjustments can be made for such **segregation distortion.** Only one affected child is sampled per family, so as not to bias the analysis by genetic correlations that occur in pedigrees (Box 3.4). A positive association between the frequency of one of the transmitted alleles and having the disease is expected to arise either because the SNP actually causes the disease, or because it is in linkage disequilibrium with the causal SNP in the population from which the families were drawn. The only assumption is that the marker and disease loci are identical by descent in the population, as a result of which the method is sensitive to genetic heterogeneity.

BOX 3.4 Family-Based Association Tests

A difficult distinction to grasp in terms of disease mapping is that between linkage mapping and linkage disequilibrium (LD) mapping, particularly where family pedigrees are the source of data. Linkage mapping uses the small amount of recombination that occurs in each generation within a pedigree to locate small chromosomal regions containing hundreds of genes. It is like recombination mapping of quantitative trait loci in model organisms, except that many pedigrees are used instead of a single cross with hundreds of progeny. LD mapping, by contrast, uses the very large amount of recombination that has occurred during the history of a population to ensure that any associations that are detected only extend over one or a few loci. (Imagine recombination mapping where you were only typing the children a few thousand generations removed from when a cross was first established: recombination would have occurred between most pairs of markers, so the resolution would be very high, but you would have to sample at an extremely high density.) Case-control population-based LD mapping was explained in Box 3.2, and family-based association tests also utilize the low LD in populations.

Family-based linkage disequilibrium mapping of genetic diseases is based on unequal transmission of alleles from parents to a single affected child in each family. If only a single affected individual is typed in each family, and all of the families are unrelated genetically, then the only sites that can be associated statistically with a disease are those that contribute directly to the disease susceptibility, or markers in linkage disequilibrium with such sites in the general population. The families must be unrelated, as this ensures that the chromosomes are sampled without biases due to shared genetic factors. As we have seen, linkage disequilibrium among sites generally extends only 100 or so kilobases along a chromosome. Consequently, the resolution of LD mapping is at the level of individual genes, and in some cases it may be possible

to resolve which SNPs within a gene are likely to affect the disease susceptibility.

There are actually several forms of family-based linkage disequilibrium tests. The approach mentioned in the text is known as the **transmission disequilibrium test,** or **TDT.** Information is extracted from the ratio of transmission of the two alleles from heterozygous parents to the affected child. Genotypes are obtained for the "triad" of two parents and their child. If both parents are homozygous (whether or not for the same allele), offspring always receive the same genotype independent of disease status, so this situation is uninformative. If, by contrast, both parents are heterozygous, then both alleles in the affected child are informative. If only one parent is heterozygous, it is always possible to deduce which allele was transmitted, so long as the genotype of the second parent is known. Even if the second parent's genotype is unknown, if the child has a different heterozygous genotype for a multiallelic marker, it is still possible to infer which allele was transmitted.

With this information, a Z-test can be designed to test the association of each allele with the disease (Spielman et al. 1993). The Z-test statistic takes the form of the difference between the observed and expected number of transmitted alleles, divided by the variance of the observed number of transmitted alleles. Suppose that in a set of triads, the A allele is transmitted b times from heterozygous parents, and the other allele(s) c times. Then:

$$Z_{TDT} = \frac{b-c}{\sqrt{(b+c)}}$$

has a two-sided p-value that is chi-square distributed. The significance can be assessed by squaring the Z_{TDT} score and comparing this value with the chi-square table with 1 degree of freedom. Corrections for multiallelic markers, as well as for situations in which there is biased segregation of alleles in both affected and unaffected offspring (segregation distortion), have been employed, as have modifications to apply the test to contrast extreme individuals for continuous (as opposed to discrete) traits (Allison 1997).

An alternative test, the **sibling-transmission disequilibrium test (S-TDT)** proposed by Spielman and Ewens (1998), uses information from one affected and one unaffected child with different genotypes in a set of unrelated families. In this case, the test statistic is

$$Z_{\text{S-TDT}} = \frac{Y - T}{\sqrt{V}}$$

where Y is the number of A alleles in the affected children in all of the families, T is half the total number of A alleles in all of the pairs of children (which is equivalent to the expected value of Y), and V is the variance of Y. Each of these quantities is easy to calculate given the observed number of homozygous and heterozygous affected and unaffected siblings. If there are r homozygotes and s heterozygotes for the allele, then $T = r + s/2$, and the variance V is the sum of the number of sibling pairs that are discordant homozygotes plus $1/4$ times the number of pairs in which only one of the siblings is heterozygous. An example of a situation in which the S-TDT may be preferable is for late-onset diseases, where parental genotypes are not always available.

To illustrate these tests, consider the data set in the table below, which shows the genotypes at a single biallelic site in two parents, an affected child, and an unaffected child for a hypothetical disease. Applying the TDT, 11 heterozygous parents in 8 of the triads transmit the A allele 9 times to affected children and the a allele twice. Note that two children provide no information because both parents were homozygous, while three children are counted twice each because both parents were heterozygous. Then:

$$Z_{\text{TDT}} = \frac{9 - 2}{\sqrt{(9 + 2)}} = \frac{7}{\sqrt{11}} = 2.11$$

Applying the S-TDT, the A allele is present 14 times in the affected siblings compared with an expected number of $(6 + 8/2) = 10$ times. The variance is $(1 \times 1 + 6 \times 0.25) = 2.5$, where the discordant homozygotes are found in the eighth family on the list. Thus,

$$Z_{\text{S-TDT}} = \frac{14 - 10}{\sqrt{2.5}} = 2.53$$

Neither of these measurements is significant with such a small sample size, but both suggest a trend toward biased segregation of the A allele. For biallelic markers, the same result is obtained for the alternate allele.

Parent 1	Parent 2	Affected	Unaffected	*b*	*c*	*r*	*s*	*Y*
AA	*Aa*	*AA*	*Aa*	1	0	1	1	2
Aa	*Aa*	*Aa*	*aa*	1	1	0	1	1
Aa	*aa*	*Aa*	*aa*	1	0	0	1	1
AA	*AA*	*AA*	*AA*	0	0	2	0	2
Aa	*Aa*	*AA*	*Aa*	2	0	1	1	2
Aa	*aa*	*Aa*	*aa*	1	0	0	1	1
aa	*Aa*	*aa*	*aa*	0	1	0	0	0
Aa	*Aa*	*AA*	*aa*	2	0	1	0	2
Aa	*AA*	*AA*	*Aa*	1	0	1	1	2
AA	*aa*	*Aa*	*Aa*	0	0	0	2	1
			Sum:	9	2	6	8	14

(A) Initial replication

Locus	T	U	Ratio	Significance
ABCC8	26	12	2.2	0.012
PPARγ	81	104	0.8	0.045
IRS1	30	26	1.2	0.30
ADRB2	96	124	0.8	NS
INS	104	115	0.9	NS
IRS1	14	21	0.7	NS
KCNJ11	138	154	0.9	NS
TNF	14	13	1.1	0.42

(B) Further replication

Study	N	Risk ratio	Significance
Sibships	1130	0.74	0.071
Case-control			
Scandinavia	481	0.88	0.10
Quebec	127	0.71	0.08
Total	2071	0.78	0.002

(C)

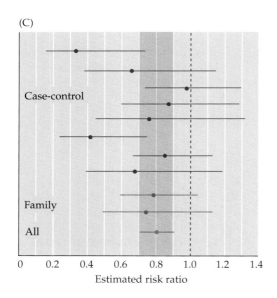

Figure 3.14 PPARγ alanine substitution and type 2 diabetes risk. Published reports had implicated common polymorphisms in 8 different genes with type 2 diabetes susceptibility. (A) Initial replication studies on a set of 333 parent-offspring trios from Scandinavia confirmed just two of these associations (T = transmitted alleles, U = untransmitted; four nonsignificant studies actually showed trends in the opposite direction to that initially reported). (B) Follow-up replication in three other cohorts confirmed only the association with the peroxisome proliferation-activated receptor-γ (PPARγ) Pro12Ala, at marginal levels. (C) Taken together, however, the consistent relative risk estimate and high frequency (~85%) of the susceptibility allele results in the SNP being associated with a population attributable risk for the disease as high as 25%. (After Altschuler et al. 2000.)

A battery of modified family-based tests are now being applied, particularly in replication studies (Jorde 2000). The combination of case-control and transmission disequilibrium tests may provide the best available evidence for association of a marker with a disease (Figure 3.14). Where an association is due to linkage disequilibrium (often reflecting proximity) rather than causation, it is likely that multiple SNPs in proximity to the locus will show association. Whether or not the strongest association is indicative of the closest linkage to the true causal SNP will depend on the precise LD structure at the locus. Since initial scans utilize SNPs at intervals of 10 kb or more, and SNPs typically occur every kilobase, a final survey of all of the SNPs in a locus is required to identify the most likely causal site.

Finally it must be noted that, given the high variance of background and environmental contributions to the effects of disease loci, there is in fact no guarantee that causal SNPs or clusters of SNPs can be identified. This fact is particularly sobering when it is realized that genes with antagonistic effects on a trait can lie adjacent to one another, and that regulatory and structural polymorphisms can have very different effects. Ultimately, suggestive statistical associations must be confirmed with biochemical, cell biological, and physiological methods.

EXERCISE 3.3 *Perform a case-control association test*

Two parallel association studies between a candidate gene and skin cancer are performed in New York City and Miami, Florida. The numbers of cases and controls who are homozygous for a 3bp insertion or deletion polymorphism or who are heterozygous are as follows:

	New York				Miami		
	Insertion	Hetero-zygous	Deletion		Insertion	Hetero-zygous	Deletion
Case	221	198	55	Case	171	236	83
Control	279	212	35	Control	189	244	77

Calculate the probability of association between the deletion polymorphism and the disease for each population separately and for the combined population, using the formula:

$$\chi^2 = N\,[N(r_1 + 2r_2) - R(n_1 + 2n_2)]^2 / RS[N(n_1 + 4n_2) - (n_1 + 2n_2)^2]$$

Discuss the possible reasons for any discrepancy in the conclusions from the two studies.

ANSWER: Following the method in Box 3.3, you should obtain the following values:

Quantity	New York	Miami	Combined
r_1	198	236	434
r_2	55	83	138
n_1	410	480	890
n_2	90	160	250
N	1,000	1,000	2,000
R	474	490	964
S	526	510	1,036

Plugging these values into the formula results in χ^2 values of 7.6, 0.8, and 6.9 for New York, Miami, and the combined population, respectively. Prima facie this suggests little effect of the polymorphism in Miami, but suggests a significant effect at $p < 0.01$ in New York and across the two populations.

(continued on next page)

Exercise 3.3 (continued)

Overall, the New York frequencies closely match Hardy-Weinberg expectations assuming the insertion frequency is 0.3, whereas the Miami figures fit an insertion frequency of 0.4. Consequently, there is evidence for allele frequency differences between the two populations, and population stratification could affect the overall conclusion. Also, skin cancer is likely to be of higher prevalence in Miami, so genotype-by-environment interactions could occur. Since the insertion polymorphism is less frequent than expected in cases than controls in New York, it could be more protective in lower radiation.

SNP Discovery

Resequencing

SNPs are discovered by comparing sequences derived from different chromosomes. The probability of detecting a SNP for which the more rare of two alleles has a frequency of p in a comparison of n individuals is given by $P = 1 - (1 - p)^{2n}$. For an intermediate frequency allele, this probability rises quickly from less than 50% for two chromosomes to more than 90% in a sample of 10 chromosomes (5 individuals). Even for rare alleles with a frequency of 0.1, a sample of 10 chromosomes gives a reasonable probability of detection, but most such alleles will be missed in a comparison of just 2 chromosomes.

SNP identification.   A number of different sources of chromosomes can be used to identify SNPs. The most obvious for many species is a comparison of the whole genome sequence with existing cDNA sequences deposited in GenBank, which typically derive from different strains. For example, at least two-thirds of all discrepancies between the Celera whole genome sequence of *Drosophila* and cDNA reports generated over two decades are true SNPs when verified in a dozen wild-type strains. A drawback of this approach is that GenBank files do not supply sequence quality scores, so there is no way to determine whether a difference is the result of a sequencing error in the original cDNA sequence.

A more conservative approach that has been reported to result in better than 80% true SNP identification is to assess EST sequence quality using a statistical assessment of the likelihood that a called sequence variant is due to a sequencing error (Marth et al. 1999). Some SNP databases now allow individual investigators to download the trace files so they can calculate phred or other quality scores for themselves, or to visually compare their own sequence traces with public sequence reads. Software such as Poly-

Bayes can be used to compare these scores with a model of error probabilities in finished genome sequence, generating a SNP probability score.

An alternative to EST sequence comparisons is to compare genomic sequences derived from different individuals. This ensures the identification of SNPs in noncoding regions of genes, which may actually account for a large proportion of quantitative trait locus effects. One source of alternative genomic sequences is scaffold fragment overlaps. Some genome sequencing projects used multiple libraries, each derived from a diploid individual, in the creation of the BAC or P1 clones that were shotgun sequenced. Typically, in regions of clone overlap, at least one clone is sequenced to a high level of accuracy, such as 1 error in 50,000 bases. If the other clone is also a finished sequence, the two contigs can be compared to identify SNPs: failure to detect any differences may be because the clones derived from the same chromosome of a single library, or because the region is monomorphic in the sample of two chromosomes.

SNP verification.   Verification of SNPs can be pursued with a variety of approaches. The most direct is to design primers for amplification of a half-kilobase including the SNP, and to resequence the fragment in a sample of at least 10 individuals (preferably up to 100 individuals), from a variety of populations. The phred supplement PolyPhred has been written specifically to identify polymorphic sites in multiple sequence alignments (Nickerson et al. 1997). Due to the expense involved in sequencing, an option is to pool up to five DNA samples prior to amplification and watch for the appearance of double peaks in the sequence trace specifically at the candidate SNP. Rare SNPs will only result in a small secondary peak, but this can be confirmed by sequencing the original samples individually.

A more radical approach is to use a completely different technology in the verification step. These techniques can include any of the genotyping methods described in the next section, or a different high-throughput sequencing method such as sequencing by hybridization. Sequencing chips are custom-fabricated oligonucleotide chips (very similar to Affymetrix oligonucleotide microarrays described in Chapter 4) that are modified for DNA-DNA hybridization (Figure 3.15; Wang et al. 1998). Once a sequence is known, a chip can be designed to detect the complete set of polymorphisms in the sequence. Each nucleotide in the sequence is the subject of eight 15-mer oligonucleotides that are synthesized by photolithography on the silicon chip. The central nucleotide of each oligo represents the four possible types, on both strands. For a 50-kb segment of the genome, 50,000 of these sets of eight oligonucleotides are arrayed on the chip and probed with pools of fluorescently tagged PCR products generated from a set of individuals to be typed, with one chip per individual. The strongest pair of signals corresponding to the two strands indicates the sequence of each site in a homozygote, or a quartet of signals of similar intensity signifies a heterozygote.

Figure 3.15 Sequencing by hybridization. Sequencing chips consist of four rows of 25-mers in which the central position is varied. Each successive column represents a shift of one nucleotide along the sequence. Hybridization of DNA from two different homozygotes for the central position results in differential hybridization, as shown. (After Wang et al. 1998.)

Sequence-Free Polymorphism Detection

Several methods have been developed that detect SNPs without sequencing. These are most useful for mapping SNPs in previously unsequenced DNA (for example, in nonmodel organisms of agronomic, ecological, or evolutionary interest) or in regions of low polymorphism (such as nonrecombining regions of sex chromosomes). These methods also have applications in mutation detection, where the goal is to identify a point mutation that distinguishes a mutagenized chromosome from the progenitor.

After mapping a mutation to the vicinity of a locus, a small number of fragments covering the region are amplified, sequentially screened for SNPs, and only those fragments that seem to be different are sequenced. Several of these methods rely on the observation that DNA tends to denature in bubbles, so that a small mismatch caused by a SNP in heteroduplex DNA will quickly propagate over several hundred bases once the two strands begin to separate. **Heteroduplex DNA** refers to double-stranded molecules formed by the reannealing of two strands derived from different alleles.

DHPLC, which stands for denaturing high performance liquid chromatography, uses the long-established HPLC technique to detect differences in stretches of DNA up to several kilobases in length (Underhill et al. 1997). The chromatography step takes several minutes per assay, but 96-well plates are accepted by robots specifically designed for DHPLC (for example, Transgenomics' Wave®) and can be left to run overnight. Crude PCR products are loaded onto the chromatography column and an organic solvent is used to elute the fragments. Because heteroduplexes have a slightly lower melting point than homoduplex DNA in which there are no mismatches, they emerge less quickly and can be detected as a double peak in the elution profile. Different haplotypes consisting of several SNPs can also be distinguished, making this method suitable for relatively high-throughput genotyping as well.

Two related procedures have been developed for the detection of allelic variants in small fragments up to several hundred bases in length. **DGGE, denaturing gradient gel electrophoresis,** is a low-tech assay that can be established in any laboratory. It uses acrylamide gels to separate the DNA fragments, and ethidium bromide or silver staining to detect the molecules. The PCR primers used to amplify the target sequence from individual genomic DNA samples include a 40-bp "GC clamp" that will not denature under the electrophoresis conditions used in the assay. The DGGE gel is poured while mixing two acrylamide solutions, one of which includes the denaturant urea, resulting in a denaturing gradient. Double-stranded heteroduplex loaded onto such a gel initially migrate into it at a uniform rate. At a particular point in the gradient, the heteroduplex target separates, resulting in a Y-shaped molecule in which the base of the Y is formed by the clamp, and the heteroduplex is suddenly retarded in its migration. Homoduplex will migrate further in the gel, so polymorphisms can be detected by the presence of upper and lower bands (left side of Figure 3.16).

SSCP, single-stranded conformation polymorphism analysis, utilizes intramolecular base-pairing inherent in all single-stranded nucleic acid molecules to detect SNPs. The target PCR product is denatured in weak alkali

Figure 3.16 DGGE and SSCP analysis.   In DGGE, duplexes stop migrating when a Y-shaped molecule is formed by denaturation up to the GC clamp, which occurs earlier on the urea gradient for heteroduplexes than for homoduplexes. In SCCP, the four classes of single-stranded molecule each have their own rate of migration. "C" and "W" stand for the Crick and Watson strands of the denatured DNA product.

solution and then loaded onto a standard acrylamide gel, but the electrophoresis is performed in a chamber that bathes the gel at a constant temperature of 4°C. Under these conditions, a large fraction of the denatured strands remain single-stranded as they enter the gel, and instead of heteroduplexes they form intramolecular secondary structures. Slight variation in the folding or stability of these structures results in differential migration through the gel (right side of Figure 3.16). The products are again detected by staining the gel with standard dyes, or by radioactive labeling. Multi-SNP haplotypes can also be detected, as each allele will produce a distinct migration pattern. The limitations of the method are that it can be persnickety to identify the correct gel conditions to achieve good separation (and not all SNPs will be resolved); only short sequences can be processed; and it is not as high-throughput for genotyping as most of the methods discussed in the next section.

TILLING, which stands for **targeting induced local lesions in genomes** (Till et al. 2003), relies on the ability of single-strand-specific nucleases to cleave heteroduplexes. PCR-amplified fragments of DNA from two different individuals are mixed together, denatured, and reannealed, so that all polymorphic sites are exposed as mismatches in the hybrid duplexes. These mismatches are recognized and cleaved by the nuclease, producing two fragments in place of one. If the two ends of the PCR product are labeled with fluorescent dyes, the location of a SNP can be inferred from the difference in length of fragments separated on an automated sequencer (Figure 3.17). The method lends itself to very high-throughput screening, and has been used both to detect induced mutations and to find polymorphisms in natural populations. For the major model organisms, PCR primer pairs for all loci are being designed to allow rapid screening for nucleotide variants at any location in the genome.

SNP Genotyping

The ability to genotype large numbers of SNPs in large numbers of individuals rapidly and cost-effectively is essential for many applications, from linkage and association mapping to mutation detection and diagnosis. As of 2004, no single method has emerged as a general panacea for the range of problems that face individual investigators, from cost and access to technology to reliability and precision. New methods are proposed and tested on a monthly basis, resulting in a bewildering range of choices, with no clear criteria for evaluating which method is most appropriate for a given study.

What is clear is that it is now feasible for single research groups, once they have invested the time and effort required to establish a method, to genotype thousands of SNPs per week, and for large consortia to genotype several hundred thousand SNPs per week. The development of microfluidic robotic workstations will mean that at least another order of magnitude improvement in throughput can be expected over the next decade—including the development of methods for genotyping specimens in the field or even at home. In this section, we simply review the basic principles of some

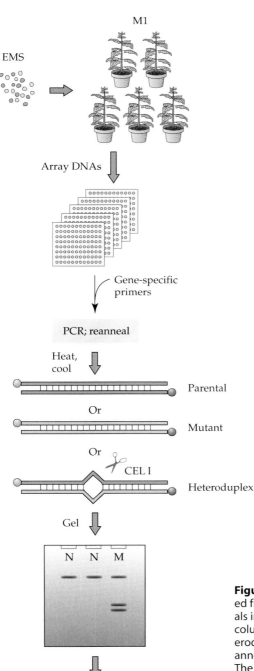

Figure 3.17 Tilling. DNA is extracted from homozygous mutant individuals in 96-well format. Pools of rows and columns are amplified by PCR, and heteroduplexes form wherever a mutant anneals to a parental-type molecule. The strands are end-labeled and after cleavage with Cel I, cleaved fragments migrate more rapidly than full-length strands on an acrylamide gel. Cross-comparison of pools leads to identification of mutant individuals.

of the major methods in use (Mir and Southern 2000), starting with procedures based on traditional molecular biological tools and equipment.

Low-Technology Methods

With the exception of the Invader assay, all current SNP genotyping methods depend on specific amplification of the DNA sequence surrounding the site to be genotyped, which is generally achieved using the PCR. By far the simplest way to detect a SNP within an amplified fragment is to monitor the cleavage of the fragment upon digestion with a 4-cutter or 6-cutter restriction endonuclease. These enzymes usually require a palindromic recognition site—that is, a site that reads the same sequence on the reverse strand, such as GATC or CCGG. Between one-fifth and one-half of all SNPs actually lie within just such a short palindromic sequence, and provided that a relevant restriction enzyme is commercially available, the SNP can be detected rapidly and for relatively little cost simply by running fragments out on agarose gels (Figure 3.18; Wicks et al. 2001). This PCR-RFLP, or Snip-SNP, method is a simplification of the **restriction fragment length polymorphism** method that was initially devised for detecting restriction site variants by whole genome Southern blots.

It is also possible to turn most SNPs into **derived cleavable amplified polymorphic sequences (dCAPS)** by clever design of the PCR primers to introduce a restriction site associated with one allele, as shown in Figure 3.19 (Neff et al. 1998). Typically the cleaved product is just 20 or so nucleotides shorter than the original amplified product, so small acrylamide gels are used to separate the fragments, which can be visualized with simple ethidium bromide staining. The method costs just cents per assay and is ideal for screening a relatively small number of sites in a large population.

A higher throughput, but often less accurate, genotyping method is **ASO,** or **allele-specific oligohybridization** (Stoneking et al. 1991). The technique is basically the same as dot-blotting (see Chapter 4), except that the probe is just 15 nucleotides long instead of a complete cloned fragment. Fifteen-mer oligonucleotides have melting temperatures in the range of 30–60° C, depending on GC content, and a single base mismatch in the middle of the probe typically results in a 3–5° change in melting temperature. The target sequence is amplified in 96- or 384-well format from a series of individuals, blotted onto a nylon filter under vacuum, and crosslinked in place by UV irradiation. The ASO probe is end-labeled, hybridized to the filter for several hours, and washed off at the predetermined stringency conditions, only leaving a signal if there is a perfect match.

Minisequencing Methods

A philosophically different approach to genotyping is to actually resequence each allele in a sample of individuals. Instead of carrying out a complete sequencing reaction, which is relatively expensive and takes hours rather

TTTTGATTTTTCAGCAGAATTTGCGGGAAACGATTCGAGTCCCAATTTTA
AATCTCAATTATTTGCAAAGGATTCTCAAATATCAGTATTAAAAACGAAA
CTTTCTGAAGTGGAAAGAAAATTTGAAAAACGTAGTCAAGATTATTACGA
GATGAAAGCTGAAAAGGAATGTTAGAGAAGAGAGTTGAAAACCAAAAG
TATCAAGTCATGAAATGGATAGTCTTCAAGAATTGAAATTAGCCAGGTAA
TTATATTAACATCTTGACGTGTTGTTTTGCATTTAATTCAAGTTT TCTAG
[A/G]CAAAAAGCACAAGATCAAAAAGAGAAAGCAGTGGAGGAGTGTAAC
ATGCATAAGAGAAAAATAGTTGGTTTGGAAGAAGAAATTCGTGCGATGGT
CGAACAGTTGAGGCTGGCAAAGTTCAATCTGAATGAGAACAAAAAAGAAT
TTGATGAGTACAAGAACAAGGCGCAGAAAATTTTGACAGCTAAAGAGAAA
CTGGTGGAGTCGCTGAAGTCAGAGGTTTGAGAGATTTGACCGAGGATTGA
TATTAAAATTCAATTCAGCAAGGAATCGGATCCAGTGATCGTCCTGTTCA

Figure 3.18 Bulked segregant mapping of mutations using Snip-SNPs. A polymorphic SNP restriction site in a PCR product, such as this XbaI site at position 20210 in clone T24B1 from *C. elegans* chromosome I, allows rapid genotyping of individuals with the two alleles. In bulked segregant analysis, pools of wild-type and mutant individuals from the F₂ population of a cross between the two parental types are genotyped at multiple markers across a region. The relative intensity of the single higher molecular weight band, and the two bands produced by digestion of each marker, provides a ratio of wild-type to mutant alleles in the populations. In this example from Wicks et al. (2001), side-by-side comparison of wild-type (left) and mutant (right) pools led to rapid localization of a Mendelian mutation in the nematode to within 100 kb.

than minutes, methods have been developed in which only one or a few bases are sequenced. Some of these are performed in microtiter plates using detectors that monitor fluorescence over a range of wavelengths and temperatures, such as the ABI Prism® 7700 Sequence Detection System, and others are being adapted for microarray formatting, or for microcapillary elec-

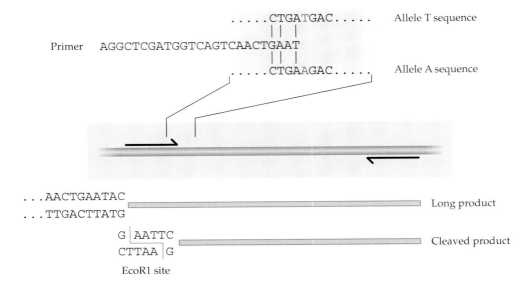

Figure 3.19 dCAPS.   A derived cleavable amplified polymorphic sequence (dCAPS) is generated by using a PCR primer that has a mismatch a few bases before the SNP. In this example, the second last base is an A instead of a C, as a result of which the sequence of the amplified fragment includes a new restriction site for EcoR1 (GAATTC) for the A allele of the SNP (shown in blue at the top). By contrast, the T allele sequence (GAATAC) does not create a restriction site. Cleavage with EcoR1 results in a shorter fragment for the A allele.

trophoresis. The cost-effectiveness of this approach is a function of the density of SNPs under consideration: If each genotype costs a couple of dollars, and there are several SNPs within less than a kilobase, complete sequencing may be more efficient. However, if the SNPs are more widely dispersed, minisequencing will be cheaper once a particular method has been established in a particular laboratory.

Single-base extension.   The conceptually most direct of these methods is **single-base extension,** or **SBE,** shown in Figure 3.20. In theory, DNA polymerase provides much higher specificity for detection of single-base differences than does differential melting temperatures of hybridized sequences. The idea behind SBE is to incorporate a single dideoxy fluorescently labeled nucleotide at the SNP position immediately adjacent to the 3' end of the genotyping primer. Since only the base complementary to the SNP will be incorporated, SBE provides a direct readout of the genotype, including differentiation of homozygotes from heterozygotes, on the basis of the wavelength of the fluorescence.

Because the reagents and protocol for SBE are so similar to that of regular automated sequencing, the challenge in making the approach broadly

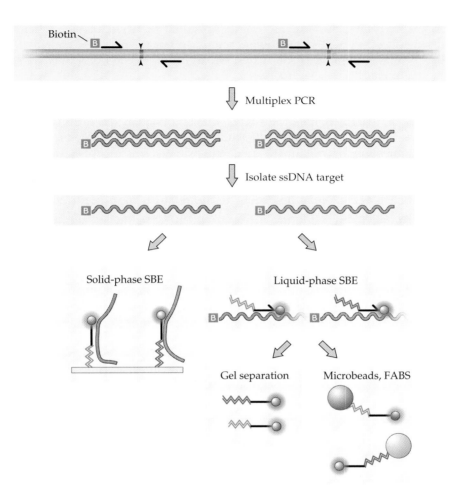

Figure 3.20 Single-base extension methods.   Multiplex PCR is performed to amplify a number of different SNP-containing fragments from genomic DNA. Methods such as trapping a biotin-labeled strand with streptavidin-coated beads in microtiterwells are then used to purify the target strand, which is hybridized to the allele-specific SBE probe. A variety of approaches can then be used to detect the incorporation of the appropriate fluorescently labeled single nucleotide. In solid-phase methods, the target is hybridized to the probe directly on a microarray, and the reaction is performed on the array. In liquid-phase methods, the SBE reaction is performed in solution. The products can then be separated by electrophoresis on the basis of size differences between the probes (possibly using new automated sequential injection techniques to stagger oligonucleotides, such as Molecular Dynamics' MegaBACE apparatus). Or they can be crosslinked to red/orange fluorescence-coded microbeads by way of a molecular "zip code" on the probe, then sorted by fluorescence-activated bead sorting. (After Chen et al. 2000.)

useful has been to develop multiplex assays (that allow simultaneous geno-typing of several SNPs in a single reaction) and/or to apply it to a microar-ray format allowing thousands rather than hundreds of individuals to be scored in parallel.

A gel-based solution to the multiplexing problem is to generate a series of genotyping oligonucleotides, each of which differs in length by a couple of nucleotides. The polymorphic targets are amplified in separate PCR reac-tions as required to generate sufficient templates, then are pooled together. After annealing to a mixture of annealing primers, single-base extension is performed in the presence of each of the dideoxy fluorescent nucleotides. The products are separated by gel or capillary electrophoresis, and the iden-tity of each SNP is read according to the relative migration rate of the oligonucleotide.

Another innovation is to couple the SBE products to color-coded microbeads by way of a molecular "zip code," and then use flow cytome-try to decode the beads (Chen et al. 2000). Bead based technologies are described in the next chapter (see Figure 4.17) in the context of gene expres-sion profiling. Companies such as Illumina® and Luminex® offer variations on the theme, and the method can be adapted for use in academic labora-tories, offering a cost-effective approach to genotyping hundreds of SNPs in large numbers of individuals.

An alternative to fluorescence detection is separation of extended prod-ucts on the basis of subtle differences in charge and size, using mass spec-trometry. Resolution by this method is achieved by inclusion of a single dideoxy nucleotide in the extension reaction. If the SNP allele is comple-mentary to the dideoxy nucleotide, the extension will terminate immedi-ately; otherwise it will extend for a few bases until the relevant base is encountered in the flanking sequence. The two different reaction products are easily distinguished by MALDI-TOF mass spectrometry as described in Chapter 5, and each separation takes just a few seconds to perform. Sequenom® offers a 384-well chip array platform for highly automated par-allel SNP genotyping by mass spectrometry (Tang et al. 1999).

Pyrosequencing.   **Pyrosequencing** is another minisequencing method that has been highly automated, as described in Ronaghi (2001) and illus-trated in Figure 3.21. The idea is to perform 96 parallel minisequencing re-actions in a microtiter plate, using fluorescence to detect incorporation of each nucleotide onto a primer in real time as nucleotides are cycled in and out of the microtiter wells. The fluorescence that is detected is actually de-rived from enzymatic cleavage of the pyrophosphate that is liberated from a dNTP when it polymerizes with a growing chain. The target poly-morphism is amplified from genomic DNA and mixed with a genotyping primer that binds a few bases upstream of the SNP. The first couple of re-action cycles allow incorporation of the common bases between the 3′ end of the primer and the SNP. In each cycle, the complementary nucleotide is added, DNA polymerase catalyzes incorporation and release of py-

(A)

(B)

Heterozygote

Nucleotide added

Figure 3.21 Pyrosequencing. (A) Pyrosequencing is based on cycling reactions, one of which generates ATP when a specific base is incorporated (top), one that converts ATP to light (center), and one that removes unincorporated nucleotides as well as excess ATP prior to the next cycle (bottom). (B) By controlling the order in which nucleotides are added, minisequencing traces known as pyrograms are produced over a matter of seconds. In this example, an A/T SNP is distinguished between homozygotes and heterozygotes. The height of each peak produced by the flash of light is proportional to the number of nucleotides incorporated.

rophosphate, and the pyrophosphate is used by luciferase to convert luciferin to oxyluciferin, releasing a detectable flash of light. The enzyme apyrase continuously removes excess unincorporated nucleotides and ATP, allowing a fresh cycle to commence upon addition of the next nucleotide after a few seconds. At the SNP, both nucleotide types are incorporated in successive reactions. On the pyrogram (Figure 3.21B), the first homozygote will produce a peak the same height as the preceding nucleotides, then a gap before the next common nucleotide is incorporated. The other homozygote will show a gap and then a peak, while heterozygotes will show two smaller peaks.

Pyrosequencing can be used to genotype several closely linked SNPs within up to 20 bases, and to distinguish complex polymorphisms involving indels and base substitutions. Unlike the other minisequencing methods, it is semi-quantitative, and so can be used to estimate allele frequencies in a mixed sample derived from tens or hundreds of individuals, with reported accuracy to less than 5%.

EXERCISE 3.4 *Designing a genotyping assay for a double polymorphism*

Design three different strategies for genotyping the two polymorphisms indicated in the following sequence:

GGTGCAGGCATGCAGAAGCGTG (G/A) CATG (T/C) GGAACAATGCTGAGTCCTAAT

ANSWER:

1. Use dCAPS finder 2.0 (**http://helix.wustl.edu/dcaps/dcaps.html**) to generate primers and restriction sites that will recognize the two sites from either direction. For example, the primer GGTGCAGGCATGCA-GAAGGGTG terminates one base before the (G/A) polymorphism, and will produce a restriction site for HphI (GGTGA) with the A allele, but no restriction product with the G allele. The primer ATTAG-GACTCAGCATTGTTGC terminates one base before the (T/C) on the reverse strand, and will produce a restriction site for CviRI (TGCA) with the T allele, but no restriction product with the C allele.

2. Use PCR to amplify the whole sequence, and then a forward primer TGCAGGCATGCAGAAGCGTG to initiate a single base extension reaction for the G/A, and a reverse primer TTAGGACTCAGCATTGTTCC for the T/C polymorphism.

3. Use a single forward primer such as GGTGCAGGCATGCAGAAGCG to initiate a 7-nucleotide pyrosequencing reaction that will type both sites in the same reaction. You would add the nucleotides in the order T, G, A, C, A, T, G, T, C, G to generate a pyrogram.

Homogeneous Fluorogenic Dye-Based Methods

Homogeneous assays are those that are carried out in a single reaction in solution. They have the advantage that the amplification and genotyping steps are coupled in a single process. There is no need to purify or transfer reagents and products, which saves labor and time and eliminates a potential source of error. All of the assays described in this section utilize fluorogenic dyes that have overlapping yet distinct spectral ranges between 500 and 650 nm and can be detected during the course of a thermal cycling reaction by a camera located beneath a microtiter plate or behind glass capillaries. Each assay uses a different "trick" to induce a change in fluorescence spectrum by the coupling or uncoupling of a "reporter" dye and a "quencher" dye. When both dyes are in close proximity, such as at either end of a 15-mer oligonucleotide, no fluorescence of the reporter dye is detected.

TaqMan.   The first of these assays to be developed was the **TaqMan 5′ exonuclease assay** (Holland et al. 1991). This assay is similar to ASO, except that the two allele-specific oligonucleotides are put in competition with one another in a single reaction, and specificity is achieved by cleavage of the reporter dye from hybridized oligonucleotides. The reaction is set up with four primers: two regular oligonucleotides flanking the SNP that prime amplification of a fragment up to 150-bp long, and two probes with or without the potential mismatch close to the 5′ end of the probe. Each probe has a reporter dye such as FAM or TET attached to the 5′ end, and a quencher dye at the 3′ end. When a Taq polymerase enzyme molecule meets a bound probe during a polymerization cycle, it digests the probe in a 5′ to 3′ direction, liberating the reporter dye to fluoresce (Figure 3.22A). If the correct conditions are chosen, the probe with the mismatch will bind to the target to a significantly lower degree, so it will not be digested by Taq (which requires a double-stranded template in order to catalyze excision). Consequently, the fluorescence spectrum due to FAM or TET is indicative of the genotype.

PE Biosystems has developed a series of kits for applying this methodology to genotyping a variety of common human polymorphisms and mutations. TaqMan probes can also be synthesized de novo for use with the ABI sequence-detection hardware and software in a wide range of applications that go beyond genotyping, as for example, in quantitation of RNA expression levels.

Molecular beacons.   Molecular beacons work on the principle that quenchers will stop absorbing fluorescence from the reporter dye without cleavage of the probe, so long as the distance between the two dyes is great enough. The beacons are essentially ASOs flanked by complementary arms that are five nucleotides long. When free in solution, these molecules fold into a fairly stable hairpin structure in which the reporter (FAM or TET, among others) is adjacent to the quencher. Different combinations of dyes can be synthesized with attachments to either the 5′ or 3′ end of the beacons. In the presence of a template, the beacon hybridizes specifically to the perfect match, separating the two dyes by 26 bases, and allowing the reporter to fluoresce (Figure 3.22B). Specificity of the hybridization is apparently increased by the extra energy required to destabilize the hairpin, increasing the discrimination over normal ASOs by several degrees. This allows the fluorescence to be measured directly during the annealing step of the PCR cycles. By using four different fluorophores, it is possible to detect both alleles of two different SNPs in a single duplex reaction (Marras et al. 1999).

Dye-labeled oligonucleotide ligation.   The dye-labeled oligonucleotide ligation assay (DOL) relies on fluorescence resonance energy transfer (FRET) to detect covalent linkage of oligonucleotides containing two different fluorophores as a result of allele-specific ligation (Figure 3.22C;

Figure 3.22 Fluorogenic dye-based genotyping methods.   (A) In the TaqMan assay, Taq polymerase enzyme digests the probe and liberates a reporter dye. Mismatch probes are not digested by Taq, so the resulting fluorescence spectrum is indicative of the genotype. (B) Molecular beacons are initially held in a stable hairpin structure. However, after hybridization to a template match, the reporter and quencher dyes are separated, allowing the reporter to fluoresce. (C) Dye-labeled oligonucleotide ligation, or DOL, relies on fluorescence resonance energy transfer (FRET) to detect covalent linkage of oligonucleotides containing two different fluorophores as a result of allele-specific ligation.

Chen et al. 1998). Ligation assays have been used in conjunction with PCR as a genotyping method for some time, but the requirement that the oligonucleotides be separated according to size on an acrylamide gel to detect the ligation event is a considerable drawback.

The principle of the DOL method is that the annealing step of the PCR reaction to amplify the target SNP region is performed at a relatively high temperature for 35 cycles, at which point the annealing temperature is dropped to allow the three genotyping probes to bind. A 5′ oligonucleotide probe binds adjacent to the SNP, and carries the donor fluorophore fluorescein. Two 3′ probes differ at the first base, which is specific for the SNP: after annealing to the template, a thermostable ligase can only ligate the two oligonucleotides together if the 5′ nucleotide of the 3′ probe is complementary to the SNP. After ligation, the donor transfers its fluorescence energy to the now nearby reporter dye (such as ROX or TAMRA) on the 3′ portion of the probe, and these dyes are stimulated to fluoresce at their own specific wavelength. In this way, ligation is accompanied by the generation of an allele-specific signal in the form of the slope of fluorescence spectra in the cycles after the annealing temperature of the reaction is dropped.

The Invader assay.   The final assay presented here has perhaps the greatest potential for genome-wide SNP screening. The appeal of the Invader assay (Lyamichev et al. 1999) derives from the feature that no amplification step is required—reagents are incubated directly with genomic DNA in a homogeneous assay that avoids PCR amplification.

The specificity and general applicability of Invader has not yet been firmly established, but one study using PCR-amplified template demonstrated better than 99% typing accuracy over 14,208 genotypes at 36 SNPs (Mein et al. 2000). The assay utilizes enzymes involved in DNA repair from thermophilic bacteria to achieve signal amplification from genomic SNPs, again in a thermal cycling reaction with detection of fluorescence in a real-time microtiter plate reader.

As shown in Figure 3.23, three oligonucleotides are employed. The first is the "invader," which binds to the target leaving a one-base overhang at the 3′ end. The second is the "probe," the 3′ end of which is complementary to the target DNA on the other side of the SNP to the invader (and to one of the two SNP types) and also includes a 5′ "flap" that does not hybridize

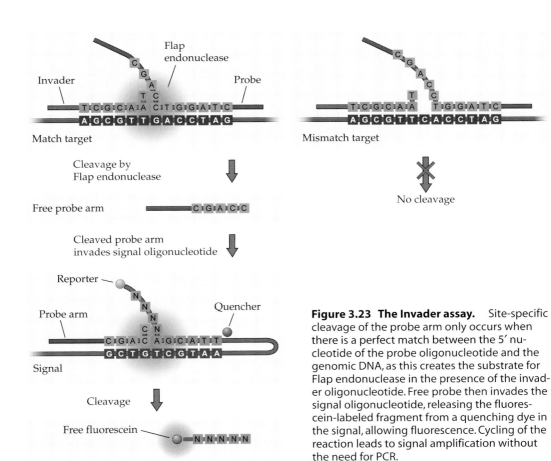

Figure 3.23 The Invader assay. Site-specific cleavage of the probe arm only occurs when there is a perfect match between the 5′ nucleotide of the probe oligonucleotide and the genomic DNA, as this creates the substrate for Flap endonuclease in the presence of the invader oligonucleotide. Free probe then invades the signal oligonucleotide, releasing the fluorescein-labeled fragment from a quenching dye in the signal, allowing fluorescence. Cycling of the reaction leads to signal amplification without the need for PCR.

to the target. If the probe matches the template at the SNP, a tertiary structure forms that is recognized by the "Flap endonuclease" enzyme, and the flap is cleaved off. If there is not a match between the probe and SNP position, no structure is formed and no cleavage occurs. This cleaved flap then becomes an invader of the "signal" oligonucleotide. The signal carries a reporter fluorogenic dye at the 5' end, a hairpin stem that is invaded by the probe arm, and a quencher dye near the point where the probe invades the hairpin. Annealing of probe to signal again produces the substrate for Flap endonuclease, cleaving off the reporter, which is now free to fluoresce. After denaturation, the probe is free to invade another signal, and so there is amplification of the signal in each cycle.

Haplotype Phasing Methods

For many applications in quantitative genetics, it is important not just to have the genotypes of individual SNPs, but also to be able to assemble the haplotype of a series of linked SNPs. Suppose an individual is heterozygous at two adjacent SNPs, yielding the genotypes A/T and A/C. This information does not tell you which sites are in phase with one another: are the haplotypes AA and TC or TA and AC? Establishment of the "linkage phase" over more than a couple of kilobases is a statistical problem discussed in Fallin and Schork (2000). Even within a short fragment, phase cannot be ascertained from the SNP genotypes alone, unless the PCR products are individually cloned prior to genotyping (in which case they must derive from the same chromosome). For SNPs that are too far apart to be amplified in the same clone (aside from X or Y chromosomal haplotypes in males), there is no direct method for establishing linkage phase. The best solution is to monitor transmission of alleles from parents to offspring. Except in the case of double heterozygotes, co-transmission allows deduction of which SNP alleles must be derived from the same chromosome. For double heterozygotes, maximum likelihood estimates (Excoffier and Slatkin 1995) that fit the probability of gametic association dependent on individual allele frequencies and haplotype frequencies in the known sample can be used.

Summary

1. SNPs are single nucleotide polymorphisms, which are the most common form of segregating molecular variation in natural populations. Depending on the species and gene, they tend to occur at a density of between one in every 30 bases to one in every kilobase.

2. Population geneticists are interested in the distribution of SNPs, including their frequency in distinct populations, effect on encoded proteins, rate of divergence between species, and degree of linkage disequilibrium and haplotype structure.

3. A large proportion of SNPs are rare (at a frequency of less than 5% of the alleles), and the overall distribution of SNPs usually accords with the predictions of neutral molecular evolutionary theory.

4. SNPs can be used to characterize population structure, infer selection and other evolutionary forces, as a tool in mapping mutations and quantitative trait loci (QTL), and for linkage disequilibrium mapping of disease genes.

5. Ultimately, most genetic variation must trace to SNPs and small indel polymorphisms, but there are complicated statistical difficulties in detecting the 10 or so SNPs that have the strongest genetic influence on any trait, among the millions of SNPs in a genome.

6. Case-control disease mapping in outbred populations refers to procedures for detecting SNPs that are more (or less) frequent in the affected than unaffected individuals. They are adversely affected by numerous factors including population stratification, admixture, variation in the level of environmental and cultural influence, influence of other loci on penetrance and expressivity, amount and distribution of linkage disequilibrium, genetic heterogeneity, and sampling artifacts.

7. Family-based association mapping procedures such as transmission-disequilibrium and sibling-disequilibrium methods are also available. Generally, more than 500 trios of parents and affected offspring must be studied to achieve statistical significance with reasonable power. Even with this study size, however, false positive and false negative results are not uncommon.

8. SNPs can be detected de novo using methods such as DHPLC and SSCP that rely on the effect of mismatches on the rate of migration of heteroduplex and homoduplex DNA molecules through some matrix. Alternatively, they can be extracted from comparison of genomic and EST sequences, or genomic sequences of multiple individuals.

9. Sequencing by hybridization is a high throughput method for detecting polymorphisms based on the specificity of hybridization to short oligonucleotides on a chip.

10. Low-technology methods for SNP genotyping include allele-specific oligonucleotide hybridization (ASO), which can be performed on a membrane filter or in microtiter plates, and Snip-SNPs or dCAPS (derived Cleavable Amplified Polymorphic Sequences), which are SNPs that affect the ability of a restriction endonuclease to cleave a short PCR product.

11. A variety of new technologies for SNP screening are available, including minisequencing and single base extension methods for microarrays, electrophoretic separation, and scoring on microbeads.

Homogeneous assays can be performed in a single tube or microtiter plate well and use fluorescent energy transfer between a donor and quencher fluorophore as the basis for discrimination of alleles.

Discussion Questions

1. Why is linkage disequilibrium so important for population and quantitative genetic analysis, and what level of sampling is required to quantify LD?

2. What are the major factors affecting the level of population structure observed for nucleotide diversity?

3. Contrast the case-control and transmission-disequilibrium strategies for mapping disease loci.

4. Why is it essential that SNP association studies be replicated before it is concluded that a particular polymorphism is associated with a disease or clinical phenotype? Does failure to replicate a finding mean that the original study was flawed?

5. If you were asked to survey the population structure of an endangered species, which of the methods for SNP genotyping would you prefer to use? Are any of these methods superior to existing methods for characterizing genetic variation, such as allozymes and microsatellites?

Web Site Exercises

The Web site linked to this book at http://www.sinauer.com/genomics **provides exercises in various techniques described in this chapter.**

1. Use GDA to compute the pattern of linkage disequilibrium in a sequence data set.

2. Carry out a hypothetical case-control association study.

3. Carry out a hypothetical transmission-disequilibrium association study.

Literature Cited

Allison, D. B. 1997. Transmission-disequilibrium tests for quantitative traits. *Am. J. Hum. Genet.* 60: 676–690.

Altschuler, D. et al. 2000. The common PPARγ Pro12Ala polymorphism is associated with decreased risk of type 2 diabetes. *Nat. Genetics* 26: 76–80.

Begun, D. and C. Aquadro. 1992. Levels of naturally occurring DNA polymorphism correlate with recombination rates in *D. melanogaster*. *Nature* 356: 519–520.

Cann, R., M. Stoneking and A. Wilson. 1987. Mitochondrial DNA and human evolution. *Nature* 325: 31–36.

Chen, J. et al. 2000. A microsphere-based assay for multiplexed single nucleotide polymorphism analysis using single base chain extension. *Genome Res.* 10: 549–557.

Chen, X., K. Livak and P.-Y. Kwok. 1998. A homogeneous, ligase-mediated DNA diagnostic test. *Genome Res.* 8: 549–556.

Clark, A. G. et al. 1998. Haplotype structure and population genetic inferences from nucleotide-sequence variation in human lipoprotein lipase. *Am. J. Hum. Genet.* 63: 595–612.

Clark, A. G. et al. 2003. Inferring nonneutral evolution from human-chimp-mouse orthologous gene trios. *Science* 302: 1960–1963.

Clegg, M. T., J. F. Kidwell, M. G. Kidwell and N. J. Daniel. 1976. Dynamics of correlated genetic systems. I. Selection in the region of the glued locus of *Drosophila melanogaster*. *Genetics* 83: 793–810.

Curran, M., I. Splawski, K. Timothy, G. Vincent, E. Green and M. T. Keating. 1995. A molecular basis for cardiac arrhythmia: HERG mutations cause long-QT syndrome. *Cell* 80: 795–803.

Devlin, B. and N. Risch. 1995. A comparison of linkage disequilibrium measures for fine-scale mapping. *Genomics* 29: 311–322.

Devlin, B. and K. Roeder. 1999. Genomic control for association studies. *Biometrics* 55: 997–1004.

Doebley, J., A. Stec and L. Hubbard. 1997. The evolution of apical dominance in maize. *Nature* 386: 485–488.

Excoffier, L. and M. Slatkin. 1995. Maximum-likelihood estimation of molecular haplotype frequencies in a diploid population. *Mol. Biol. Evol.* 12: 921–927.

Falconer, D. S. and T. F. C. Mackay. 1996. *Introduction to Quantitiative Genetics*, 4th Ed. Longman, Essex, U.K.

Fallin, D. and N. Schork. 2000. Accuracy of haplotype frequency estimation for biallelic loci, via the expectation-maximization algorithm for unphased diploid genotype data. *Am. J. Hum. Genet.* 67: 947–959.

Frary, A. et al. 2000. *fw2. 2*: A quantitative trait locus key to the evolution of tomato fruit size. *Science* 289: 85–88.

Freimer, N. and C. Sabatti. 2004. The use of pedigree, sib-pair, and association studies of common diseases for genetic mapping and epidemiology. *Nat. Genetics.* 36: 1045–1051.

Graur, D. and W.-H. Li. 2000. *Fundamentals of Molecular Evolution*. Sinauer Associates, Sunderland, MA.

Halushka, M. K. et al. 1999. Patterns of single-nucleotide polymorphisms in candidate genes for blood-pressure homeostasis. *Nat. Genetics* 22: 239–247.

Hartl, D. and A. Clark. 1998. *Principles of Population Genetics*, 3rd Ed. Sinauer Associates, Sunderland, MA.

Hästbacka, J. et al. 1994. The diastrophic dysplasia gene encodes a novel sulphate transporter: positional cloning by fine structure linkage disequilibrium mapping. *Cell* 78: 1073–1087.

Hedrick, P. W. 1985. *Genetics of Populations*. Jones and Bartlett Publishers, Boston.

Hill, W. G. and A. Robertson. 1968. Linkage disequilibrium in finite populations. *Theoret. Appl. Genet.* 38: 226–231.

Holland, P., R. Abramson, R. Watson and D. Gelfand. 1991. Detection of specific polymerase chain reaction product by utilizing the 5′-3′ exonuclease activity of *Thermus aquaticus* DNA polymerase. *Proc. Natl Acad. Sci. (USA).* 88: 7276–7280.

Horikawa, Y. et al. 2000. Genetic variation in the gene encoding calpain-10 is associated with type 2 diabetes mellitus. *Nat. Genetics* 26: 163–175.

International HapMap Consortium. 2003. The International HapMap Project. *Nature* 426: 789–796.

International SNP Map Working Group. 2001. A map of human genome sequence variation containing 1.42 million single nucleotide polymorphisms. *Nature* 409: 928–933.

Ioannidis, J., E. Ntzani, T. Trikalinos and D. Contopoulos-Ioannidis. 2001. Replication validity of genetic association studies. *Nat. Genetics* 29: 306–309.

Jorde, L. 2000. Linkage disequilibrium and the search for complex disease genes. *Genome Res.* 10: 1435–1444.

Kreitman, M. 2000. Methods to detect selection in populations with applications to the human. *Annu. Rev. Genomics Hum. Genet.* 1: 539–559.

Krings, M. et al. 2000. A view of Neandertal genetic diversity. *Nat. Genetics* 26: 144–146.

Lander, E. and D. Botstein. 1989. Mapping Mendelian factors underlying quantitative traits using RFLP linkage maps. *Genetics* 121: 185–199.

Lewontin, R. C. 1964. The interaction of selection and linkage. I. General considerations; heterotic models. *Genetics* 49: 49–67.

Lewontin, R. C. 1995. The detection of linkage disequilibrium in molecular sequence data. *Genetics* 140: 377–388.

Lewontin, R. C. and K. Kojima. 1960. The evolutionary dynamics of complex polymorphisms. *Evolution* 14: 450–472.

Long, A. and C. Langley. 1999. The power of association studies to detect the contribution of candidate genetic loci to variation in complex traits. *Genome Res.* 9: 720–731.

Long, A. D., S. Mulleney, T. F. C. Mackay and C. H. Langley. 1996. Genetic interactions between naturally occurring alleles at quantitative trait loci and mutant alleles at candidate loci affecting bristle number in *Drosophila melanogaster. Genetics* 144: 1497–1510.

Lyamichev, V. et al. 1999. Polymorphism identification and quantitative detection of genomic DNA by invasive cleavage of oligonucleotide probes. *Nat. Biotechnol.* 17: 292–296.

Lynch, M. and B. Walsh. 1998. *Genetics and Analysis of Quantitative Traits.* Sinauer Associates, Sunderland, MA.

Marras, S., F. R. Kramer and S. Tyagi. 1999. Multiplex detection of single-nucleotide variations using molecular beacons. *Genet. Anal. Biomed. Eng.* 14: 151–156.

Marth, G. T. et al. 1999. A general approach to single-nucleotide polymorphism discovery. *Nat. Genetics* 23: 452–456.

Mein, C. et al. 2000. Evaluation of single nucleotide polymorphism typing with Invader on PCR amplicons and its automation. *Genome Res.* 10: 330–343.

Mir, K. and E. Southern. 2000. Sequence variation in genes and genomic DNA: Methods for large-scale analysis. *Annu. Rev. Genomics Hum. Genet.* 1: 329–360.

Neff, M. M., J. D. Neff, J. Chory and A. E. Pepper. 1998. dCAPS, a simple technique for the genetic analysis of single nucleotide polymorphisms: experimental applications in *Arabidopsis thaliana* genetics. *Plant J.* 14: 387–392.

Nickerson, D. A., V. Tobe and S. Taylor. 1997. PolyPhred: Automating the detection and genotyping of single nucleotide substitutions using fluorescence-based resequencing. *Nucl. Acids Res.* 25: 2745–2751.

Nickerson, D. A. et al. 1998. DNA sequence diversity in a 9.7-kb region of the human lipoprotein lipase gene. *Nat. Genetics* 19: 233–240.

Nielsen, D. and B. S. Weir. 1999. A classical setting for associations between markers and loci affecting quantitative traits. *Genetical Research* 74: 271–277.

Parker, H. G. et al. 2004. Genetic structure of the purebred domestic dog. *Science* 304: 1160–1164.

Pritchard, J., M. Stephens, N. Rosenberg and P. Donnelly. 2000. Association mapping in structured populations. *Am. J. Hum. Genet.* 67: 170–181.

Reich, D. E. et al. 2001. Linkage disequilibrium in the human genome. *Nature* 411: 199–204.

Rockman, M. V. and G. A. Wray. 2002. Abundant raw material for *cis*-regulatory evolution in humans. *Mol. Biol. Evol.* 19: 1991–2004.

Rockman, M. V., M. W. Hahn, N. Soranzo, D. B. Goldstein and G. A. Wray. 2003. Positive selection on a human-specific transcription factor binding site regulating IL4 expression. *Current Biology* 13: 2118–2123.

Ronaghi, M. 2001. Pyrosequencing sheds light on DNA sequencing. *Genome Res.* 11: 3–11.

Rosenberg, N., J. K. Pritchard, J. L. Weber, H. M. Cann, K. K. Kidd, L. A. Zhivotovsky and M. W. Feldman. 2002. Genetic structure of human populations. *Science* 298: 2381–2385.

Sasieni, P. D. 1997. From genotypes to genes: Doubling the sample size. *Biometrics* 53: 1253–1261.

Spielman, R. and W. J. Ewens. 1998. A sibship test for linkage in the presence of association: The sib transmission/disequilibrium test. *Am. J. Hum. Genet.* 62: 450–458.

Spielman, R., R. McGinnis and W. Ewens. 1993. Transmission test for linkage disequilibrium: The insulin gene region and insulin-dependent diabetes mellitus (IDDM). *Am. J. Hum. Genet.* 52: 506–516.

Stephens, C. A. et al. 2001. Haplotype variation and linkage disequilibrium in 313 human genes. *Science* 293: 489–493.

Stoneking, M., D. Hedgecock, R. Higuchi, L. Vigilant and H. Erlich. 1991. Population variation of human mtDNA control region sequences detected by enzymatic amplification and sequence-specific oligonucleotide probes. *Am. J. Hum. Genet.* 48: 370–382.

Tang, K., D. J. Fu, D. Julien, A. Braun, C. R. Cantor and H. Coster. 1999. Chip-based genotyping by mass spectrometry. *Proc. Natl. Acad. Sci. (USA)* 96: 10016–10020.

Thornsberry, J., M. Goodman, J. Doebley, S. Kresovich, D. Nielsen and E. Buckler IV. 2001. *Dwarf8* polymorphisms associate with variation in flowering time. *Nat. Genetics* 28: 286–289

Till, B. J. et al. 2003. Large-scale discovery of induced point mutations with high-throughput TILLING. *Genome Res.* 2003 13: 524–530.

Underhill, P. A. et al. 1997. Detection of numerous Y chromosome biallelic polymorphisms by denaturing high-performance liquid chromatography. *Genome Res.* 7: 996–1005.

Vogel, T., D. Evans, J. Urvater, D. O'Connor, A. Hughes and D. Watkins. 1999. Major histocompatibility complex class I genes in primates: co-evolution with pathogens. *Immunol. Rev.* 167: 327–337.

Wang, D. G. et al. 1998. Large-scale identification, mapping, and genotyping of single-nucleotide polymorphisms in the human genome. *Science* 280: 1077–1082.

Weir, B. S. 1996 *Genetic Data Analysis II*. Sinauer Associates, Inc. Sunderland, MA.

Wicks, S., R. Yeh, W. Gish, R. Waterston and R. Plasterk. 2001. Rapid gene mapping in *Caenorhabditis elegans* using a high density polymorphism map. *Nat. Genetics* 28: 160–164.

Zeng, Z.-B. 1994. Precision mapping of quantitative trait loci. *Genetics* 136: 1457–1468.

Zondervan, K. T. and L. R. Cardon. 2004. The complex interplay among factors that influence allelic association. *Nat Rev Genet*. 5: 89-100.

4 Gene Expression and the Transcriptome

After genome sequencing and annotation, the next major branch of genome science is analysis of the transcriptome, namely documenting gene expression on a genome-wide scale. The **transcriptome** is the complete set of transcripts and their relative levels of expression in a particular cell or tissue type under defined conditions. Several technologies have been developed for parallel analysis of the expression of thousands of genes, among which *cDNA microarrays* and *oligonucleotide arrays* are the best known. These methods are most suitable for contrasting expression levels across tissues and treatments of a chosen subset of the genome, but they do not provide data on the absolute levels of expression. A third method called *serial analysis of gene expression* (SAGE) relies on counting of sequence tags to estimate absolute transcript levels, but is less suited to experimentation with large numbers of samples.

In addition to describing these three methods, this chapter also discusses techniques for verifying differential gene expression on a gene-by-gene basis and outlines some of the applications of comparative expression analysis. Since transcription is only one level of gene regulation, transcript levels do not necessarily accurately indicate protein expression or activity. Methods for characterization of the proteome—the structure and expression of the proteins encoded in the genome—are described in Chapter 5.

Parallel Analysis of Gene Expression: Microarrays

The basic procedure of microarray analyses is to deposit a very small amount of DNA corresponding to each one of a collection of thousands of genes (the "probes") onto a few square centimeter array surface, and then to interrogate these probes by hybridization to a "target" of mRNA that has been labeled with a fluorescent dye (Duggan et al. 1999). The amount of target that sticks to each spot of probe is proportional to the abundance of the transcript in the sample, and is detected as the intensity of fluorescent signal. A change in abundance is measured as an increase or decrease in the signal, relative either to the signal from a control reference sample (a "ratio"), or to the signals from the other probes on the array (a "relative intensity"). Estimates of the absolute abundance of each transcript in each sample are only approximate, so most inferences involve comparison of expression levels across treatments.

Applications of Microarray Technology

Gene expression profiling is much more than simply a method for rapidly finding genes that are differentially used under particular circumstances. Used properly, microarrays are a powerful tool for generating and testing hypotheses, classifying specimens, annotating genes, studying developmental and evolutionary processes, and performing clinical assessments. There are four basic steps in all microarray analyses, as shown in the form of a flow diagram in Figure 4.1: experimental design, technical performance, statistical analysis, and data mining. The first of these is arguably the most important, as the starting point for any gene expression profiling experiment should always be the precise formulation of a biological question. This will guide the implementation of an efficient and informative experimental design. Consultation with a statistician on issues such as the levels of replication that will be required to detect an effect or perform a contrast, and how to lay out the hybridizations, is always advisable, since the way an experiment is performed can have a large effect on statistical power and hence the conclusions that are reached.

There are at least half a dozen different common applications of microarray methodology, some of which are explored in the context of case studies at the end of this chapter. **Detection of candidate genes** remains a major objective of many studies, based on the supposition that if a gene is transcribed under one set of conditions but not another, then its expression is likely to be required for, or at least contributes to, any biological differences between the conditions. Examples include implicating genes in disease states such as cancer or adverse immune responses, finding target genes for a regulatory factor, and describing the genetic response to drug exposure. An analogous usage is in **annotation of gene function,** generally following the principle of guilt-by-association. In any list of differentially expressed genes, there are likely to be several that encode proteins with unknown functions, and these can be tentatively ascribed a molecular or cellular function based on their pattern of shared regulation with genes whose function is known.

Microarrays can also be used to dissect genetic mechanisms in a hypothesis-driven manner. An important application is **definition of genetic pathways,** by a combination of precise temporal profiling and focused bioinformatic analysis. Time-series can be used to order the activation and repression of transcription, suggesting which genes regulate the expression of other genes. Inclusion of mutants and experimental manipulations in the

Experimental design
Frame a biological question
Choose a microarray platform
Decide on biological and technical replicates
Design the series of hybridizations

Technical performance
Obtain the samples
Isolate total RNA
Label cDNA or mRNA
Perform the hybridizations
Scan the slides or chips

Statistical analysis
Extract fluorescence intensities
Normalize data to remove biases
t-tests for pairwise comparisons
ANOVA for multifactorial designs

Data mining
Cluster analysis and pattern recognition
Study lists of gene ontologies
Search for regulatory motifs
Design validation and follow-up experiments

Figure 4.1 Flow diagram for gene expression profiling.

analysis allows causation to be distinguished from correlation. Investigation of the distribution of conserved sequence motifs in putative control regions of genes can lead to **dissection of regulatory mechanisms** that link pathways and networks of genes. In Chapter 6, we discuss how the systems biology approach combines microarray analyses with proteomics, functional genomics, and metabolic profiling to arrive at a more complete description of complex cell biological processes.

Quantitative approaches are increasingly important in transcriptomics, both in basic research and its translation into applications in agriculture and medicine. **Quantification of transcriptional variance** is an essential tool in evolutionary and quantitative genetics, allowing investigators to partition the differences among individuals, populations, and species under a variety of different environmental conditions. Rather than simply describing how a treatment affects a tissue, inclusion of individual variation in the experimental design enables assessment of how individuals respond differently to a treatment. Animal and plant breeders can attempt to select for specific phenotypes associated with this variation at the transcriptional level. Similarly, clinical applications of **molecular phenotyping** are being developed with regard to prognosis of cancer progression, classification of infection or toxin exposure, and prediction of pharmacological responses.

Experimental Design

The first experimental design decision is which microarray platform to use. As described below, there are currently three major types of platform: cDNA and long-oligonucleotide microarrays, and short-oligonucleotide gene chips. Other technologies based on beads and three-dimensional arrays are just emerging. Major considerations involved in selecting a platform include cost (from $50 per homemade cDNA array to over $500 for commercial whole-genome oligonucleotide arrays), coverage (from several hundred genes to multiple probes for over 20,000 annotated genes), availability (from having to make your own to having numerous options for model organisms), and quality (the relatively high cost of commercial arrays is generally offset by their very high repeatability and uniformity). There are now numerous options for custom synthesis of microarrays for new organisms, or for representing a user-defined fraction of the transcriptome, with the result that microarray technology is accessible to most biologists.

The second essential consideration is level of replication, with choices generally representing a balance between available funding and experimental objectives. In the early days of microarray analysis, it was rare to see more than two replicates of any sample in an experiment, and most inferences were based on large—more than twofold—changes in expression observed in multiple samples. It has since been recognized that subtle differences down to 1.2-fold or even less can be detected with moderate replication levels of 6–10 hybridizations per sample.

In addition, the fundamental distinction has been drawn between technical and biological replicates (Churchill 2002). Technical replicates are repeated samples of the same biological material, for example RNA preparations, labeling reactions, hybridizations, and duplicate spots of the same probe on each array. Biological replicates are independent samples of similar material—for example, different individuals with the same genotype, leaves of a plant, trials of a treatment, or populations of a species. Because of cost and practicality, it is rarely possible to include all levels of replication in an experiment, so choices must be made based on which contrasts are of most interest. In some cases, pooling of samples allows the technical or biological source of variation to be included in the analysis without being measured explicitly. Examples include performing two independent labeling reactions and combining them before hybridization (to reduce noise in the stoichiometry of labeling), and mixing a population of individuals to gain a better representation of the response in a population.

The third major design issue is which samples to contrast on which arrays. For short-oligonucleotide "Affymetrix" gene chips, this is not an

Figure 4.2 Common experimental designs for microarray analysis.

issue, since only one sample is hybridized to each chip. However, most other microarrays involve competitive hybridizations of two samples labeled with two different-colored dyes, and in these cases there are three basic types of experimental design to choose from, as shown in Figure 4.2.

Reference sample designs contrast each experimental sample against a common reference sample, which is generally designed to include an average or intermediate level of transcript for every gene on the microarray. This will often be a mixture of each of the experimental samples, or some easily obtained tissue that can be prepared repeatably.

Though popular, reference designs are wasteful of expensive materials and generally less statistically powerful than designs that employ more replicates of just the experimental samples. **Loop designs** are optimal where there are multiple biological samples of one or two treatment classes (Churchill 2002). A simple approach with two replicates of each sample is to contrast A→B, B→C, C→D, …, G→A where the tail of the arrow represents one dye, and the head the other. In such a design, each sample is "balanced" with respect to the number of labeling reactions with each dye. Complete loop designs utilize multiple replicates so that each sample is contrasted at least once with each other sample. An alternative approach modeled after classical agricultural field trials is the **split-plot design** involving multiple factors (for example sex, drug, and population) in which the number of contrasts within and between factors is varied based on considerations of statistical power (Jin et al. 2001). The more direct contrasts there are, the more power there is for that contrast, but it is still possible to draw inferences involving contrasts that are not made directly. A fourth type of approach is not to adopt a specific design, either because of lack of attention during planning or because an experiment grew or changed unexpectedly after it was commenced. It is generally possible to extract information from any design, but statistical power is strongly influenced by the way an experiment is set up.

EXERCISE 4.1 *Design a microarray experiment*

Suppose that you are setting up an experiment designed to see how gene expression in the liver of four strains of mice (A, B, C, and D) is affected by treatment with a particular drug, with the complication that you suspect that the two sexes respond differently. You have enough money to perform 16 two-color microarrays. Describe how you would set up the experiment in three ways: (a) using a reference sample; (b) maximizing your ability to find significant genes for each drug for a particular sex and strain combination; (c) maximizing your ability to detect interactions between the treatments by contrasting female against male, drug against control, and strain against strain as much as possible.

ANSWER: There are 16 different samples to be contrasted: 4 strains × 2 treatments (drug vs. control) × 2 sexes (male vs. female). For design (a) it is simply a matter of setting up one array with each sample hybridized against the reference (which might, for example, be a mixture of each of the 16 experimental samples). Most often the reference is labeled consistently with just one dye. None of the samples are replicated.

For design (b) you would be well-advised to use a split-plot strategy, in which case you will be able to have four replicates of each sample, in eight mini-experiments as follows:

EXERCISE 4.1   TABLE 1

	Red dye	Green dye
Array 1:	Female A, Drug	Female A, Control
Array 2:	Female A, Control	Female A, Drug
Array 3:	Female B, Drug	Female B, Control
Array 4:	Female B, Control	Female B, Drug
Array 5:	Male A, Drug	Male A, Control
Array 6:	Male A, Control	Male A, Drug
Array 7:	Male B, Drug	Male B, Control
Array 8:	Male B, Control	Male B, Drug
Array 9:	Female C, Drug	Female C, Control
Array 10:	Female C, Control	Female C, Drug
Array 11:	Female D, Drug	Female D, Control
Array 12:	Female D, Control	Female D, Drug

EXERCISE 4.1 TABLE 1 (CONTINUED)

	Red dye	**Green dye**
Array 13:	Male C, Drug	Male C, Control
Array 14:	Male C, Control	Male C, Drug
Array 15:	Male D, Drug	Male D, Control
Array 16:	Male D, Control	Male D, Drug

Note that each pair of arrays involves the same set of treatments, in replicate, with the dyes flipped. The figure illustrates that every array is Male against Male (squares) or Female against Female (circles) for each strain, and every array contrasts Drug (shaded) against Control (unshaded).

For design (c) you would choose a loop strategy in which each possible combination of strains is contrasted against each other, while all arrays contrast Drug against Control and Male against Female (see the figure at the end of this exercise). An example setup follows, but note that with three treatments it is not always possible to completely balance the design, so in this case Arrays 8 and 16 are same-sex contrasts:

EXERCISE 4.1 TABLE 2

	Red dye	**Green dye**
Array 1:	A, Drug, Female	A, Control, Male
Array 2:	A, Control, Male	B, Drug, Female
Array 3:	B, Drug, Female	D, Control, Male
Array 4:	D, Control, Male	C, Drug, Female
Array 5:	C, Drug, Female	C, Control, Male
Array 6:	C, Control, Male	D, Drug, Female
Array 7:	D, Drug, Female	B, Control, Male
Array 8:	B, Control, Male	A, Drug, Male
Array 9:	A, Drug, Male	D, Control, Female
Array 10:	D, Control, Female	D, Drug, Male
Array 11:	D, Drug, Male	A, Control, Female
Array 12:	A, Control, Female	C, Drug, Male
Array 13:	C, Drug, Male	B, Control, Female
Array 14:	B, Control, Female	B, Drug, Male
Array 15:	B, Drug, Male	C, Control, Female
Array 16:	C, Control, Female	A, Drug, Female

(continued on next page)

Exercise 4.1 (continued)

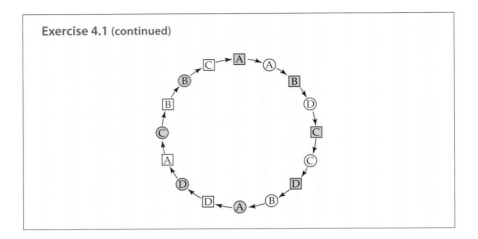

Microarray Technologies

cDNA microarrays.   The technology of cDNA microarrays, which quickly gained usage in academic settings, is a conceptually simple method for monitoring the relative levels of expression of thousands of genes simultaneously (Schena et al. 1995). PCR-amplified cDNA fragments (ESTs) are spotted at high density (10–50 spots per mm²) onto a microscope slide or filter paper, and probed against a fluorescently or radioactively labeled target (Figure 4.3). The platform is very flexible, since users print their own microarrays and are free to add or subtract clones as they refine their collection of ESTs. In recent years, similar genomic DNA microarrays have also been introduced, using PCR products generated with gene-specific primers that amplify predicted genes or gene segments from genomic DNA.

For the choice of which clones to array is influenced by the level of annotation of the cDNA library from which they were extracted and by the aims of the experiment. Ideally, each EST should be sequenced, and should represent a unique gene or alternative splice variant, in which case the collection is called a **unigene set.** Since cDNA libraries are tissue-specific, multiple libraries must be sampled to produce an unbiased set of clones that is likely to be representative of genes expressed under different conditions. It is possible to generate a cDNA microarray simply from randomly picked, unsequenced, clones. This leads to overrepresentation of a small subset of highly expressed genes on the microarray; however, molecular subtraction techniques can be used to reduce the redundancy. It is common practice to resequence the clones of interest at the end of an experiment to confirm their identity and ensure that no mistakes were made in the printing of the array.

For many organisms, the first generation of unigene sets was assembled by identifying unique clones in EST databases. Multigene families may be represented by several different clones, but in these cases there is some potential for cross-hybridization to occur, meaning that the same clone is recognized by transcripts from different genes. This problem can be avoided

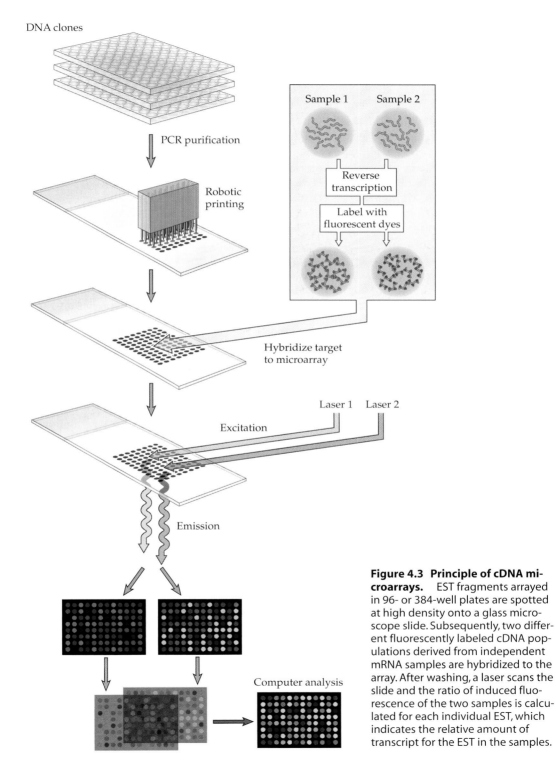

DNA clones

PCR purification

Robotic printing

Sample 1   Sample 2

Reverse transcription

Label with fluorescent dyes

Hybridize target to microarray

Laser 1   Laser 2

Excitation

Emission

Computer analysis

Figure 4.3 Principle of cDNA microarrays.   EST fragments arrayed in 96- or 384-well plates are spotted at high density onto a glass microscope slide. Subsequently, two different fluorescently labeled cDNA populations derived from independent mRNA samples are hybridized to the array. After washing, a laser scans the slide and the ratio of induced fluorescence of the two samples is calculated for each individual EST, which indicates the relative amount of transcript for the EST in the samples.

to some degree by choosing probes that correspond to a non-conserved portion of the gene. Since cDNA microarrays typically have an upper limit of 15,000 elements (and often include fewer than 5,000 elements), they are unable to represent the complete set of genes present in higher eukaryotic genomes. Consequently, arrays have been developed that are specific for a certain developmental stage or tissue (for example, gastrula or leaf), or class of gene (transcription factor or receptor).

The most expensive and time-consuming step in cDNA microarray analysis is the amplification of the EST set. While each spot on a microarray should contain just 10 ng of DNA, it is delivered as a few nanoliters of 5 mg/ml DNA solution. PCR-amplified fragments have been found to yield much stronger signals than concentrated plasmid DNA. PCR amplification is performed in 96- or 384-well plate format, typically in 100-μl volumes, and the products must be precipitated and re-suspended in a special spotting solution. Variation in clone length complicates amplification, and presumably affects signal strength, so clones in the 1–2 kb range are used preferentially.

Once amplified, the fragments may be spotted onto either a coated glass microscope slide or a nitrocellulose or nylon membrane. Membranes are most suited to applications where radioactivity is used to label the cDNA, while glass only supports fluorescence-based detection. Slides can be prepared in-house by coating them with a polylysine solution, but the uniformity of this surface is not guaranteed and a number of commercial suppliers have developed alternative coatings. Most solid support surfaces bond covalently with the sugar-phosphate backbone of spotted DNA after crosslinking by ultraviolet irradiation or baking at the end of the printing step.

Printing is performed with a robot that picks up samples of DNA from a 384-well microtiter plate and deposits aliquots sequentially onto a field of 100 or so slides. The spacing between spot centers is specified from 120–250 μm according to the density required (thus the robot must be accurate within <20 μm with respect to where it deposits each spot). The entire microarray usually covers an area 2.5×5.0 cm, though shorter grids can be printed when fewer clones are to be represented.

Commonly used print heads have from 4 to 32 individual printing pins spaced several millimeters apart. The pins pick up DNA solution from nearby wells on the master plate. At a rate of a few deposits per second, it takes up to 2 days for the robots to prepare 100 microarrays containing at least 5,000 clones. Since only a few microliters of PCR product are used in each print run, each amplified batch of clones is sufficient for multiple print runs and hence at least 1,000 slides.

Several types of printing pin are in use (Figure 4.4). Most rely on capillary action to transfer solution from the pin to the slide. This method has the virtue that spot size can be controlled somewhat by adjusting the amount of time the pin touches the slide; with spot diameters close to 50 μm, extremely high-density microarrays can be produced. The pins are expensive and fragile, they sometimes have a tendency to deliver doughnut-shaped spots (as the DNA spreads away from the tip), and can be prone to

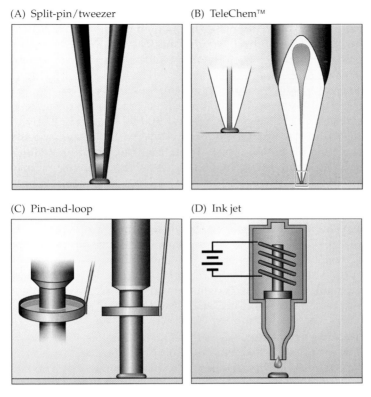

Figure 4.4 Types of printing pins. (A) Tweezer or split-pin designs transfer low nanoliter amounts of DNA to the array by capillary action as the tip strikes the solid surface. (B) TeleChem™ tips and pins apply small droplets by contact between the pin and substrate. (C) The pin-and-loop design picks up the DNA in a small loop, and a pin stamps solution on a slide at a uniform density. (D) Ink jets spray picoliter droplets of liquid under pressure.

clogging. An alternative innovation is the pin-and-loop system, in which the DNA is picked up as a meniscus in a small loop, and a pin of a given diameter literally stamps the solution onto the slide. Ink-jet printers promise highly reproducible and efficient spot deposition, but have nevertheless not seen wide usage.

Long-oligonucleotide microarrays. Long oligonucleotides in the range of 50–70 bases have gained in popularity as probes since the late 1990s. The principal differences relative to cDNA arrays are that the user is spared the pain of having to PCR-amplify thousands of clones (instead, oligonucleotides are ordered from a company), and a single strand of DNA of uniform length is deposited, affording greater control over hybridization specificity. Once the strand of interest has been identified in a cDNA or genomic sequence, freely available algorithms (for example,

OligoArray; Rouillard et al. 2002) can be used to design long oligonucleotides with the desired base composition, which promotes hybridization at the same temperature for all probes. Checks are also performed to eliminate the potential for cross-hybridization, and in general over 90% of genes yield suitable probes.

Spotting is performed with the same types of robotic arrayers as described above. A concern is that if the probe DNA crosslinks to the glass through its backbone, then only 20–30 bases of each individual molecule are likely to be exposed to the labeled target, which may be too short to ensure specific hybridization. Modified 5′ amino groups have been used to promote linkage of the oligonucleotides to an aldehyde on coated slides, so that the whole molecule is exposed.

Alternatively, the oligonucleotides can be synthesized in situ on the slide surface (Hughes et al. 2001). Agilent Technologies will synthesize arrays containing up to 44,000 long oligonucleotides (60-mers) using a combination of ink-jet printing to deliver the appropriate nucleotide in 60 consecutive steps, and phosphoramidite chemistry to ensure the incorporation of a single base in each step. For a design fee, a list of gene accession numbers can be turned into a custom microarray. Hybridizations are performed using essentially the same methods as for cDNA microarrays.

Short-oligonucleotide gene chips.   The third general approach to parallel analysis of gene expression is the use of short oligonucleotide microarrays, originally developed and marketed under the trademark Affymetrix GeneChip® (Figure 4.5; Lockhart et al. 1996; Lipshutz et al. 1999). The unit of hybridization here is a series of 25-mer oligonucleotides designed by a computer algorithm to represent known or predicted open reading frames. Each gene is represented by between 10 and 20 different oligonucleotides to control for variation in hybridization efficiency due to factors such as GC content; this variation can be exacerbated due to the short length of the probes. The possibility of cross-hybridization with similar short sequences in transcripts other than the one being probed is controlled for by including a mismatch control adjacent to each oligonucleotide that has a single base change at the center of the oligonucleotide; this control should not hybridize under high-stringency conditions.

With Affymetrix chips, the level of expression of each gene is calculated as an "average" of the differences between the perfect match and mismatch using a procedure (currently MAS 5.0) provided by Affymetrix with their scanning software. It should be noted that the utility of the mismatch control is not universally agreed upon, and it has been suggested that better analytical performance is achieved by ignoring the mismatch data, and instead adopting statistical models based on intensity values for each of the perfect match probes (Li and Wong 2001; Chu et al. 2002). There is typically a high variation in fluorescence intensity of the 10 or more oligonucleotides that represent each gene, which implies that choosing a different set of nucleotides may give a slightly different absolute measure of abundance. However, as with the other microarray technologies, gene chips are more

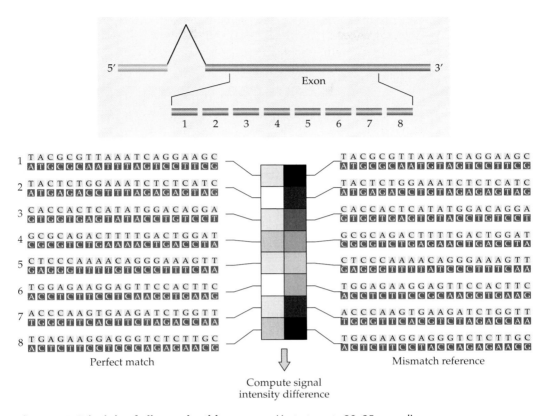

Figure 4.5 Principle of oligonucleotide arrays.   Up to twenty 20–25-mer oligonu-cleotide sequences are designed from an exon and printed on a chip adjacent to a mis-match oligonucleotide with a single base change (red) at the central position. When la-beled RNA is hybridized to the chip, each oligonucleotide will produce a slightly different signal intensity. These are processed in conjunction with mismatch data to calculate a value for the expression level of the transcript, based approximately on the average difference between the perfect match and mismatch intensities.

useful for contrasting the same gene across treatments than for comparing expression levels of genes within a single treatment.

    High-density short oligonucleotide arrays are constructed on a silicon chip by photolithography and combinatorial chemistry (Figure 4.6; McGall et al. 1996). For a 25-mer, 100 sequential nucleotide-addition reactions are performed across the surface of the chip in 25 cycles of A, T, G, and C. In each cycle, a localized flash of light "deprotects" the growing nucleotide chain just on that portion of the chip where the next nucleotide should be added. When a solution containing the nucleotide is added, a single nucleotide adds on to each deprotected chain, after which the remainder are washed off before the next cycle begins. The localization of the light is achieved either by inserting a mask between the light and the chip, using

(A)

(B)

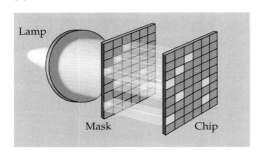

Figure 4.6 Construction of oligonucleotide arrays.   Oligonucleotides are synthesized in situ on the silicon chip, as described in the text. (A) In each step, a flash of light "deprotects" the oligonucleotides at the desired location on the chip; then "protected" nucleotides of one of the four types (A, C, G, or T) are added so that a single nucleotide can add to the desired chains. The light flash (B) is produced by photolithography, using a mask to allow light to strike only the required features on the surface of the chip.

technology developed by the microprocessor industry, or using miniature mirrors to focus a laser beam on each spot with technology developed by Nimblegen. Several hundred thousand oligonucleotides with their mismatch controls can be synthesized rapidly on thousands of identical chips, with extraordinarily high-quality reproducibility.

Both oligonucleotide and cDNA microarray technologies have their advantages and disadvantages. In brief, oligonucleotide arrays can accommodate higher densities of genes, including predicted genes not represented in cDNA libraries; have lower variability from chip to chip; incorporate mismatch controls; can be used by researchers without access to microarray construction facilities; and lend themselves to data comparison across research groups. Meanwhile, cDNA methods can be applied to any organism, irrespective of the status of genome sequencing efforts; are considerably cheaper once the platform has been developed, and so offer higher levels of replication; and are more flexible in design. They also rely on hybridization over kilobases rather than tens of bases, which may reduce cross-hybridization artifacts and the effect of sequence polymorphism on hybridization.

Labeling and hybridization of cDNAs.   There are two common protocols for production of labeled target, both of which can be modified in several ways. These are direct labeling of cDNA prepared from mRNA by reverse transcription, and linear amplification of labeled mRNA (Figure 4.7A,B). Total RNA is usually used as the template, rather than polyA fraction, which is generally less than 1% of total cellular RNA. If a radioactive label

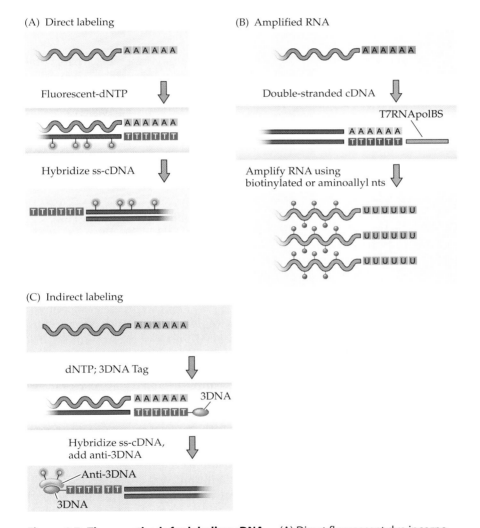

Figure 4.7 Three methods for labeling cDNA. (A) Direct fluorescent dye incorporation during reverse transcription. (B) In aRNA synthesis, amplified RNA is produced by in vitro transcription from a T7 promoter at the 5′ end of the primer used in cDNA synthesis. Nucleotides are incorporated that have attached biotin groups, or can be crosslinked to fluorescent dyes through aminoallyl linkages. If biotin is used, as is the case with AffymetrixTM gene chips, it is recognized by a fluorescent streptavidin-phycoerythrin compound on the array. (C) Genisphere® 3DNA submicro labeling, in which the 3D reagent is attached to an oligonucleotide that is complementary to the 5′ end of the primer used in cDNA synthesis, then hybridizes to it on the microarray.

is used (^{33}P, ^{35}S or ^{3}H), it is incorporated directly on one of the nucleotides. More commonly, fluorescent dyes such as Cy3 and Cy5 conjugated to the base of a nucleotide are incorporated during the reverse transcription reaction. These dyes are bulky and do not incorporate efficiently, so an alternate approach is to use nucleotides with a small aminoallyl group that can be crosslinked to the dyes after the cDNA has been synthesized. Synthesis of cDNA is usually primed with polyT or with random hexamers, the latter being essential for bacteria, which do not have polyadenylated mRNAs.

Several micrograms of template RNA are required for direct incorporation protocols. This places a prohibitive constraint on experiments using biopsies, small organisms, or fractionated cell types. The linear amplification approach (Eberwine 1996) circumvents this problem by running off several copies of complementary RNA from each template using T7 RNA polymerase that initiates synthesis from an oligonucleotide attached to the polyT primer (Figure 4.7B). In this way, 100 ng of template can be converted into micrograms of labeled aRNA, sufficient for a few hybridizations. Some protocols call for an extra step in which aRNA is reverse transcribed into more stable acDNA. As with direct labeling reactions, the dyes can be incorporated as part of the nucleotides, or attached later to an aminoallyl group. For work with Affymetrix GeneChips, biotinylated nucleotides are incorporated into the aRNA, and these are recognized after the hybridization reaction by streptavidin conjugated to fluorescent phycoerythrin.

A third labeling approach shown in Figure 4.7C is to use normal nucleotides throughout, but detect hybridized target through coupling of fluorophores to a substrate attached to the polyT primer. In the Genisphere® 3DNA Submicro system, the sensitivity of detection can be controlled by adjusting the amount of fluorescent anti-3DNA reagent that recognizes this substrate.

Where radioactive detection is used, the same filter must be stripped and re-probed to provide second or multiple samples. Unlike glass microarrays, which are single-use, filters can be re-probed half a dozen times. However, there is a risk that washing is incomplete, or that it affects subsequent hybridizations and introduces the potential for experimental artifacts. While radioactivity is less expensive than fluorescent dyes, exposure to radiation can add up quickly, placing a working constraint on use of radioactive probes.

Hybridization is performed in special humidified chambers. Since the aim is to detect expression of specific transcripts, hybridization and washes are carried out under high-stringency conditions that minimize any propensity for cross-hybridization between similar genes. Visualizing the hybridized target on a microarray can be performed by laser-induced fluorescence imaging using either a confocal detector or a CCD camera (Figure 4.8). Because fluorescence is quenched with time and exposure to light, errors at this step (due, for example, to improper setting of the laser—too much activation can saturate the signal, too little results in loss of signal) can have a profound effect on the conclusions reached. It has also been found

(A)

(B)

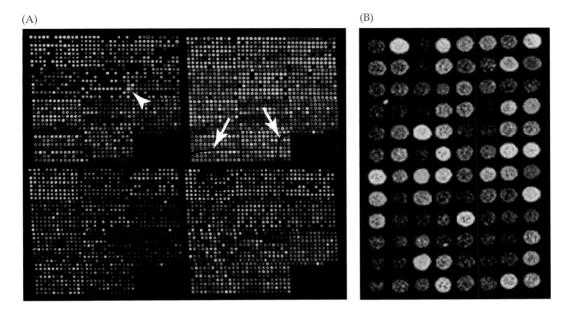

Figure 4.8 Microarray images produced with a pin-and-loop arrayer.   (A) Two common undesirable features are indicated, namely high local background (arrowhead) and scratches (two arrows) that would suggest "flagging" of the associated spots. (B) A close-up of a portion of the array demonstrates the uniformity of relative hybridization within each spot and differences in the red:green ratio for each clone.

that differential oxidation of fluorophores in the presence of high levels of atmospheric ozone can affect the ratio of signals from two dyes, suggesting that steps should be taken to control the laboratory environment.

Statistical Analysis of cDNA Microarray Data

Ratios and reference samples.   The standard approach to analyzing microarray data is to compute the ratio of fluorescence intensities for two samples that are competitively hybridized to the same microarray. One sample acts as a control, or "reference" sample, and is labeled with a dye (often Cy3) that has a different fluorescence spectrum from the dye (Cy5) used to label the experimental sample. In theory, three- and four-sample competitive hybridizations can be performed, but there are concerns that the fluorescence spectra of multiple dyes are overlapping, so signals may be partially confounded.

Soon after the development of cDNA microarray technology, a convention emerged that twofold induction or repression of an experimental sample, relative to the reference sample, were indicative of a meaningful change in gene expression. This convention does not reflect standard statistical definitions of significance (Cui and Churchill 2003). Nevertheless, genes that show such a response in two or more samples from a series of experiments, or a mean change above some arbitrary threshold, are typically chosen for further analysis. This often has the effect of selecting the top 5% or so of the clones present on the microarray. Adjustment of the fold-difference threshold results in a tightening or loosening of this fraction.

There are sound theoretical and experimental reasons for adopting ratios as the standard for comparison of gene expression (Eisen et al. 1998).

- First, the procedure emphasizes the fact that microarrays do not provide data on absolute expression levels. Formulation of a ratio captures the central idea that it is a *change in relative level of expression* that is biologically interesting. A corollary is that a higher fluorescence intensity for one spot on the array does not necessarily mean that a particular gene is expressed at a higher level than genes that produce lesser fluorescence signals. This is because fluorescence intensity is a function of the length of the EST, the amount of label incorporated into the cDNA during reverse transcription, the efficiency of hybridization, and the concentration of DNA prepared for the particular clone, among other factors.

- Second, competitive hybridization removes variation among arrays from the analysis. If signal is compared from two different arrays, it is possible that a difference in fluorescence intensity is due not so much to differential gene expression as to differences between the arrays—such as the absolute amount of DNA spotted on the arrays, or local variation introduced either during slide preparation and washing, or during image capture. As described below, such variation can be accounted for by replication, but many experiments involve large numbers of time points or other treatments, and due to the expense of the method many studies adopt sample sizes of just one or two per treatment.

There are pitfalls in analyzing simple ratios, and data must be transformed to remove obvious biases, as shown in Exercise 4.2 and in Figure 4.9. Commercial imaging software offers a range of different options for data manipulation, and uses different default procedures that may complicate comparison of data sets generated by multiple research groups.

The nature of the reference sample can also impact the analyses, because if the transcript abundance is too high or low relative to the experimental sample, power to discriminate differential expression is reduced. For this reason, many authors choose to use a pooled aliquot of all of their experimental samples as their reference, which should average representation of each transcript class. Alternatively, a standard reference that can be prepared easily by different groups facilitates more direct comparison of different experiments. A third option is to label organismal DNA or a pool of oligonucleotides uniformly representing each gene on the microarray.

Data normalization. All microarray experiments must be "normalized" to ensure that biases inherent in each hybridization are removed. This is

Red	Green	Difference	Ratio (G/R)	Log$_2$ Ratio	Centered R
16,500	15,104	−1,396	0.915	−0.128	−0.048
357	158	−199	0.443	−1.175	−1.095
8,250	8,025	−225	0.973	−0.039	0.040
978	836	−142	0.855	−0.226	−0.146
65	89	24	1.369	0.453	0.533
684	1,368	529	2.000	1.000	1.080
13,772	11,209	−2,563	0.814	−0.297	−0.217
856	731	−125	0.854	−0.228	−0.148

Figure 4.9 Simple normalization of microarray data.   The difference between the raw fluorescence intensities is a meaningless number. Computing ratios (green/red) allows immediate visualization of which genes are higher in the red channel than the green channel, but logarithmic transformation of this measure on the base 2 scale results in symmetric distribution of values. Finally, normalization by subtraction of the mean log ratio adjusts for the fact that the red channel was generally more intense than the green channel, and centers the data around zero.

true whether ratios or raw fluorescent intensities are adopted as the measure of transcript abundance. The most obvious bias is due to an overall difference in mean sample fluorescence intensity, either as a result of more mRNA being used as the template for one of the dyes, or a difference in sensitivity or power associated with detection of the fluorescence signal. For this reason, raw ratios are usually centered to produce a mean ratio of one, after first log-transforming them on the base 2 scale. This procedure produces symmetry of relative increases and decreases in expression about zero: twofold induction (a ratio of 2.0) or repression (a ratio of 0.5) transform to values of +1 or −1, respectively. It also tends to ensure that the frequencies of the log ratios are closer to a normal distribution than are raw ratios, which facilitates statistical analysis.

Recognizing that the logarithm of a ratio is equivalent to the difference in the logarithms (that is, log[A/B] = log[A] − log[B]), it is also permissible to directly normalize each channel separately. Subtraction of the mean log fluorescence intensity for the channel from each value transforms the measurements such that the abundance of each transcript is represented essentially as a fold increase or decrease relative to the sample mean, namely as a relative fluorescence intensity. Some authors choose to take the further step of ensuring that the variance of each channel is also constant, meaning that the dispersion about the mean is the same for all hybridizations. This is done by dividing the relative fluorescence intensities by their standard deviation so that the log-transformed values have a mean of zero and a variance of one. This preserves the rank order of expression and removes any systematic biases in the efficiency of labeling for classes of treatment, but may remove some true biological signal.

BOX 4.1 Microarray Image Processing

After labeled probe is hybridized to a microarray, several steps must be taken in order to transform the fluorescence intensity associated with each probe into a measure of transcript abundance. Most of these steps are automated by software provided with commercial scanners, but because each program performs the transformations in a slightly different manner, and since repeated measures from the same image can give slightly different results, it is worth considering just how these images are processed. There are essentially four steps: (1) image acquisition, (2) spot location, (3) computation of spot intensities, and (4) data reporting.

The raw image of a microarray scan is usually a 16-bit TIFF file that is a digital record of the intensity of fluorescence associated with each pixel in the array, represented as a number between 0 and 65,536 (i.e., 2^{16}). Higher resolution can be achieved by decreasing the pixel size (for example, 0.1 mm^2 results in approximately 130 pixels covering a spot 150 μm in diameter) or by storing the data in 32-bit format. However, the files are extremely memory-hungry (a normal scan usually requires 40 Mb of disk space), and the sources of experimental error are greater than those associated with image resolution.

The image is usually captured after first performing a pre-scan, both to confirm that the hybridization worked and to estimate the appropriate gain on the laser in order to capture as much information as possible without saturating the signal. If the gain is set too high, all high-intensity spots will converge on the same upper value, leading either to loss of data if both channels (dyes) are similarly affected, or unwanted bias if only one channel is saturated. If the gain is set too low, information at the low end of the scale is lost in the background.

Because dyes quench with time, and possibly at different rates, it is not a good idea to repeatedly scan the same array. Red-green color images of spots such as Figure 4.8 are actually false-color represen-tations of underlying digital values; single channels are normally visualized in black and white, as shown in the figure here.

Once the image has been captured, the individual spots must be located. This is most simply achieved by laying a grid over the image that places a square or circle around each spot. In ScanAlyze, which is free software for Microsoft Windows® available from **http://rana.lbl.gov/ EisenSoftware.htm** (Eisen et al. 1998), the grid is produced by specifying the number of rows and columns and the spacing of the centers of each spot (which will be the same as the spacing used by the arraying robot). Subsequently, a circle with the same diameter as the average spot is drawn around the grid centers. For example, an 8×12 grid with circles 150 μm in diameter spaced at 200 μm will overlay each of 96 spots, with 50 μm between the edges of adjacent circles. Since there are always imperfections in the spacing of spots of perhaps up to 10 μm, the spots must be re-centered by deforming the grid so as to maximize the coverage of the spots by the circles. This is done semiautomatically by most software, focusing on subsets of the total microarray, using visual confirmation that each round of centering (which takes less than a second) improves the fit. At the same time, flaws on the microarray due to dust specks, coverslip movement, and blotches of unwashed dye can be flagged to exclude the underlying data. The whole process can take an experienced analyst half an hour or more per microarray.

Next the spot intensities are calculated. The simplest approach is to compute the mean intensity for each pixel *within* the circle surrounding a spot, and subtract from this number the mean intensity of the *background* pixels immediately surrounding the spot. Since background intensity can be strongly affected by dust specks that increase the signal, some users prefer to subtract the median background pixel intensity.

The spots produced by capillary transfer often have a donut shape, or are otherwise

Visualization of single-channel microarrays. The microarray on the left has poor-quality spots; it is difficult to pinpoint the appropriate background or spot intensity values. In the array on the right, however, the uniformity of the spots makes this task much more obvious, and repeated measures are more likely to be the same.

uneven in intensity, due to spread of the DNA solution to the perimeter of the spot. For this reason, a more accurate way of reading spot intensity is to draw a histogram of pixel intensities throughout the square grid around a spot. As shown in the right-hand figure, the distribution will usually be bimodal, with one peak associated with the background and the other with the desired hybridization signal. Subtracting the background from the signal peak values may give the most robust measure of fluorescence. The variance in pixel intensity also supplies a measure of spot quality that can be used to flag data for exclusion.

Data is usually reported as a tab-delimited text file in columns that associate the particular measures of spot intensity with row and column identifiers and spot quality values. Most software supports data normalization to remove overall biases associated with the amount of cDNA and quality of the labeling reaction, but these manipulations can also be performed with familiar software, such as Microsoft Excel® or any number of statistical packages.

A final concern is to align spot numbers with clone identity. Again, the process is usually automated, but is not necessarily trivial since the spotting process results in different juxtaposition of clones on the array than those in adjacent wells of the microtiter plates.

Further linkage of the data to genome databases that allow users to call up information on the function and sequence of interesting clones requires merging the output of the data analysis with relational databases. As described later in this chapter, protocols built on XML-based languages that will allow storage and retrieval of microarray data from public databases are under development.

A different approach to normalization is to regress either the ratios or raw values onto a standard curve made of a set of control probes. These can match a series of spiked-in transcripts from a different organism that cover the full range of expression levels but do not hybridize to any of the probes of interest. Alternatively, the user can define a set of probes that they expect not to be differentially expressed across treatments, or choose a set of probes that empirically have invariant expression levels in the experiment. There is no one correct procedure for data normalization, and in practice it may be a good idea to use several methods and compare the results.

EXERCISE 4.2 *Calculate which genes are differentially expressed*

Data for the following 10 genes are obtained from a microarray, where each number represents the fluorescence intensity for the Cy3 or Cy5 channel after background subtraction. Calculate which genes are at least twofold different in their abundance on this array using two different approaches: (a) by formulating the Cy3:Cy5 ratio, and (b) by calculating the difference in the log base 2 transformed values. In both cases, make sure that you adjust for any overall difference in intensity for the two dyes, and comment on whether this adjustment affects your conclusions.

EXERCISE 4.2 TABLE 1

	Cy3	Cy5
Gene 1	1,378	1,678
Gene 2	331	797
Gene 3	1,045	1,001
Gene 4	3,663	2,557
Gene 5	259	133
Gene 6	1,882	2,264
Gene 7	604	738
Gene 8	2,611	2,671
Gene 9	528	1,092
Gene 10	458	593

ANSWER:

EXERCISE 4.2 TABLE 2

	Ratios method				Log transformation method				
	Cy3	Cy5	Ratio	Adj. ratio	$\log_2$ (Cy3)	$\log_2$ (Cy5)	Adj. Cy3	Adj. Cy5	Differ.
Gene 1	1,378	1,687	0.817	0.856	10.43	10.72	0.62	0.71	−0.09
Gene 2	331	797	0.415	0.435	8.37	9.64	−1.43	−0.37	−1.07
Gene 3	1,045	1,001	1.044	1.094	10.03	9.97	0.23	−0.04	0.26
Gene 4	3,663	2,557	1.433	1.502	11.84	11.32	2.03	1.31	0.72
Gene 5	259	133	1.947	2.041	8.02	7.06	−1.79	−2.95	1.16
Gene 6	1,882	2,264	0.831	0.871	10.88	11.15	1.08	1.14	−0.06
Gene 7	604	738	0.818	0.858	9.24	9.53	−0.57	−0.48	−0.09
Gene 8	2,611	2,671	0.978	1.025	11.35	11.38	1.55	1.38	0.17
Gene 9	528	1,092	0.484	0.507	9.04	10.09	−0.76	0.09	−0.85
Gene 10	458	593	0.772	0.810	8.84	9.21	−0.96	−0.79	−0.17

Using the ratios method, without adjustment for overall dye effects, genes 2 and 9 appear to have Cy3/Cy5 < 0.5, suggesting that they are differentially regulated. No genes have Cy3/Cy5 > 2. However, the average ratio is 0.95, indicating that overall fluorescence is generally 5% greater in the Cy5 channel. One way to adjust for this is to divide the individual ratios by the average ratio, which results in the adjusted ratio column. This confirms that gene 2 is underexpressed in Cy3, but not gene 9, whereas gene 5 may be overexpressed.

Using the log transformation method, you get very similar results. Base 2 is used because it is intuitive to think of a twofold change as less than −1 or greater than 1. The adjusted Cy3 and Cy5 columns indicate the difference between the $\log_2$ fluorescence intensity and the mean $\log_2$ intensity for the respective dye, and hence express the relative fluorescence intensity, relative to the sample mean. The difference between these values gives the final column on the right-hand side, indicating that genes 2 and 5 may be differentially expressed by twofold or more. By contrast, if you just subtract the raw $\log_2$ values, you will see that gene 9 appears to be underexpressed in Cy3, but gene 5 appears to be slightly less than twofold overexpressed.

Most imaging software also provides options for dealing with fluctuation in signal intensity across a spot, as well as for subtraction of the mean or median local background fluorescence intensity, as described in Box 4.1. More complex transformations have been proposed to deal with systemic artifacts that for unknown reasons tend to bias ratios as a function of the overall intensity of fluorescence in both channels (Quackenbush 2002). A popular one is the Lowess procedure, which reduces any relationship between the ratio and intensity on each array that is due to print tip effects or differences in the efficiencies with which the dyes may label low-abundance transcripts.

It is generally assumed that microarrays provide accurate readouts of gene expression over three orders of magnitude. This is approximately the range of pixel intensities that scanners can resolve above background. Most scanners produce TIFF output files, with pixel intensities ranging from 0 to 65,536 (namely 2^{16}, the range of 16-bit data) and background values typically between 50 and 200. While direct measures of transcript levels support the supposition that transcription is regulated within such a narrow range for the majority of genes, departures from linearity of the abundance-fluorescence intensity relationship are not accounted for routinely in microarray analyses.

Significance testing. Rather than simply adopting a twofold cutoff, most gene expression studies now require an assessment of the statistical significance of the difference between samples (Draghici 2003). The theoretical basis of robust statistical methods based on formulating a Student's *t*-test is described in Box 4.2. For a single treatment, the *t*-test evaluates whether the ratio of experimental to reference sample is different from unity for each gene. For two treatments, a slightly different form of *t*-test evaluates whether the mean ratios are different from one another. In either case, the key is to judge the apparent difference in transcript abundance relative to the variance among replicates. Significance Analysis of Microarrays (SAM) software available at http://www-stat.stanford.edu/~tibs/SAM (Tusher et

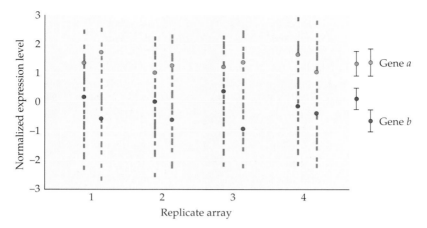

Figure 4.10 Analysis of variance (ANOVA) for gene expression data.   Each column of points shows the normalized level of expression of a set of genes relative to the sample mean. Contrasting the relative expression levels for two genes across four replicate arrays shows that there is no difference between the red and green labeled samples for gene *a*, whereas gene *b* is less strongly expressed relative to the sample mean in the green samples.

al. 2001) builds in some modifications to standard *t*-tests specifically for microarray data. It also allows the user to assess correlations with covariates such as clinical status.

Where there are multiple levels of each factor or multiple factors in the experimental design, analysis of variance (ANOVA) methods similar to those adopted by quantitative and agricultural geneticists more than 50 years ago can be used (Kerr et al. 2000; Wolfinger et al. 2001). Formulation of *F*-ratios is more efficient than comparison of a series of pairwise *t*-tests, and correlations between variables are accounted for by fitting a combined statistical model. ANOVA is a robust procedure for partitioning sources of variation—for example, testing whether or not the variation in gene expression is less within a defined subset of the data than it is in the total data set. Statistical testing requires moderate levels of replication (between 4 and 10 replicates of each treatment), but the extra effort and expense of this can be counteracted by the elimination of the reference sample. As the sample size of an experiment increases, the power of the statistical tests increases.

Analysis of variance of microarray data commonly proceeds in two steps. First, the raw fluorescence data is log-transformed on the base-2 scale, and the arrays and dye channels within each array are normalized with respect to one another. Tests for overall differences among samples (perhaps due to subtle differences in the amount of mRNA, or proportion of total RNA that is polyadenylated) can also be performed at this step. Removing the dye and array effects leaves normalized expression levels of each EST clone relative to the sample mean.

Subsequently, a second model is fit with these relative expression levels associated with each gene. As shown in Figure 4.10, the question is posed as to whether the mean difference between treatments is significant given the variability across arrays. This is similar to formulating a *t*-test, but extra adjustments are made to account for small correlations between the two measurements from each spot, and to control for dye effects. Furthermore, where multiple different treatment factors (e.g., sex, drug, and cell type) are being contrasted in the same experiment, it is possible to ask whether, in addition to the individual treatment factors, interactions between treatments affect the gene expression levels.

Statistical power is a major issue with all microarray experiments. Given the variation from experiment to experiment, multiple hybridizations must be performed to detect a change of expression of 1.2-fold with confidence. Such changes may be among the most significant biological effects (for example, if they involve the level of expression of key regulatory kinases). A volcano plot of significance against magnitude of effect (Figure 4.11) may assist in the choice of genes for more detailed analysis. On this plot, genes

Figure 4.11 Volcano plot of significance against effect. Based on 24 replicate arrays contrasting male against female adult *Drosophila*, the *x*-axis shows the difference in normalized log-transformed expression level between the two treatments, and the *y*-axis the significance of the difference as the negative logarithm of the *p*-value. On this scale, more significant effects (smaller *p*-values) are at the top of the plot, and the nominal $\alpha = 0.05$ takes a value of 1.3. If fold change is used as the measure of significance, genes are selected that lie to the left or right of the two vertical lines (–1 and 1), which represent the cutoff for twofold difference in expression. If significance is chosen, genes are selected that lie above a horizontal line that represents a chosen significance threshold, for example $p = 0.00001$ (–log$p = 5$). The two regions marked A represent genes with a large-fold change and high significance; region B indicates genes with high significance but only a small difference; genes in the two C regions show large but insignificant differences; and region D genes do not differ by either criteria. (After Jin et al. 2001.)

BOX 4.2 Basic Statistical Methods

Most scientific experiments and surveys in some way aim to compare numerical values between two or more groups of interest. A molecular biologist may want to compare the expression level of a gene under two different cellular environments, or a human geneticist may want to know if five different human populations have the same average cholesterol level. The field of statistics allows for such comparisons to be made in a meaningful way.

Suppose two different laboratories carry out the same experiment: both labs measure the average weight of two strains of white mice. Lab A weighs four mice from each strain and finds that the average weight in strain 1 is 18.0 grams, while the average of the four mice from strain 2 is 21.0 grams. Lab B weighs 125 mice from each strain and finds averages of 20.2 grams for strain 1 and 20.1 grams for strain 2. The results from Lab A suggest that the strain 1 mice weigh an average of 3 grams less than the strain 2 mice. In contrast, the data from Lab B imply that the strain 1 mice weigh an average of 0.1 grams more than strain 2 mice. How should one evaluate such data?

Clearly, there are differences in the amounts of information in the two experiments. Lab B weighed many more mice than did Lab A, so intuitively we place more value on the result from Lab B. The statistical methods of confidence intervals and significance tests allow for quantitative answers to the question, "Do the average weights of strains A and B differ, and if so, by how much?"

Estimating Unknown Means: Confidence Intervals

Many experiments ask questions about the values of unknown mean values. In the mouse example, there are actually two unknown means, one for each strain of mouse. When we collect *sample* data from a *population*, we most likely will compute the sample mean $\overline{X} = \sum x_i / n$ as an estimate of the true population mean. When reporting such a result, one should also recognize that if the experiment were to be repeated, a different value of $\overline{X}$ would be found. Thus, it is important to provide an indica-

tion of the *variability* of the estimate. A *confidence interval* computed from sample data provides a range of values likely to contain the true population mean, and simultaneously accounts for the effects of sample size and individual-to-individual variation. When estimating a population mean, the formula for a confidence interval is:

$$\overline{X} \pm t_{n-1} s / \sqrt{n}$$

The value of $\overline{X}$ is the single best *point estimate* of the unknown mean. The remainder of the formula creates a symmetric interval around $\overline{X}$ to account for individual variation (the sample standard deviation, s), sample size (n) and the desired level of confidence (t_{n-1}). The calculation is illustrated below using the first column of data from Table 1. The confidence interval is interpreted as a range of values having a 95% chance of containing the true (population) mean value. For the sample calculation, there is a 95% chance that the mean weight of strain 1 mice falls between 17.37 and 22.63.

Table 1

Strain 1	Strain 2	Strain 3
21	23	24
18	19	21
23	21	25
20	20	22
18	21	24

For example, compute the confidence interval for the mean weights of the 5 mice in Strain 1 that are shown in the first column of data.

$$\overline{X} = \frac{1}{5}(21 + 18 + 23 + 20 + 18) = \frac{1}{5}(100) = 20$$

$$s^2 = \frac{\sum x_i^2 - \frac{1}{n}\left(\sum x_i\right)^2}{n-1}$$

$$= \frac{(21^2 + \cdots + 18^2) - \frac{1}{5}(100)^2}{4}$$

$$= \frac{2018 - 2000}{4} = 4.5$$

$$s = \sqrt{s^2} = \sqrt{4.5} = 2.12$$

The value of t_4 is found by using a table of the t distribution with 4 degrees of freedom (df). For a 95% confidence interval, the value is 2.776. The final confidence interval becomes:

$$20 \pm 2.776 \times 2.12 / \sqrt{5}$$
$$= 20 \pm 2.63 = (17.37, 22.63)$$

Evaluating Hypotheses about Means: Significance Tests

The fundamental steps of the scientific method include developing hypotheses, designing and carrying out an experiment to test that hypothesis, and then either rejecting or accepting the hypothesis based on the experimental data. A *significance test* (or *hypothesis test*) is a formalization of these steps for use with numerical data. We will consider a test of the *null hypothesis* that the mean of a population is equal to μ_0, a value motivated by the topic being studied. The significance test proceeds first by calculating a *test statistic*, t. The test statistic is then used to calculate a *p-value*, which is used to decide whether the null hypothesis should be rejected (small *p*-values, often 0.05 or less) or accepted (larger *p*-values, often 0.05 or larger). The formula for the test statistic for testing the hypothesis that the mean of a population is equal to μ_0 is:

$$t = \frac{\overline{X} - \mu_0}{s / \sqrt{n}}$$

The *p*-value can be approximated using a table of the t distribution with $n - 1$ degrees of freedom (df). For positive values of t, the desired value is the probability of a value of t greater than the one observed. The calculation is illustrated using the male data from Strain 1 in Table 1. Suppose Strain 1 has been maintained in a different lab for many generations where it is known to have an average weight of 18 grams, and you want to ensure that the mean has not changed. In this setting, the null hypothesis would be that the mean weight of the male mice in Strain 1 is 18 grams, and the computed *p*-value tells us the probability of getting a sample mean of 20 or higher if the true population mean really is 18 grams.

Strain 1
$\overline{X}_1 = 20$ $t = \dfrac{20 - 18}{2.12 / \sqrt{5}} = 2.11$
$s_1 = 2.12$

Using a table of the t distribution with 4 df, the *p*-value associated with this test statistic is found to be between 0.05 and 0.1. There is some evidence that the mean has increased from 18 grams, but since the *p*-value is greater than 0.05, the evidence is not strong enough to reject the null hypothesis of mean 18 grams. That is to say, if we were to sample a new set of five individuals from this same population of mice, it would not be surprising for them to have an average weight of 20 grams or higher.

Comparing Three or More Unknown Means: One-way ANOVA

It might be the case that a scientist has three or more populations represented in the data and wants to know if there are any differences among the means of those populations. To address such a question, he or she would take a sample from each of the populations and apply a technique known as *analysis of variance*, more commonly called ANOVA. ANOVA leads to significance tests of the null hypothesis (which here states that there are no differences among the population means). This approach is much more efficient than performing three separate *t*-tests comparing each pair of means, and only gets more useful as the number of populations increases.

The calculations for ANOVA are simple, but a bit tedious. Define X_{ij} to be the jth observation from the sample taken from the ith of P populations. Furthermore, let $X_{i\cdot}$ be the sum of all n_i observations in the sample from population i, and let $X_{\cdot\cdot}$ and $\overline{X}$ be the grand sum of all N observations from all samples and the sample mean of those N values, respectively. With these values defined, we first compute the *total sum of squares*,

$$SS\text{Tot} = \sum_{i=1}^{P} \sum_{j=1}^{n_i} \left(X_{ij} - \overline{X}\right)^2 = \sum_{i=1}^{P} \sum_{j=1}^{n_i} X_{ij}^2 - \frac{1}{N} X_{\cdot\cdot}^2$$

The left-hand portion of the equation indicates that the total sum of squares is simply the sum of squared deviations of individual values from the overall mean. The right-hand portion is an algebraically equivalent expression that is much faster to calculate. Next, compute the *treatment sum of squares*,

$$SS\text{Trt} = \sum_{i=1}^{P} \sum_{j=1}^{n_i} \left(\overline{X}_i - \overline{X}\right)^2 = \sum_{i=1}^{P} \frac{X_{i\cdot}^2}{n_i} - \frac{1}{N} X_{\cdot\cdot}^2$$

(Continued on next page)

BOX 4.2 (continued)

Table 2

Source	df	SS	MS	F
Treatments	$dfT = p - 1$	SSTrt	MSTrt $= SST/dfT$	$F = MS$Trt$/MSE$
Error	$dfE = i - p$	SSE	$MSE = SSE/dfE$	
Total	$i - 1$	SSTot		

From the left-hand equation we see that the treatment sum of squares is the sum of squared deviations of the P sample means from the overall mean. Again, the right-hand portion is a faster computational form.

Finally, we must compute the *error sum of squares*,

$$SSE = SS\text{Tot} - SS\text{Trt}$$

The three sums of squares, along with some other useful values, are usually arranged in an ANOVA table in the form of Table 2.

The value of F serves as a test statistic for the null hypothesis that all populations have equal means. A table of the F distribution, with dfT numerator degrees of freedom, and dfE denominator degrees of freedom, is necessary to calculate a p-value. Large values of F suggest that the means are not all equal to one another.

We illustrate this below using:

Strain 1	Strain 2	Strain 3	
$\overline{X}_1 = 20$	$\overline{X}_2 = 20.8$	$\overline{X}_3 = 23.2$	$\overline{X} = 21.3$
$X_{1.} = 100$	$X_{2.} = 104$	$X_{3.} = 116$	$X_{..} = 320$

Referring to a table of the F distribution with $dfT = 2$ and $dfE = 12$, and an F-value of 4.43, we find that the p-value for the test is between 0.05 and 0.10. There is some, but not especially strong, evidence that the three strains of mice do not all have equal means. Subsequently, so-called *post-hoc* tests can be used to assess which subsets of the strains are different from the others. In this case there is some evidence that strain 3 differs from strains 1 and 2.

Following slight modifications, both pairwise t-tests and ANOVA are useful for assessing significance in microarray analyses. There is some debate as to the most appropriate ways to implement the methods and interpret the output from tests on tens of thousands of genes. ANOVA can also be extended into more complex two- or multifactorial models. These procedures allow the investigator to assess the significance of interaction effects. In the introductory example above, we might want to know whether the difference in conclusions from the two laboratories is itself significant, namely whether there is a laboratory by strain interaction effect: perhaps strain 1 does not grow so well in Lab A for some reason. Combining all of the data from many microarrays of different types provides increased statistical power for hypothesis testing, but also raises computational and statistical issues that are the subject of ongoing research.

$$SS\text{Tot} = \left(21^2 + 18^2 + 23^2 + \cdots + 22^2 + 24^2\right) - \frac{1}{15}(320)^2 = 6892 - 6826.67 = 65.33$$

$$SS\text{Trt} = \left(\frac{100^2}{5} + \frac{104^2}{5} + \frac{116^2}{5}\right) - 6826.67 = 6854.4 - 6826.67 = 27.73$$

$$SSE = SS\text{Tot} - SS\text{Trt} = 65.33 - 27.73 = 37.60$$

$$MS\text{Trt} = SS\text{Trt} / 3 - 1 = 27.73 / 2 = 13.87$$

$$MSE = SSE / 15 - 3 = 37.60 / 12 = 3.13$$

$$F = \frac{MS\text{Trt}}{MSE} = \frac{13.87}{3.13} = 4.43$$

in the lower left and right sectors (C) represent potential false positives, which show a large difference that is not significant. Those in the top center (B) are potential false negatives if fold change is the criterion for acceptance, since the effect is relatively small but is highly significant.

EXERCISE 4.3 *Evaluate the significance of the following gene expression differences*

Five replicate microarrays were performed contrasting an experimental treatment with a reference sample. Three of the genes gave the following results (after adjustments for dye effects). For which genes is there strong evidence for elevated gene expression in the experimental sample?

EXERCISE 4.3 TABLE 1

	Gene 1		Gene 2		Gene 3	
	Exp.	Ref.	Exp.	Ref.	Exp.	Ref.
Array 1	606	287	1,519	1,007	771	586
Array 2	441	198	3,783	1,166	738	502
Array 3	702	366	1,496	1,562	699	493
Array 4	597	255	3,472	1,773	800	625
Array 5	888	402	2,374	2,029	854	718

ANSWER: The average ratios of experimental:reference are 2.16, 1.77, and 1.33 for genes 1, 2, and 3, respectively. At first glance, these values suggest that gene 1 and perhaps gene 2 have elevated expression. However, the variance of the expression ratios for gene 2 is much higher than that of the other two genes: note that in the second replicate array the level for the experimental sample was more than three times the reference sample, but in the third replicate the measurement for the reference sample actually had the higher value. We can test for gene expression differences using the *t*-test described in Box 4.2. Begin by computing the experimental:reference ratio for each replicate array. The null hypothesis of interest is that the mean of the ratios is equal to 1 (i.e., no change in expression level). The *t*-statistics for each of the three genes are 7.3, 0.8, and 3.0, respectively, each with 4 degrees of freedom, leading to *p*-values less than 0.05 for genes 1 and 3, but not for 2. Thus, we conclude that the level of expression is increased only in genes 1 and 3. You should compute confidence intervals for the expression ratios to arrive at an estimate of the possible extent of the increases in gene expression.

Bioconductor. Freely distributed software for statistical analysis of microarray data is available from a number of sources, including the Bioconductor Project (http://www.bioconductor.org). Software from this site is written in the open source statistical programming language *R*, and implements procedures that are similar to those that can be found in commercial products such as SAS. Implementation requires more knowledge of statistics and computer science than most molecular biologists possess, but the basic skills can be learned quickly through collaboration. Various commercial packages for microarray analysis also offer statistical algorithms, but these tend to be more directed toward identifying the genes that have extreme values in a single hybridization, and toward data-mining procedures, as discussed in the next section.

A particularly useful program in Bioconductor is the *q-value* program for assessing the false discovery rate (FDR) in microarray experiments (Storey and Tibshirani 2003). FDR has emerged as a useful procedure for conservative selection of genes based on significance testing. This is because thousands of contrasts are performed, and significance thresholds must be adjusted to reflect this fact. In a set of 10,000 probes, approximately 500 should exceed the nominal 0.05 p-value by chance, and appear as false positives, so a much more stringent p-value cutoff is advised. Traditionally, the multiple comparison problem is addressed by Bonferroni adjustment, namely by dividing the testwise significance level by the number of tests to arrive at an experimentwise significance level. With 10,000 tests, this is usually too stringent and results in exclusion of many false negatives. The q-value is a measure of the proportion of genes at any significance level that are expected false positives: A q-value of 0.05 implies that only 5% of the genes at that level are likely to be false positives. Intuitively, if in a set of 10,000 probes, 5 are expected to be positive at $p = 0.005$, but 100 are observed significant at this level, then 95 of these are probably true positives. The use of FDR allows the investigator to select a set of genes with a level of confidence that is most appropriate for their purposes.

Given a list of significant genes, it is usually useful to know whether particular categories of gene are under- or overrepresented relative to all of the genes on the microarray. For example, if 10% of the probes represent mitochondrial proteins, but 20% of the differentially expressed genes are of this type, this difference in percentage may itself be significant. A useful Web-based tool for assessing this question is DAVID/EASE software (http://david.niaid.nih.gov/david/ease.htm), which converts lists of accession numbers for a wide range of organisms into gene ontology classes and then tests for the significance of changes in abundance of GO categories (Hosack et al. 2003).

Microarray Data Mining

In attempting to make biological sense of microarray data, it is useful to convert strings of hundreds of thousands of numbers into a format that the human brain can process. Invariably, this entails graphical representation, either in the form of line drawings or color-coding, that places genes into clusters with similar expression profiles (Eisen et al. 1998). Clustering implies co-regulation, which in turn may imply that the genes are involved in a similar biological process. Consequently, in addition to describing how indi-

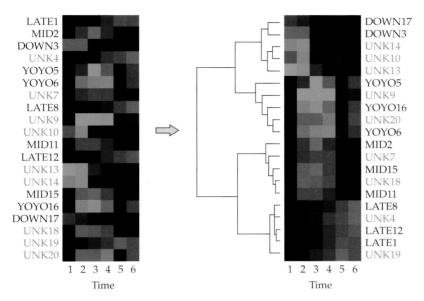

Figure 4.12 Hierarchical clustering of gene expression. An initially disordered set of gene expression profiles (left) can be converted into an immediately intelligible set of clusters by hierarchical clustering and rendering of the profiles in color, as in this hypothetical TreeView representation of a time-series with 20 genes (right). The observation that the genes of the DOWN, YOYO, MID, and LATE classes cluster together suggests that the unknown (UNK) genes may have functions of the respective groups in which they cluster.

vidual genes respond to certain treatments, microarray analysis describes the level of coordinate regulation of gene expression on the genome-wide scale. Since the clustering process groups unknown genes with annotated genes, it can lead to the formulation of hypotheses concerning the possible function of the unknown genes.

Researchers recognized early on that color-coding provides a direct (though not particularly precise) means to immediately identify co-regulated genes. Raw fluorescence intensities are transformed into false-color representations according to the convention that relatively high ratios of expression of experimental-to-reference sample are coded red and low ratios are coded green,* with the brightness of the color proportional to the magnitude of the differential expression (Figure 4.12). A ratio of 1 is black. (Because most color-blind people cannot contrast red and green, some authors are now using yellow and blue, respectively.) Alignment of all the genes on an array one above the other with the experimental contrasts across the figure allows the researcher to see patterns, especially if some sort of clustering algorithm has been used to filter the sample.

---

*Note that there is no relationship between this use of color and the red/green/yellow coloring of microarray images such as that shown in Figure 4.8.

BOX 4.3 Clustering Methods

The rapid advances in technology for simultaneously measuring levels of gene expression at many loci have presented a number of challenging data analysis problems for bioinformaticians. One of the most common tasks is the identification of "clusters" of genes that share an **expression profile.** Experiments of this sort collect expression data from G genes using E experiments. A typical example is measuring expression from many genes at a number of time points (e.g., the expression data from $G = 2000$ genes at $E = 8$ time intervals). Intuitively, the object is to identify groups of genes that appear to undergo coordinated changes in expression level, either positive or negative, as indicated in Figure 4.12.

Fortunately, methods for defining such clusters have been around for a number of years in different contexts. Thus, while new approaches for clustering that take explicit account of the nuances of gene expression data are being created, most widely used algorithms are simple modifications of traditional statistical methodologies. Most approaches fall into one of two categories. *Bottom-up* clustering methods begin with each gene in its own cluster. Clusters are then recursively clustered based on similarities, creating a hierarchical, treelike organization. (Indeed, many of the same methods are used for both gene expression clustering and phylogeny reconstruction.) *Top-down* methods begin by selecting a predetermined number of clusters. Genes are then assigned to those clusters to minimize variation within clusters and maximize variation between them.

We will demonstrate a variation of the popular clustering algorithm proposed by Eisen et al. (1998). Like other bottom-up methods, it consists of three distinct steps: (1) construct a matrix of similarity measures between all pairs of genes; (2) recursively cluster the genes into a treelike hierarchy; and (3) determine the boundaries between individual clusters.

We denote the (normalized) measurement of expression for gene g in experiment e as x_{ge}. For each of the $[G(G-1)]/2$ pairs of genes i and j, compute the correlation coefficient r_{ij} between the E experimental measurements for the two genes:

$$r_{ij} = \frac{1}{E} \sum_{e=1}^{E} \left(\frac{x_{ie} - \overline{x}_i}{s_i} \right) \left(\frac{x_{je} - \overline{x}_j}{s_j} \right)$$

where $\overline{x}_i$ is the average expression level for gene i over the E experimental conditions and s_i is the standard deviation of those same E measurements. Genes with similar expression profiles will have values of r_{ij} near unity.

With the matrix of correlations in hand, we proceed to the clustering portion of the algorithm. The process is easily described in the following recursion:

1. Find the pair of clusters with the highest correlation and combine the pair into a single cluster.

2. Update the correlation matrix using the average values of the newly combined clusters.

3. Repeat steps 1 and 2 $G-1$ times until all genes have been clustered.

This algorithm is simple enough to provide a numerical example. Suppose the initial correlation matrix for $G = 5$ genes is:

	2	3	4	5
1	0.3	0.2	0.8	0.1
2		0.9	0.1	0.8
3			0.2	0.7
4				0.1

The highest correlation is the 0.9 observed between genes 2 and 3, so we combine them into a cluster and recompute the correlation matrix:

	23	4	5
1	0.25	0.8	0.1
23		0.15	0.75
4			0.1

Note that the "correlation" between the clusters 23 and 4, for example, is the average of the correlations going into the new cluster, $r_{23,4} = \frac{1}{2}(r_{24} + r_{34}) = \frac{1}{2}(0.1 + 0.2) = 1.5$. We continue this process, clustering 1 with 4, then 23 with 5. The resulting hierarchy takes the form:

2   3 5   1   4

The final step in the clustering process is to determine the boundaries of individual clusters. In the above example, do 2 and 3 form a cluster to the exclusion of 5, or is there a single 235 cluster? Eisen at al. (1998) made no formal attempt to address this question, relying instead on defining clusters based on shared biological function. For instance, if genes 2, 3, and 5 were all heat-shock proteins, it might be sensible to include all of them in a single cluster. More recently, authors such as Hastie et al. (2000) and Kerr and Churchill (2001) have used metrics based on principles of ANOVA or the bootstrap procedures used in phylogenetic analysis to define well-supported clusters.

Clustering can be performed in a supervised or unsupervised manner, as described in Box 4.3. A common initial approach is to use correlation coefficients to sort the genes that share the most similarity in expression profile across a range of conditions into their own clusters. Such hierarchical clustering is an excellent method for identifying groups of co-expressed genes, and is incorporated into freeware such as Cluster/TreeView (http://rana.lbl.gov/EisenSoftware.htm), several programs available through Bioconductor, as well as in most commercial packages.

Hierarchical clustering can be performed both on the genes and the treatments, allowing detection of patterns in two dimensions. In cases where the treatments represent a series of drugs or mutations or tissue types rather than a temporal sequence, this two-dimensionality can be an extremely powerful mechanism for identifying similarities in genome-wide responses. It is important to recognize that there are actually numerous different types of hierarchical clustering algorithms that may or may not give the same tree structure, and that there is often no solid statistical support for the clustering produced by these methods. Also, ambiguities can arise as a result of the somewhat arbitrary orientation of branches stemming from each node.

A more sophisticated approach to clustering is to specify the number of clusters that are desired in advance, and then force the data to conform to this structure. Self-organizing maps (SOMs) and *k*-means clusters are similar procedures that automate this task (Tamayo et al. 1999). In *k*-means clustering, all genes are initially assigned at random to one of *k* clusters. The mean value for each treatment in each cluster is then computed, and each gene is reassigned to the cluster to which it shows the closest similarity. This procedure is reiterated until a stable structure is achieved. Self-organizing maps are also assembled by an iterative procedure that moves the centroid of each cluster in each step. The choice of the number of clusters can be made according to biological criteria, or may depend on a post-hoc comparison of the results of analyses run with different numbers of clusters. Since the objective is to identify coordinately regulated genes, the precise statistical significance of clusters is less important than failing to see a potentially significant relationship. False-positive groupings will presumably be invalidated by subsequent analyses—for example, if co-expressed genes are later found to be expressed in completely different cell types under circumstances that make no biological sense.

An alternative to clustering methods that rely on pairwise comparisons of values within each treatment is to ask how expression of each gene relates to the major sources of variation in the total sample. In **principal component analysis** (**PCA**, also known as **singular value decomposition**), the major axes of variation among the treatments are identified and each gene is assigned a value representing how that component of the variation contributes to its profile of expression (Holter et al. 2000). These axes will typically contain contributions from multiple treatments, so the clustering is sensitive to interaction as well as to primary effects. For example, in a study contrasting the effects of sex, cancer type, and chemotherapeutic agent, it may be that an interaction between sex and cancer type is the major source of gene expression variation, followed by a major effect of chemotherapy. Plotting the principal component scores of all genes on these two axes will identify clusters of genes that respond in like manner to the major sources of variation. An implication of this

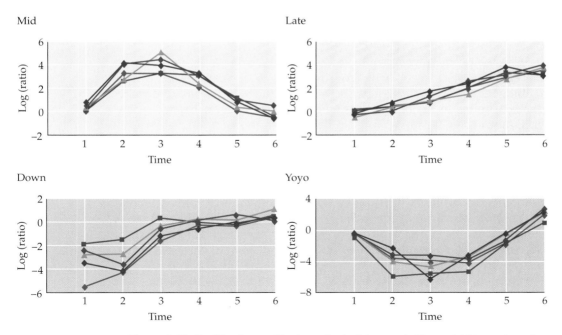

Figure 4.13 Profile plots.   The hypothetical data set in Figure 4.12 is re-plotted, showing the actual ratios at each time point. In this case, *k*-means clustering recognizes the same clusters as hierarchical linkage clustering.

approach is that it does not confine genes to a single cluster, since different genes may cluster on different pairs of axes.

Cluster analysis is a useful procedure for seeing trends in data, and can be supplemented by parallel plots of the profile of level of expression of genes in the cluster across samples, as shown in Figure 4.13. Actually, there is no reason why genes cannot belong to multiple clusters, reflecting differential regulation in different tissues and under different circumstances. In the final analysis, however, the biological meaning of clusters is assessed by their predictive power and their capacity for generating hypotheses. These features, in turn, are a function of the ability to align cluster data with gene ontology and other genomic annotation, including linkage to literature databases (Jenssen et al. 2001).

EXERCISE 4.4 *Perform a cluster analysis on gene expression profiles*

Suppose that while performing a microarray analysis of gene expression at four successive time points, you are particularly interested in whether any of the following five genes are co-regulated. By estimating the correlation between each profile, draw a plausible cluster diagram indicating which genes are most closely co-regulated, and comment on the pattern you observe.

EXERCISE 4.4 TABLE 1

	T1	T2	T3	T4	T5
Gene A	10.2	13.6	15.1	14.9	12.1
Gene B	13.7	13.5	12.9	13.3	13.8
Gene C	8.6	11.2	12.7	12.8	12.5
Gene D	13.3	13.3	11.1	9.6	9.6
Gene E	12.2	14.5	11.1	11.2	13.7

ANSWER: Use software such as the Correlation Function Data Analysis Tool in Microsoft Excel (or compute the correlations by hand—see Box 4.3) to show that the correlation between each expression profile is as follows:

EXERCISE 4.4 TABLE 2

	Gene B	Gene C	Gene D	Gene E
Gene A	−0.82	0.82	−0.43	−0.36
Gene B	—	−0.49	0.14	0.65
Gene C	—	—	−0.80	−0.18
Gene D	—	—	—	0.34

(continued on next page)

Exercise 4.4 (continued)

This matrix shows that genes A and C are most alike, and that B and E also have somewhat similar profiles. By contrast, genes A and B are strongly negatively correlated, as are genes C and D. Consequently, a plausible clustering of the profiles looks like:

However, the strong negative correlation between A and B, and between C and D, might imply that the two major clusters of genes are actually regulated by some of the same factors, just with opposite effects. If you plot the expression values of each gene across the time course, you will see that whereas A and C increase from T1 to T3, B and D tend to decrease. Clustering by the absolute value of the correlation coefficients results in quite a different picture of the relationships among the genes.

ChIP Chips and Gene Regulation

In addition to finding differentially regulated genes, microarrays can be used to help dissect the mechanisms by which differential transcription is achieved. There are two aspects to this problem: identifying the genes that are targets for specific transcription factors, and finding the transcription factor binding sites in the regulatory regions of those genes. The two pursuits can reinforce one another, and are most successful when combined with experimental manipulation to confirm the interactions.

An indirect approach to identifying target genes is to compare gene expression between wild-type and mutant organisms that either have a deletion of a transcription factor gene, or ectopically express it under control of a tissue-specific or inducible promoter. Differential regulation need not imply that the gene regulation is direct, since an intermediate factor may lie in the regulatory pathway.

A more direct approach is to isolate fragments of DNA that bind to the transcription factor, and determine which gene(s) those fragments regulate. The genomic technique of **chromatin immunoprecipitation microarrays** (**ChIP chips**; Lieb 2003) has been developed for this purpose, as diagrammed in Figure 4.14A. Protein complexes are crosslinked to chromatin by light chemical treatment, and then an antibody specific for the transcription factor of interest is used to purify the protein-DNA complexes by affinity chromatography (see Figure 5.5). The DNA is then liberated, and competitively hybridized with whole genomic DNA to a DNA microarray containing probes for the regulatory regions of genes.

ChIP chip has been used to systematically explore the design of regulatory pathways in budding yeast, *Saccharomyces cerevisiae* (Lee et al. 2002). Target genes for each of the 106 known transcription factors were deter-

(A)

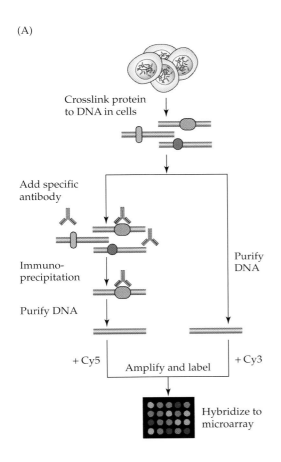

Figure 4.14 Chromatin immunoprecipitation and regulatory pathways. (A) ChIP chips involve capture of DNA crosslinked to a transcription factor protein, which is then hybridized to a genomic DNA microarray. Fragments that produce a stronger signal in the immunoprecipitated fraction than whole gDNA identify likely targets of the transcription factor. (B) Comprehensive assembly of lists of targets of each of the 106 known yeast transcription factors allowed computational reconstitution of regulatory pathways in yeast. These networks adopt five common forms, as shown, with an estimate of the number of examples of each. (After Lee et al. 2002.)

(B)

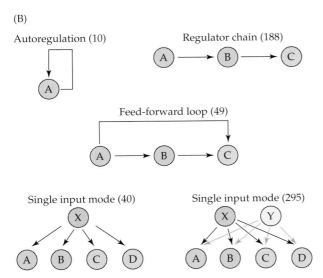

mined and ordered into five types of network shown in Figure 4.14B. These include autoregulation (where a protein regulates expression of its own gene); simple two- or three- (or more) step regulator chains in which gene A regulates gene B, which regulates gene C; feed-forward loops, in which the first gene also regulates the last gene; and branching networks with single or multiple inputs.

Binding sites tend to be less than 12 bp in length, so are too short to detect by standard local alignment search algorithms, and considerable attention is being given to development of new bioinformatic approaches to binding site detection. The Gibbs sampling approach is described in some detail in Box 4.4. Other methods include regression of motif counts on transcript abundance (Bussemaker et al. 2001), and searching for clusters of biochemically characterized binding sites upstream of genes. Such procedures generally result in good success in finding novel target genes, but remain subject to high rates of detection of false targets and failure to detect known targets. Incorporation of empirical and biophysical data into search algorithms, in software such as Ahab: http://gaspard.bio.nyu.edu/Ahab.html, is likely to increase the performance of regulatory motif discovery (Rajewsky et al. 2002).

SAGE and Microbeads

Serial Analysis of Gene Expression

Serial analysis of gene expression, commonly known as **SAGE,** is a method for determining the absolute abundance of every transcript expressed in a population of cells. The SAGE protocol is based on the serial sequencing of 15-bp tags that are unique to each and every gene (Velculescu et al. 1995). These gene-specific tags are produced by an elegant series of molecular biological manipulations and then concatenated for automated sequencing, with each sequencing reaction generating up to 50 high-quality tags. At least 50,000 tags are required per sample to approach saturation, the point where each expressed gene is represented at least twice in the tag collection. Since this costs upward of $2,000 per sample in even the best-equipped facilities, the method is not appropriate for the types of replicated contrasts that microarrays are well suited for. The advantage of SAGE is that the method is unbiased by factors such as reference samples, hybridization artifacts, or clone representation, and it gives what is believed to be an accurate reading of the true number of transcripts per cell.

The procedure used to generate tags is diagrammed in Figure 4.15. Starting with less than a microgram of polyA mRNA, a double-stranded cDNA library is constructed using biotinylated polydT oligonucleotides to prime the reverse transcription reaction. This cDNA is then cleaved with an "anchor enzyme"—typically NlaIII, a four-cutter restriction enzyme that cleaves DNA an average of once every 400 bases. Streptavidin-coated magnetic beads are then used to purify the 3′ end of each cDNA from the anchoring site to the polyA tail, which binds to the streptavidin through the biotin

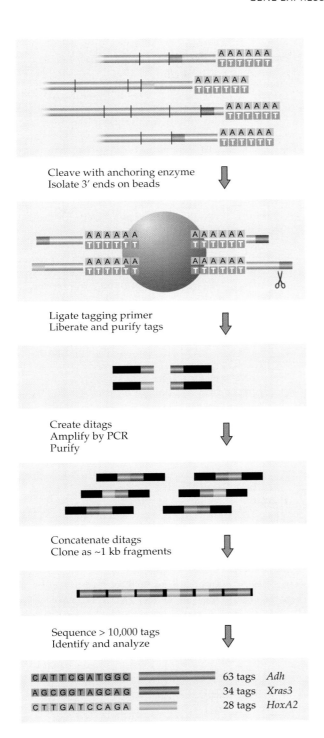

Cleave with anchoring enzyme
Isolate 3′ ends on beads

Ligate tagging primer
Liberate and purify tags

Create ditags
Amplify by PCR
Purify

Concatenate ditags
Clone as ~1 kb fragments

Sequence > 10,000 tags
Identify and analyze

C A T T C G A T G G C	63 tags	*Adh*
A G C G G T A G C A G	34 tags	*Xras3*
C T T G A T C C A G A	28 tags	*HoxA2*

Figure 4.15 Principle of SAGE. In serial analysis of gene expression, the absolute abundance of every transcript expressed in a cell population is determined by serial sequencing of 15-bp, gene-specific tags produced by molecular manipulations and concatenated for automated sequencing. The text describes the method in detail.

BOX 4.4 Motif Detection in Promoter Sequences

Given a set of co-regulated genes, it is generally assumed that transcription of all or most of the genes is likely to be regulated by a common transcription factor or factors. Many of these factors act by binding to specific DNA elements upstream of the promoters. Consequently, there ought to be conserved regulatory DNA sequences in these promoter regions. In some cases, such elements have been identified by biochemical and/or molecular genetic procedures (for example gel-shift assays) in which the transcription factor protein is shown to bind to the DNA sequence. In general, however, most regulatory sequences remain to be identified, and bioinformatic strategies are being employed to help find them. There are two major obstacles: *statistical* and *biological* constraints.

The statistical problem is that regulatory elements tend to be short (between 6 and 10 bases in length, reflecting the length of helical groove that the proteins recognize through specific hydrogen bonding) and somewhat variable. The expected frequency of any n-base sequence is one in every 4^n nucleotides, which for a 7-base element is once every 16 kb. Thus, even in a microbial genome, every short sequence will be represented thousands of times, so the problem of identifying elements that are overrepresented in the promoters of just a few percent of the genes (in each case in just a handful of copies) is acute. Adding to this the possibility that one or two of the bases diverge from the canonical motif only exacerbates the problem.

The biological constraint is that transcriptional regulation is complex. The same pattern of transcription can be generated by different transcription factors, so not all genes with the same profile will share a particular motif. Also, the same transcription factor can be involved in the generation of a wide variety of expression profiles, sometimes because the response it elicits is concentration-dependent (the same factor can be an activator at some concentrations, a repressor at others), sometimes because of variation in the spacing between the motif

and the promoter, and often as a function of the co-factors it associates with on the DNA. Thus, regulatory element function is heavily context-dependent, and while it may be possible to identify conserved core elements, there is no one-to-one correspondence between the presence of that core and transcriptional output.

Importantly, the binding sites that confer important biological responses may often be the ones with the lowest affinity for the transcription site (which can be more sensitive to slight changes in protein concentration), and hence the ones that diverge most from the consensus sequence. Furthermore, in higher eukaryotes, regulatory sequences tend to be dispersed over tens and even hundreds of kilobases, and can be positioned downstream of, within, or upstream of genes. This greatly increases the amount of space that must be searched in order to find these elements.

One common statistical strategy for identifying motifs in unaligned promoter sequences is based on a method called **Gibbs sampling.** The idea is to iterate randomly through (ungapped) alignments of the promoter regions, searching for alignments that result in the identification of blocks of conserved residues of some prespecified length w. In essence, the result is an optimal local multiple sequence alignment.

Discussion of the exact details of the Gibbs sampling method require statistical concepts beyond the scope of this text. However, the flavor of the method is easily demonstrated by example. Suppose the following sequences are promoter regions of a cluster of co-regulated genes (clearly, the sequences are shorter than what we would find in practice):

ACCGTGGTGT

TGGCACAAGC

GCCGATAGTC

AGTGGCGAAC

CCTGTGGTCA (sequence Z)

Initialize the iteration by selecting a random starting sequence (the last one, for our example) and designating it as sequence Z. The idea of each step in the iteration is to use the remaining sequences to find the location of the shared regulatory motif in Z. Pick a random initial alignment for the remaining sequences and starting point for the motif of width $w = 4$.

★ ★ ★ ★

ACCGTGGTGT

TGGCACAAGC

GCCGATAGTC

AGTGGCGAAC

For each position i, where $i = 1..w$, in the alignment within the motif, tabulate the nucleotide frequencies q_{ij}:

	1	2	3	4
A	0.25	0.25	0.00	0.50
C	0.25	0.50	0.00	0.25
G	0.50	0.00	1.00	0.25
T	0.00	0.25	0.00	0.00

Compute the corresponding nucleotide frequencies p_j for the pool of sites outside the pattern:

$$p_A = 5/24 = 0.21 \qquad p_C = 6/24 = 0.25$$
$$p_G = 7/24 = 0.29 \qquad p_T = 6/24 = 0.25$$

Select a starting point, x, for the motif in sequence Z:

★ ★ ★ ★

CCTGTGGTCA

Calculate the probability of the pattern using the values from the profile (q's):

$$Q(x) = 0.5 \times 0.25 \times 1.0 \times 0.25 = 0.00625$$

and also the probability of the pattern using the "background" values from sequence regions outside the profile (p's):

$$P(x) = 7/24 \times 6/24 \times 7/24 \times 7/24 = 0.006203$$

The ratio of the two probabilities, $R(x) = Q(x)/P(x) = 1.01$, is indicative of how likely it is that sequence Z has an example of the motif beginning at position x, so we select as the location of the motif the position with largest $R(x)$. The current sequence Z is added to the alignment, with a new sequence designated as Z for the next

cycle of the iteration. Placement of the motif in sequence Z becomes more and more refined each iteration, and the complete sequence alignment (i.e., the placement of the motif's starting point in each sequence) eventually converges.

The idea driving Gibbs sampling methods is that the better the description of the profile probabilities (q's), the more accurately the position of the motif in sequence Z can be identified. In its early stages, the locations of the motifs are chosen essentially at random. Frequencies in the p's and q's are similar, and, consequently, values of $R(x)$ hover near 1. As the iterations continue, however, some of the motif locations are placed correctly by chance, leading to more pronounced differences between the profile frequencies in q and the background frequencies in p, higher ratio values, and better placement of the motif in that iteration's sequence Z.

For the example data, the alignment eventually converges to:

★ ★ ★ ★

ACCGTGGTGT

TGGCACAAGC

GCCGATAGTC

AGTGGCGAAC

CCTGTGGTCA

with q's:

	1	2	3	4
A	0.00	0.20	0.00	0.00
C	0.00	0.00	0.00	0.40
G	0.00	0.80	1.00	0.00
T	1.00	0.00	0.00	0.60

and p's:

$$p_A = 9/30 = 0.30 \qquad p_C = 11/30 = 0.37$$
$$p_G = 8/30 = 0.27 \qquad p_T = 2/30 = 0.07$$

For further details, including methods for selecting the optimal value of the motif width w, see Lawrence et al. (1993) and Roth et al. (1998); the latter describes the popular AlignACE software.

Iteration proceeds by dropping one sequence from the analysis and recomputing the weight matrix and alignment. At the

(Continued on next page)

BOX 4.4   *(continued)*

end of the process, a motif is identified that is over-represented in the sample, so long as some pre-set likelihood cut-off threshold is exceeded. Subsequently, further criteria might be specified, such as the location and orientation of the motif relative to the promoter, whether it is palindromic, and how specific it is relative to other sequences in the genome that were not included in the initial collection. Cross-species contrasts might also be performed on the supposition that functional elements are likely to be conserved, and the availability of complete genome sequences will greatly facilitate such contrasts (McGuire et al. 2000). A useful program for performing this type of search online is the AlignACE program at http://atlas.med.harvard.edu/cgi-bin/alignace.pl.

Other approaches to motif recognition have been considered. A hidden Markov model (HMM) approach is implemented in the MEME program described in Bailey and Gribskov (1998) and available at http://meme.sdsc.edu/meme/website. An alternative to searching for motifs in a cluster of genes is to ask whether a previously identified motif has a different frequency among clusters. Jakt et al. (2001) applied this approach to characterize and refine motifs present in several types of transcription profile observed during the cell cycle of yeast. A list of all of the matches and their respective likelihood scores in the complete set of promoters is assembled,

and the number of matches in each cluster at several specificity levels is compared with the number expected in a sample of the same size, computed from a hypergeometric distribution. This approach shows not just which clusters are enriched for a particular motif, but also whether the enrichment is only for perfect matches or includes more divergent motifs.

A different approach again is to avoid clustering, and instead just regress the presence of every possible combination of nucleotide sequences up to 7 or 8 bases, on the relative gene expression level difference between a single time point or condition and a reference point (Bussemaker et al. 2001). That is to say, a linear model is set up for each gene in which the relative expression level is fit as a function of the number of times a particular precise motif is detected in the promoter region. Multiple regression can also be applied to fit the effects of several motifs simultaneously, and to test for synergistic or antagonistic interactions between them. Applied to a single time point of synchronized yeast cells, the method identified 11 motifs involved in initial cell cycle progression, including the previously identified stress response, SFF, MCB, and M3a/b elements, as well as two apparently new motifs. Undoubtedly, a range of novel approaches to association of gene expression and regulatory motif finding await development.

---

group. An adapter oligonucleotide is then directionally ligated to the anchoring sites. This adapter contains a recognition site for a type II restriction endonuclease, BsmFI, which liberates the tags adjacent to the anchoring site by cleaving 15 bp into the cDNA fragment. The adapter-tag fragments are then ligated tail-to-tail to form ditags, which are amplified by PCR using primers complementary to the 5′ end of the adapter. These PCR products are then purified by extraction from an acrylamide gel, and again cleaved with the anchoring enzyme NlaIII, which liberates the adapters and leaves 30-bp ditags. These ditags are purified once more, then ligated end-to-end under controlled conditions that produce concatenated chains that range in length from a few hundred base pairs to a few kilobases. The approximately 1-kb fraction is excised from a gel one last time, cloned into a sequencing vector, and several thousand clones are sequenced directly.

Tag identification is automated, and for model organisms the gene corresponding to each tag can be identified immediately and unambiguously. Since each tag commences with the sequence of the anchoring site, SAGE analytical software identifies tags with the correct length and spacing and filters out artifacts due to end-filling, cloning, and PCR errors. A list of each unique tag and its abundance in the population is assembled, and where possible, the tags are annotated with whole genome and/or EST information.

While most 15-bp sequences only appear once or at most a few times even in mammalian genomes, an extra stringency step that facilitates gene identification is that the tag must include the 3′-most anchoring site in a predicted transcript. A fraction of genes will have multiple tags due to alternative splicing near the 3′ end, or use of alternative polyadenylation sites; but for the most part these can be identified. Artifacts can also arise due to sequencing errors or to mis-incorporation of bases during the cDNA synthesis and ditag PCR amplification steps. Tag sequences also provide an additional source of information for the identification of novel genes, where the tag aligns with the 3′ end of a predicted but as yet unannotated transcript. For the most part, even closely related members of gene families have unique tags. It should be recognized that in some organisms, levels of nucleotide polymorphism are sufficiently high that a nontrivial fraction of tags will be strain-specific.

Transcript abundance measured by SAGE can either be expressed in relative terms or converted to an estimate of the number of transcripts per cell (Figure 4.16). Typically, across all eukaryotic cell types, fewer than 100 transcripts account for 20% of the total mRNA population, each being present in between 100 and 1,000 copies per cell. These include transcripts that encode ribosomal proteins and other core elements of the transcriptional and translational machinery, histones, and some taxon-specific genes such as Rubisco in plants and polyadenylated mitochondrial transcripts in *Drosophila*. A further 30% of the transcriptome comprises several hundred intermediate-frequency transcripts with between 10 and 100 transcripts per cell. These include housekeeping enzymes, cytoskeletal components, and some unusually abundant cell type–specific proteins.

The remaining half of the transcriptome is made up of tens of thousands of low-abundance transcripts, some of which may be expressed at less than one copy per cell, and many of which are tissue-specific or induced only under particular conditions. Thus most of the transcripts in a cell population contribute less than 0.01% of the total mRNA. If one-third of a higher eukaryotic genome is expressed in a particular tissue sample, then somewhere in the neighborhood of 10,000 different tags should be detectable. Taking into account that half of the transcriptome is relatively abundant, at least 50,000 tags must be sequenced to approach saturation.

For comparative purposes, SAGE is most useful for identifying genes that show large increases in one or more samples, or are completely missing from

(A)

(B)

Figure 4.16 SAGE analysis of yeast and colorectal cancer transcriptomes.
(A) Almost 50% of yeast transcripts are expressed at less than 10 copies per cell, whereas 17% of the total mRNA is due to a small number of transcripts at greater than 100 copies per cell. (B) A comparison of over 60,000 tags each from both normal colon epithelium and from a colorectal tumor sample revealed 83 candidate genes that are either up- or downregulated at least tenfold in the tumor. (A after Velculescu et al. 1997; B after Zhang et al. 1997.)

others. Because even a single tag is good evidence that a gene is expressed in a tissue, SAGE also performs better than microarray analysis with respect to determining which genes are expressed at low levels, since background signals of similar intensity to true low-abundance signals often cloud microarray analysis. However, failure to observe a tag is poor evidence that a gene is not transcribed, even if the collection is approaching saturation (two or more tag sequences) for most transcripts.

Quantitative comparison of SAGE samples is not always easy to interpret. A tag present in four copies in one sample of 50,000 tags and two copies in another sample, may actually be twofold induced in the first sample; but such a difference is also expected to arise by random sampling. Even the contrast of 20 tags to 10 is not obviously significant given the large number of comparisons that are performed, though it is at least suggestive.

A great advantage of SAGE is that the method is unbiased by experimental conditions, so direct comparison of data sets is possible. As a consequence, small changes in tag abundance can also be compared across a large number of samples generated by different groups. Web-based tools for performing such comparisons online, as well as facilitating download of complete data sets, are well developed and accessible through SAGEnet (http://www.sagenet.org).

Microbead-Based Expression Profiling

Several new technologies have been proposed for measuring transcript abundance, some of which utilize 3μ **microbeads** instead of a flat surface as the substrate for hybridization. The idea behind **b**ead **a**rrays for the **d**etection of **g**ene **e**xpression (**BADGE;** Yang et al. 2001) is to capture transcripts on color-coded beads that can be interrogated in a highly parallel manner as shown in Figure 4.17. Oligonucleotides complementary to the labeled target nucleic acid are coupled to a batch of microbeads that have been uniquely coded with a combination of a small number of colored dyes. After hybridization, the identity of each bead is determined at one end of the fluorescence spectrum, and then the amount of target assessed at a different wavelength. Instead of streaming each bead past a detector, Illumina's BeadArray™ technology allows tens of thousands of beads to be processed in parallel on etched fiberoptic bundles.

Single-Gene Analyses

One of the primary objectives of all comparative gene expression studies is to identify a small subset of genes, further study of which may illuminate understanding of the biological basis of differences in the samples under study. Whether a change in gene expression causes or merely correlates with a response, the first step in further study is to use a different experimental procedure to confirm that the gene really *is* differentially expressed. Several methods for studying expression one gene at a time have been developed since the early days of molecular biology; two of these are described in this section.

Northern Blots

Northern blots are the simplest procedure used to determine whether a gene is expressed in a sample, but they are semi-quantitative, so they really only confirm twofold or greater differences in gene expression. Poly-A mRNA is isolated from the tissue of interest and the transcripts are separated according to length on a denaturing agarose gel. The mRNA molecules are then transferred to a nylon or nitrocellulose filter by capillary or

(A)

(B)

(C)

Figure 4.17 Microbead-based gene expression profiling. Luminex® Bead arrays work on the principle that tens of thousands of color-coded beads, each tagged with a unique anti-sense oligonucleotide, provide biodetectors for transcript abundance. As a bead passes through two laser beams (A), the color of the bead is read at one wavelength, and the intensity of fluorescent dye-labeled cDNA at another. In this example, the transcript (IL4) that hybridizes to the yellow bead (B) is expressed at twice the level of IL7, which hybridizes to the orange bead (C). (After an online tutorial at **http://www.luminexcorp.com**).

electrolytic blotting, fixed in place by ultraviolet crosslinking, and hybridized to a labeled short DNA probe synthesized from a cloned fragment of the gene of interest. The label can be radioactive, chemiluminescent, or histochemical, and will produce a band on the blot wherever hybridization occurs, in proportion to the amount of mRNA in the sample. If the transcript is alternatively spliced, multiple bands corresponding to the different-length transcripts will be detected.

In order to be quantitative, the intensity of signal from the gene being probed must be compared with an internal control for the amount and quality of the mRNA in different lanes on the blot. "Housekeeping" genes (such as ribosomal proteins or particular actin subunits that are thought not to fluctuate significantly in expression level across treatments) are used to reprobe the blot. The specific gene signal is then normalized to the control signal. In cases where a single transcript is produced or information about alternative splicing is not of interest, the gel electrophoresis step can be omitted, in which case mRNA samples are blotted directly onto the filter in an array of small dots or slots. Such dot/slot blots enable simultaneous comparison of expression levels from tens or even hundreds of samples and only take hours, rather than a couple of days, to perform.

Quantitative PCR

An increasingly popular method for quantifying individual gene expression is **quantitative reverse-transcription PCR,** also known as **Q-PCR or Q-RT-PCR.** Standard PCR (the polymerase chain reaction) is not generally quantitative, because the end product is observed after the bulk of the product has been synthesized, at a point where the rate of synthesis of new molecules has reached a plateau. Consequently, small differences in the amount of target at the start of the reaction are masked. As with the measurement of biochemical reaction rates, quantitative measures of nucleic acid polymerization must be made during the linear phase of the reaction. The number of cycles taken to attain the linear phase of increase in product biosynthesis is regarded as a quantitative indicator of the amount of template in the RNA population. To be truly quantitative, in Q-PCR the reaction rate is compared with that of a similar template (ideally, one with the same base composition, such as the same gene with a small internal deletion) that is spiked into a dilution series. This establishes the sensitivity of the reaction to the number of molecules of template.

Real-time measurement of product accumulation provides the most sensitive assay; otherwise, multiple reactions must be set up in parallel, with replicates, and stopped at chosen times. Commercial PCR machines for Q-PCR utilize a transparent reaction cuvette or capillary exposed to a fluorescence detector. The detector measures the signal from a dye that only fluoresces when intercalated with double-stranded nucleic acid. Readings are taken just prior to the denaturation step between 80° and 85°, since primer-dimers and nonspecific products denature below this temperature and so contribute minimally to the fluorescence signal.

The Q-PCR reaction is usually primed with cDNA synthesized by one cycle of reverse transcription, following which gene-specific primers are used to amplify the locus of interest. With each cycle, the fluorescence measures the number of molecules of double-stranded DNA that have been synthesized. If a reaction is primed with 10 molecules of the template, the rate of observable product accumulation will be delayed relative to one primed with 1,000 molecules of template, because it takes at least half a dozen cycles to produce sufficient specific product to begin seeing a signal (Figure 4.18).

An alternative to using fluorescent dyes is Applied Biosystem's TaqMan® assay. This method relies on the property of **fluorescence resonance energy transfer (FRET),** in which the fluorescence from one dye is transferred to and thereby quenched by a nearby dye that emits at a different wavelength.

(A)

Reverse transcriptase

Amplification by PCR
Add Syber Green I

Figure 4.18 Quantitative RT-PCR.
(A) mRNA is reverse transcribed and the transcript of interest is amplified using a gene-specific 5′ primer. (B) In the presence of a fluorescent dye such as Syber Green I, amplification can be monitored in real time. (C) The crossing point of the linear phase of the reaction with background fluorescence is used to establish a standard curve from which the relative amount of unknown product can be determined. (After Rasmussen et al. 1998.)

(B)

(C)

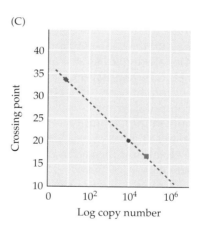

The two dyes are incorporated at either end of an internal primer that hybridizes to the template cDNA. As the PCR reaction proceeds, a 5′-to-3′ exonuclease activity of the polymerase digests this FRET primer, freeing the reporter dye into solution so that its fluorescence emission is no longer quenched. Thus, the observed fluorescence intensity increases as more PCR product is made, allowing quantification of transcript levels. (A different application of TaqMan® assay in genotyping is discussed in Chapter 3).

Properties of Transcriptomes

In this section, we describe some applications of gene expression profiling on a genome-wide basis. Literature that includes microarray data is increasing at an exponential rate as the method becomes incorporated as a tool in genetic dissection, so this section is not, nor is it intended to be, a comprehensive survey. The intention is merely to illustrate how the procedure is being used to investigate a wide range of biological processes.

Microbial Transcriptomics

Microarrrays have been used extensively both for gene expression profiling and comparative genomics in a wide range of microbes. In bacteria as diverse as *E. coli*, extremophiles, and industrial fermenters, transcription profiling has been employed to study the organization of biochemical pathways, adaptive responses to antibiotics, and the mechanistic basis of phenomena such as biofilm formation. Hybridization of bacterial DNA to microarrays is being employed to characterize strain diversity and the species content of environmental samples, and as a rapid detection technology for infectious diseases and food-borne pathogens. Similarly, fungi of various sorts and protozooan parasites have received much attention aimed toward dissecting cellular processes and identifying target genes for use in immunization and drug development.

Many of the first cDNA and oligonucleotide microarray studies were done using the budding yeast *Saccharomyces cerevisiae*, starting with an analysis of transcriptional changes that occur during the haploid cell cycle (Spellman et al. 1998). One of the advantages of yeast as an experimental system is that the early completion of the genome sequence quickly led to the annotation of the entire complement of just over 6,200 genes, which is small enough to allow complete representation of the open reading frames on a single microarray. Comparison of gene expression profiles after synchronization of cell growth resulted in the identification of hundreds of genes that are transcribed during, and presumably are required for procession through, each stage of mitosis.

Similarly, sporulation has been studied in great detail in *S. cerevisiae* (Figure 4.19; Chu et al. 1998). In this yeast equivalent of meiosis, diploid cells are converted to haploid gametes in response to starvation on nitrogen-deficient medium. Decades of genetic and molecular analysis had already identified 50 or so genes that were induced in four temporal waves during sporulation. cDNA microarray analysis identified more than 1,000 genes that showed a more than threefold change for at least one of six time points, or an average 2.2-fold change across the entire time course. In addition, global analysis allowed the early phase to be subdivided into four more subtle groupings, including a set of 52 immediate but transiently induced metabolic genes that are likely to be responding to nitrogen starvation rather than initiation of sporulation per se. Intriguingly, only a relatively small fraction of cell-cycle regulated genes are regulated in a similar fashion in the fission yeast, *Schizosaccharomyces pombe*, implying considerable divergence in all but the core functions that coordinate meiosis in divergent yeasts (Rustici et al. 2004).

A very different application of cDNA microarrays is in experimental evolution. Yeast cells cultured in glucose-limiting media adapt by reducing their dependence on glucose fermentation and switching to oxidative phospho-

Figure 4.19 Microarray analysis of sporulation in budding yeast. Large fractions of the yeast genome are either repressed or induced in response to sporulation. Microarray analysis allows clustering of suites of genes that share profiles, such as those assembled for six genes of each subclass charted on the right. (After Chu et al. 1998.)

rylation as a more efficient means of generating energy. Three replicate chemostat experiments starting from the same isogenic culture were allowed to evolve for over 250 generations, in which time six selective sweeps led to profound metabolic adaptation (Ferea et al. 1999). Remarkably, gene expression profiling demonstrated that 3% ($^{184}/_{6,124}$) of all genes showed an average twofold change in gene expression across all three replicate evolved strains, in most cases including similar responses in at least two of the strains. This corresponded to about one-third of all genes that showed at least a twofold change in one strain, indicating that despite the small number of adaptive mutations responsible for the response to glucose starvation, hundreds of genes responded in a coordinated manner.

About half of the genes that changed in at least two populations were previously characterized, and most of these have roles in respiration, fermentation, and metabolite transport (Figure 4.20). Furthermore, many of the genes showed parallel changes as a result of physiological adaptation to glucose depletion (the "diauxic shift"), indicating that long-term evolution occurred in a somewhat predictable direction that mirrored expectations derived from prior knowledge of metabolism. In a conceptually similar study of adaptation of the pathogenic yeast *Candida albicans* to fluconazole exposure, at least two different global mechanisms of response

Figure 4.20 Gene expression changes in yeast.   In this diagram of yeast central metabolism, the boxes represent transcripts encoding enzymes that are repressed (green) or induced (red) at least twofold, either after a shift to glucose-limiting media (the diauxic shift; triangles in the upper left half of boxes) or after 250 generations of adaptive evolution (triangles in lower right of boxes). Both conditions lead to reduced glucose fermentation and increased oxidative phosphorylation through the TCA cycle. (After DeRisi et al. 1997; Ferea et al. 1999.)

were observed (Cowen et al. 2002). Similar evolved expression profiles were observed in drug-resistant isolates sampled directly from HIV patients, highlighting how microarray analyses may be used to address the pressing biomedical problem of antibiotic resistance.

Figure 4.21 The compendium approach.   This technique is used to identify genes and treatments that act on similar pathways. Profiling a large number of mutations or treatment conditions identifies clusters of genes that are co-regulated across a range of conditions; for example, here genes in clusters A1 and A2 have opposite effects, while those in B and C display specific features of interest. Simultaneous hierarchical clustering by treatment (across the top of the figure) also groups conditions that lead to similar overall transcription profiles, allowing researchers to generate hypotheses as to the functions of the mutant genes, drugs, or other environmental agents that led to these perturbations in gene expression.

Numerous studies have examined the effect of gene knockouts on the yeast transcriptome, leading to insights into the targets of regulatory gene function. A particularly impressive application of this strategy is the "compendium of expression profiles" approach (schematized in Figure 4.21), in which over 300 different strains and conditions were contrasted (T. R. Hughes et al. 2000). The data set included 11 inducible transgenes, 13 drug treatments, and 276 deletion mutants, 69 of which removed unclassified ORFs. Twenty to thirty percent of all deletion mutants had more than 100 genes significantly induced or repressed relative to the control, highlighting the ubiquity of coordinate gene regulation in yeast. Interestingly, the fraction of unclassified ORF deletions that had a large effect on the transcriptome was considerably less than that of the already characterized genes, consistent with the idea that these "orphan" genes have minor roles or are important only under extreme or unusual growth conditions. Hierarchical classification of the 300 samples and 6,000 genes proved to be a powerful mechanism for clustering genes with unknown or poorly studied functions into processes shared by other genes, such as cell wall biosynthesis, steroid metabolism, mitochondrial function, mating, and protein synthesis.

Parasite gene expression profiling is exemplified by analysis of the causative agent of malaria, *Plasmodium falciparum*. The genome of this protozooan contains 5,400 predicted genes, most of which have no known function. Strikingly, 60% of the genes are expressed while the organism infects human red blood cells, in a tightly orchestrated program that follows the functional needs of the parasite (Bozdech et al. 2004). Following invasion, the organism upregulates protein synthesis and metabolic activity, and thereafter waves of genes related to DNA replication, growth, plastid formation, the next cycle of invasion, and motility are turned on. One application of this type of analysis is in the identification of candidate vaccine targets expressed on the cell surface.

Cancer and Clinical Applications

Both SAGE and microarray technology have been employed extensively to characterize the human transcriptome in diverse cell types. Estimates from SAGE analysis of the number of genes expressed in any given cell type fall in the range of 15,000–25,000 (Table 4.1). A "minimal" transcriptome consisting of at least 1,000 genes expressed in all cell types has also been suggested. The greatest variety of gene expression is apparently found in brain tissue, perhaps reflecting the complexity of neuronal types. Aside from assembling atlases of gene expression, numerous studies have begun to characterize differences in gene expression in cancer cells as well as differences associated with other human diseases.

Microarray analyses of cancer have been carried out with the following objectives:

- Enhanced classification of cancer types, including identifying cell-type of origin
- Characterization of expression profiles that may help predict therapeutic response
- Clustering genes in order to generate hypotheses concerning their mode of action in carcinogenesis
- Identification of novel gene targets for chemotherapy

TABLE 4.1 *Estimates of Transcript Diversity in Human Tissues*

Tissue[a]	Total SAGE tags	Number of unique tags	Estimated number of genes[b]
Colon epithelium	98,089	12,941	20,000
Keratinocytes	83,835	12,598	22,000
Breast epithelium	107,632	13,429	20,500
Lung epithelium	111,848	11,636	16,000
Melanocytes	110,631	14,824	21,000
Prostate	98,010	9,786	15,000
Monocytes	66,673	9,504	20,000
Kidney epithelium	103,836	15,094	22,500
Chondrocytes	88,875	11,628	19,000
Cardiomyocytes	77,374	9,449	18,000
Brain	202,448	23,580	23,500

[a]Each tissue was represented by at least two libraries.
[b]An estimate of the number of genes is provided for comparison assuming that 200,000 tags were screened for each tissue.
Data from Velculescu et al. (1999).

The usefulness of microarrays for cancer classification has been demonstrated both through studies of cancer cell lines and cancer biopsies. One early microarray study of 9,700 cDNAs in 60 cancer cell lines of diverse origin clearly demonstrated that samples of similar cancer types—for example, neuroblastomas, melanomas, leukemias, or colon or ovarian cancers (though not necessarily breast and lung carcinomas)—tend to share gene expression profiles. These cancer-type specific profiles in turn may in part reflect differences retained from their tissue of origin (Figure 4.22A; Ross et al. 2000). Hierarchical clustering identifies groups of genes that provide signature profiles for each cancer subtype and hence identify markers that may prove useful in clinical diagnosis, and suggest functions for previously uncharacterized genes.

Biopsy gene expression profiles can also be clustered relative to one another and to normal tissues to subclassify cancer types. Breast cancers in particular are not always readily classified using classical histological markers, but genome-wide comparisons have the power to resolve the likely cell-type of origin (Perou et al. 2000). While leukemias can be broadly classified by histology, until the advent of microarrays no markers were available that predicted either mortality or therapeutic response. Expression profiles have now been identified that cluster leukemias into groups that correlate with long-term prognosis, as shown in Figure 4.22B and C (Alizadeh et al. 2000). Metastatic and solid tumors can be distinguished on the basis of transcription of an extensive set of genes that regulate passage through the cell cycle, and ongoing studies focus on the quantitative correlation of specific

Figure 4.22 Molecular pharmacology of cancers. (A) Hierarchical clustering of tumor biopsy expression profiles reveals that different tumors can be identified on the basis of type-specific expression profiles. (B) Similarly, clustering of distinct cancer types, such as diffuse large B-cell lymphomas (DLBCL), uncovers the existence of novel molecular subtypes (GC B-like in gold and Activated B-like in blue) that may be predictive of survival probability, as indicated by the standard Kaplan-Meier plots shown in (C). (A after Ross et al. 2000; C after Alizadeh et al. 2000.)

genes with proliferation rates. Similarly, molecular signatures in breast cancers predict metastatic potential and responsiveness to hormonal or chemotherapy, particularly if coupled with genotype information on the status of susceptibility loci such as BRCA1 (van't Veer et al. 2002). However, there are numerous issues to be resolved concerning the power and repeatability of predictors based on microarrays, and the standards required for regulatory implementation are different from those used during exploratory research (Petricoin et al. 2002).

Microarrays are also being used in the context of infectious disease, the immune system, and disorders from asthma to autoimmunity. For example, profiling lesions due to infection with leprosy revealed how *Mycobacterium* affects the cytokine balance in T-cells in a manner that correlates with the manifestation of disease (Bleharski et al. 2003). Applications range from finding biomarkers for infection and exposure to toxins or drugs, to learning how infection with one agent may affect tolerance to another, assessing contamination in blood products, and monitoring the effects of systemic parasite infection.

The potential impact of model systems biomedical research is illustrated by a study of a murine model for the human neurodegenerative diseases Sandhoff and Tay-Sachs (Wada et al. 2000), as outlined in Figure 4.23. Sandhoff and Tay-Sachs diseases are associated with the accumulation of particular gangliosides and lipids in neurons due to the absence of lysosomal β-hexosaminidase function. Comparison of spinal cord samples from normal and mutant mice on cDNA microarrays demonstrated intense upregulation of the inflammatory response in four-month-old mice, the time at which the disease manifests. This pathology arises from the presence in the brain stem of unusually high numbers of microglial cells, which have been shown to alleviate lipid storage disorders by supplying the necessary enzymes to overloaded neurons, and by phagocytosing dying neurons. Enzyme-deficient microglia cannot perform this service, but do secrete cell-death promoting cytokines that lead to massive neurodegeneration. Encouragingly, this effect can be counteracted by bone marrow transplantation, which eventually leads to the production of normal circulating microglia.

Development, Physiology, and Behavior

Gene expression profiling has been adopted by biologists working with model systems to address a wide range of questions in development, physiology, and behavior (Reinke and White 2002). One immediate application is in annotating gene expression as development proceeds in *Drosophila*, nematodes, and plants. The life cycle of a fruitfly consists of four distinct phases: *embryogenesis*, in which the basic features of the body pattern are established; a series of larval *instars* in which the precursor imaginal tissues of the future adult fly form; the *pupa*, in which the larva is dissolved and the adult assembled; and the *adult*. One comprehensive study of 66 sequential time points (Arbeitman et al. 2002), summarized in Figure 4.24, revealed that 86% of the 4,028 genes studied change in expression during develop-

Mouse model

β-hexosaminidase knockout

Microarray analysis + histology

Activated genes in microglia

▶ BLC chemoattractant
▶ Cathepsin S, Cathepsin C
▶ Serine protease inhibitor 2–1
▶ Complement component 1
▶ Macrophage expressed 1
▶ Galectin 3
▶ Prostaglandin D2 synthase
▶ Annexin A3
▶ Benzodiazepine receptor
▶ MHC 2Q7, MHC 2L
▶ Osteopontin
▶ CD68, CD37, CD9
▶ F4/80 antigen-like EGF module

Hypothesis

Infiltration of degenerating neurons by enzymatically inactive microglia amplifies the inflammatory response, leading to heightened neuronal cell death

Test

Bone marrow transplant with *HexB+* cells ameliorates neurodegeneration

Figure 4.23 Microarray analysis and the etiology of Sandhoff and Tay-Sachs diseases. In a mouse model, microarray analysis in conjunction with histological studies linked intense upregulation of the inflammatory response to the presence of unusually high numbers of microglial cells in the brain stem. Enzyme-deficient microglia were then implicated in the etiology of two human neurodegenerative disorders. (Data from Wada et al. 2000.)

(A)

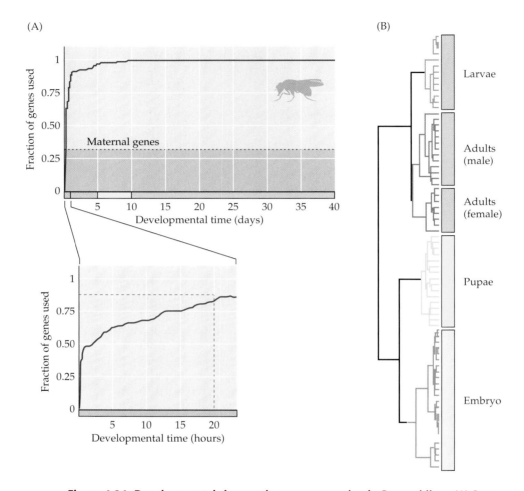

Figure 4.24 Developmental changes in gene expression in *Drosophila*. (A) Over 80% of all genes are transcribed during embryogenesis, starting with 30% of genes as maternal transcripts, followed by a gradual accumulation of usage over the next 24 hours. (B) Clustering of profiles across the complete developmental time-course reveals that transcription in adults is globally closer to that observed in larvae than in pupae and embryos. (After Arbeitman et al. 2002.)

ment. Remarkably, many genes are expressed in two waves: embryonic resembles pupal expression, and larval resembles adult. In some striking cases, such as myogenesis, microarrays reveal how analysis of RNA from whole organisms can detect transcriptional events in cellular subpopulations (White et al. 1999). It has even been possible to detect induction of a gene in a single pair of cells in the brain.

With this baseline, it is now possible to identify early and late transcriptional targets of single mutations, and to assemble pathways and networks of gene expression. This approach supplements classical genetic and cell biological studies, providing a genome-wide perspective that can suggest hypotheses that can be tested by directed studies of specific genes. As well as providing insight into organ specification, sex determination, embryonic and appendage patterning, and hormonal regulation of metamorphosis, microarrays have been used to dissect physiological responses to aging, diet (notably, caloric restriction), and sexual behavior.

Microarrays have also been used to identify genes associated with wood biosynthesis in forest trees grown on soils of varying quality; to characterize responses to viral and nematode pathogenesis; and to dissect nodule induction by symbiotic nitrogen-fixing bacteria. Dairy farmers are interested in characterizing resistance and susceptibility to mastitis, and sheep farmers would like to know more about the properties of wool follicles. In each of these cases, simple expression profile comparisons are just the first step in analysis, but careful experimental design can be employed to hone the tool for an extraordinarily wide range of applications.

Studies oriented toward nervous system function also cover a diversity of questions, including psychogenomics (the genomic basis of psychological disorders such as Alzheimer's disease and epilepsy), neuropharmacogenomics (the effect of drugs on specific parts of the brain), and behavioral genomics (the molecular basis of behavioral differences). For example, Whitfield et al. (2003) generated a molecular profile of transcript abundance in the brains of honeybees that successfully predicted whether 57 out of 60 bees were foragers or nurses. Since adult bees tend to transition from hive work to foraging with age, they used principal components analysis to compare the contributions of the different factors to overall expression variance, and found that behavior has twice the impact as their age does on transcription in the brain.

Evolutionary and Ecological Functional Genomics

A new discipline of evolutionary and ecological functional genomics is emerging, building heavily on comparative genomics and transcription profiling (Feder and Mitchell-Olds 2003). Some applications, such as studying interactions between biotic and abiotic factors, for example plant-bacterial symbiosis or insect dessication resistance, focus on finding crucial genes and biochemical or cellular pathways. Many others require complex experimental designs involving multiple levels of several factors—perhaps young and old males and females of a dozen different ecotypes. In these cases, the statistical tools used to analyze data tend to be more directed toward the precise estimation of variance components and the avoidance of false negatives at the expense of occasional false positives. Much recent focus has been on integration of population and quantitative genetic methods with gene expression profiling.

Comparative approaches promise to shed much light on the relationship between transcriptional divergence, genetic divergence, and phenotypic divergence. Among closely related species, it is not unusual to see significant differences in the transcript levels of as many as 20% of the genes

(A) Brain

(B) Liver

(C)

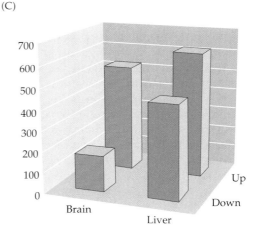

Figure 4.25 Divergence in gene expression in primates. When the relationship between the transcriptome in the brain (A) or liver (B) of individual primates is drawn according to the average distance between transcript abundance, it is observed that the branch leading to humans is twice as long in the brain as it is in the liver. This suggests an accelerated rate of transcriptional divergence in the brain, although counts of the total number of differentially expressed genes suggests greater divergence in the liver (C). (After Enard et al. 2002; Hsieh et al. 2003.)

expressed in a tissue. A landmark comparison of expression profiles in humans, chimps, and an orangutan (Enard et al. 2002) suggests that more transcripts have evolved in the liver than the brain, but that there tends to be greater divergence between humans and other primates in those genes that are different (Figure 4.25). Nevertheless, scaling by the amount of variation among individuals within a species, most divergence at the transcriptional level is probably not due to positive selection, and may be better explained by genetic drift (Hsieh et al. 2003; Khaitovich et al. 2004).

Gene expression variation among individuals within species is often likely to highlight genes that may contribute to quantitative traits. In a contrast of muscle from a northern and a southern population of *Fundulus* fish, more divergence was seen between latitudes than between two sibling species in the southern region, implicating these muscle genes in adaptive divergence (Oleksiak et al. 2002). In the progeny of crosses between phenotypically divergent parents, simultaneous expression profiling and tracking of genetic markers can be used both to map genes whose expression correlates with a trait of interest, as well as genes that regulate transcript abundance. The latter are called **expression quantitative trait loci (eQTL),** and have been reported in relation to obesity in mice, lignin content in eucalyptus, and variation in natural isolates of yeast. Some early conclusions from these studies (Figure 4.26) are that eQTL often co-localize with the genes they regulate (implying regulatory polymorphism), much of the transcriptome is affected by relatively large-effect eQTL, and the same eQTL may regulate many target genes, whether or not they encode transcription factors (Schadt et al. 2003; Yvert et al. 2003).

Many more applications are sure to emerge in the next few years. Combined with protein expression profiling (as described in the next chapter), biologists' perspective on the genetic basis of variation, divergence, and pathology is itself evolving rapidly.

Gene Expression Databases

Gene expression profiling has been criticized for its focus on data generation rather than hypothesis testing, but practitioners point out that, first, hypotheses tend to flow from large data sets and, second, whole-genome approaches often engender a shift in perspective that can spur new research directions. The successful use of microarrays thus depends on making the data available to as many potential users as possible, which in turn requires the establishment of Internet-based databases. There are three features of database establishment that need to be addressed: importing data in a standard format into the database, facilitating queries, and generating novel data-mining tools.

A consortium of users known as the Microarray Gene Expression Data Society (MGED; http://www.mged.org) have agreed on community standards and proposed a set of **m**inimal **i**nformation for the **a**nnotation of **m**icroarray **e**xperiments (**MIAME**). The minimal features of an experiment that are required for annotation are:

- Definition of the clones represented on the array (the "platform")
- Precise description of the samples and controls (reference samples)
- Characterization of the hybridization and labeling protocols
- The researchers' names and institutional addresses
- Data values associated with each point on the array

Different software associated with scanners output data in different formats, and use different criteria for calculating spot and background intensities, so users are asked to reformat their output to a standard set of criteria. The alternative of depositing TIFF files with the raw data and allowing centralized extraction of intensity values is unrealistic due to the extremely large volumes of data that would have to be stored. Nevertheless, individ-

(A)

(B)

Figure 4.26 Expression quantitative trait loci (eQTL). (A) If two parents with divergent levels of transcript abundance for a gene (red, high; green, low) are crossed, it is generally expected that their F_1 progeny will have an intermediate level of transcription (yellow). Abundance will vary in F_2 individuals, and may be associated with the genotype at a number of loci sampled throughout the genome. For example, the first locus shows a dominant effect of the A allele, since both AA and AC genotypes have relatively low transcription in a sample of 5 individuals, whereas CC have high abundance. (B) Mapping of such markers onto the genome often indicates that the expression QTL maps to the same location of the gene that encodes the transcript, suggesting that a cis-acting regulatory polymorphism affects the transcript abundance. (After Schadt et al. 2003; Yvert et al. 2003.)

Select expression data set	Example
Type of data	Microarray, SAGE, etc.
Organism	*Drosophila melanogaster, Mus musculus*
Tissue	Embryo, heart
Treatment	25°C, nicotine ingestion

Select measures	Example
Specific data set	Laboratory; platform; samples
Items	Intensity; normalized expression; ratio
Subset of ORFs	Gene families; search by ontology

Select query terms	Example
QC filters	At least 4 replicates
Fold change	>2×
Significance of difference	$p < 0.0001$
Clustering algorithm	Hierarchical, *k*-means, etc.

Figure 4.27 Flow diagram of microarray database queries.

ual users may opt to store their own data locally in a format that allows others to visualize actual spot profiles.

Ideally, databases would also allow direct contrast of experiments conducted with each of the three established methodologies (Aach et al. 2000). The major obstacle here is that SAGE reports the number of tags observed in a sample, short oligonucleotide arrays report estimates of expression level derived by subtraction of the mismatch control, and cDNA microarrays only report relative expression levels. Differences among platforms—namely, the set of clones on the array—do not necessarily present novel computational problems, but do demand careful database management. A further concern is quality control, which entails assuring that data meet minimal standards of consistency using parameters such as approximate linearity of the relationship between the two dye signals. Whether or not such measures can deal with acknowledged but unavoidable sources of variance, such as humidity during array printing or the practices of individual researchers, is unclear.

The simplest form of query is just to call up raw ratio measures from single experiments. More sophisticated protocols allow comparisons across experiments. Typically, a query entails the three steps shown in Figure 4.27: (1) selection of a set of arrays or treatments to contrast, (2) selection of a set of genes of interest, and (3) definition of the criteria by which a change in expression is regarded as meaningful and/or significant. Tools for doing this have been developed as part of the cancer genome anatomy project, initially for comparison of EST clusters in different cancer types, and for online comparison of SAGE data from human data sets. Sophisticated data-mining tools that use statistically robust approaches and facilitate online contrasts of data from different experimental groups—and potentially systems and even organisms—are under development. In the meantime, the Expression Connection of the *Saccharomyces* Genome Database represents the state-of-the-art in how gene expression can be profiled over the World Wide Web, and the Gene Expression Omnibus (GEO) at the NCBI and ArrayExpress at the European Bioinformatics Institute have become the central repositories for expression data.

Summary

1. Gene expression profiling can be performed with a variety of technologies for arraying nanomolar amounts of gene-specific probes. cDNA microarrays consist of PCR-amplified EST fragments arrayed on a glass microscope slide, while GeneChips consist of a series of 25-mer oligonucleotides synthesized directly on a silicon chip. A hybrid technology consists of long oligonucleotides (60- to 80-mers) spotted or synthesized in situ on glass slides using ink-jet printers.

2. Most cDNA microarray applications use two different fluorescent dyes so that differences among replicate arrays are controlled for by computing the ratio of gene expression in the experimental versus a reference sample. Typical analyses report those genes showing two-fold or greater differences in gene expression in one or more samples.

3. Statistical procedures are being developed to enhance data extraction from microarrays, recognizing that dyes, pins, and transcript abundance can all bias ratio measurements. Image analysis also affects measurement, since the pixel intensity across a spot is not uniform, and background fluorescense intensity varies.

4. A variety of clustering methods have been adopted to identify groups of co-regulated genes, including hierarchical clustering, self-organizing maps, and principal component analysis.

5. Similarity of expression profiles is often regarded as evidence for similarity of function of a cluster of genes, allowing putative annotation of the function of unknown genes. Though not definitive, this type of analysis is at least sufficient to generate hypotheses that can be tested by more traditional molecular biological approaches.

6. Investigation of the mechanisms that are responsible for co-regulation of sets of genes is facilitated by detailed monitoring of temporal expression profiles in mutant backgrounds. Bioinformatic methods can be used to identify shared upstream regulatory sequences that may lead to the isolation of transcription factors that mediate particular expression profiles.

7. Serial analysis of gene expression (SAGE) is an alternative method for characterizing transcript abundance in cells. It relies on sequencing large numbers of gene-specific tags.

8. Northern blots and quantitative reverse transcription-PCR (Q-PCR) can be used to confirm expression differences detected on microarrays or by SAGE.

9. The complete set of yeast ORFs can be represented on a single microarray, facilitating a wide range of studies from characterization of the cell cycle, sporulation, response to nutrient starvation, and adaptation in long-term culture.

10. Microarray analysis has revealed that cancer cells tend to retain transcriptional features of the normal cell from which they derive, allowing molecular phenotyping of cancers. In addition, gene expression profiles may be diagnostic of the metastatic potential of tumors, and may thus play a role in design of treatment protocols.

11. Applications of microarrays in human disease research range from characterization of transcriptional changes in diseased tissue to toxicology, immunology, and pharmacology. Characterization of clusters of induced or repressed genes may generate hypotheses as to the etiology of disease.

12. As costs decrease and the availability of array technology expands, gene expression profiling will be increasingly important in the developmental biology of model organisms, as well as the ecology and evolutionary biology of a wide range of parasitic and infective species.

Discussion Questions

1. Compare and contrast the advantages and disadvantages of cDNA and oligonucleotide-based gene expression profiling. What might the effect of cross-hybridization be with each method?

2. Why did the majority of early studies of transcriptional differences adopt fold change as the criterion for detecting changes in gene expression, rather than a measure of statistical significance?

3. Discuss the potential pitfalls in the "guilt-by-association" approach to assigning probable gene function, in which similarity of expression profile is used to infer similarity of function.

4. How might transcriptional profiling contribute to evolutionary and ecological genomics?

5. An emerging application of transcriptomics is in combining compendia of expression profiles with compendia of responses to drugs or other environmental agents, essentially searching for correlations

among vast volumes of data. How does this work and what sorts of corroboratory data will be required to test any hypotheses that are generated?

Web Site Exercises

The Web site linked to this book at http://www.sinauer.com/genomics **provides exercises in various techniques described in this chapter.**

1. Analyze two 4,000-gene microarray experiments using ScanAlyze.

2. Compare a variety of methods for computing ratios and normalizing the microarray data.

3. Perform an Analysis of Variance of a three-treatment experiment.

4. Use the xProfiler at NCBI online.

5. Compare several methods for clustering the data in Exercise 3.

Literature Cited

Aach, J., W. Rindone and G. Church. 2000. Systematic management and analysis of yeast gene expression data. *Genome Res.* 10: 431–445.

Alizadeh, A. A. et al. 2000. Distinct types of diffuse large B-cell lymphoma identified by gene expression profiling. *Nature* 403: 503–511.

Arbeitman, M. N. et al. 2002. Gene expression during the life cycle of *Drosophila melanogaster*. *Science* 297: 2270–2275.

Bailey, T. L. and M. Gribskov. 1998. Methods and statistics for combining motif match scores. *J. Comp. Biol.* 5: 211–221.

Bleharski, J. R. et al. 2003. Use of genetic profiling in leprosy to discriminate clinical forms of the disease. *Science* 301: 1527–1530.

Bozdech, Z., M. Llinás, B. L. Pulliam, E. D. Wong, J. Zhu and J. L. DeRisi. 2004. The transcriptome of the Intrerythrocytic developmental cycle of *Plasmodium falciparum*. *PLoS Biology* 1: 85–100.

Bussemaker, H. J., H. Li and E. Siggia. 2001. Regulatory element detection using correlation with expression. *Nat. Genetics* 27: 167–174.

Chu, S. et al. 1998. The transcriptional program of sporulation in budding yeast. *Science* 282: 699–705.

Chu, T. M., B. Weir and R. D. Wolfinger. 2002. A systematic statistical linear modeling approach to oligonucleotide array experiments. *Math. Biosci.* 176: 35–51.

Churchill, G. A. 2002. Fundamentals of experimental design for cDNA microarrays. *Nat Genet.* 32 (Suppl): 490–495.

Cowen, L. E., A. Nantel, M. Whiteway, D. Thomas, D. Tessier, L. M. Cohn and J. B. Anderson. 2002. Population genomics of drug resistance in *Candida albicans*. *Proc. Natl Acad. Sci. (USA)* 99: 9284–9289.

Cui, X. and G. A. Churchill. 2003. Statistical tests for differential expression in cDNA microarray experiments. *Genome Biol.* 4: 210.

DeRisi, J., V. Iyer and P. O. Brown. 1997. Exploring the metabolic and genetic control of gene expression on a genomic scale. *Science* 278: 680–686.

Draghici, S. 2003. *Data Analysis Tools for DNA Microarrays.* Chapman and Hall/CRC, Princeton, NJ.

Duggan, D., M. Bittner, Y. Chen, P. Meltzer and J. Trent. 1999. Expression profiling using cDNA microarrays. *Nat. Genetics* 21 (Suppl.): 10–14.

Eberwine, J. 1996. Amplification of mRNA populations using aRNA generated from immobilized oligo(dT)-T7 primed cDNA. *Biotechniques* 20: 584–591.

Eisen, M. B., P. Spellman, P. Brown and D. Botstein. 1998. Cluster analysis and display of genome-wide expression patterns. *Proc. Natl Acad. Sci. (USA)* 95: 14863–14868.

Enard, W. et al. 2002. Intra- and interspecific variation in primate gene expression patterns. *Science* 296: 340–343.

Feder, M. E. and T. Mitchell-Olds. 2003. Evolutionary and ecological functional genomics. *Nat. Rev. Genet.* 4: 651–657.

Ferea, T., D. Botstein, P. O. Brown and F. Rosenzweig. 1999. Systematic changes in gene expression patterns following adaptive evolution in yeast. *Proc. Natl Acad. Sci. (USA)* 96: 9721–9726.

Hastie, T. et al. 2000. "Gene shaving" as a method for identifying distinct sets of genes with similar expression patterns. *Genome Biol.* 1: research 003.1–003.21

Holter, N., M. Mitra, A. Maritan, M. Cieplak, J. Banavar and N. Federoff. 2000. Fundamental patterns underlying gene expression profiles: simplicity from complexity. *Proc. Natl Acad. Sci. (USA)* 97: 8409–8414.

Hosack, D. A., G. Dennis Jr., B. T. Sherman, H. C. Lane and R. A. Lempicki. 2003. Identifying biological themes within lists of genes with EASE. *Genome Biology* 4: R60.

Hsieh, W.-P., T.-M. Chu, R. D. Wolfinger and G. Gibson. 2003. Mixed-model re-analysis of primate data suggests tissue and species biases in oligonucleotide-based gene expression profiles. *Genetics* 165: 747–757.

Hughes, J. D., P. Estep, S. Tavazoie and G. M. Church. 2000. Computational identification of *cis*-regulatory elements associated with functionally coherent groups of genes in *Saccharomyces cerevisiae. J. Mol. Biol.* 296: 1205–1214.

Hughes, T. R. et al. 2000. Functional discovery via a compendium of expression profiles. *Cell* 102: 109–126.

Hughes, T. R. et al. 2001. Expression profiling using microarrays fabricated by an ink-jet oligonucleotide synthesizer. *Nat. Biotechnol.* 19: 342–347.

Jakt, L. M., L. Cao, K. Cheah and D. Smith. 2001. Assessing clusters and motifs from gene expression data. *Genome Res.* 11: 112–123.

Jenssen, T.-K., A. Lægreid, J. Komorowski and E. Hovig. 2001. A literature network of human genes for high-throughput analysis of gene expression. *Nat. Genetics* 28: 21–28.

Jin, W., R. Riley, R. Wolfinger, K. White, G. Passador-Gurgel and G. Gibson. 2001. Contributions of sex, genotype and age to transcriptional variance in *Drosophila. Nat. Genetics* 29: 389–395.

Kerr, M. K. and G. Churchill. 2001. Statistical design and the analysis of gene expression microarray data. *Genet. Res.* 77: 123–128.

Kerr, M. K., M. Martin and G. Churchill. 2000. Analysis of variance for gene expression microarray data. *J. Comput. Biol.* 7: 819–837.

Khaitovich, P. et al. 2004. A neutral model of transcriptome evolution. *PLoS Biol.* 2(5): E132.

Lawrence, C. E., S. Altschul, M. Boguski, J. Liu, A. Neuwald and J. Wootton. 1993. Detecting subtle sequence signals: A Gibbs sampling strategy for multiple alignment. *Science* 262: 208–214

Lee, T. I. et al. 2002. Transcriptional regulatory network in *Saccharomyces cerevisiae. Science* 298: 799–804.

Lieb, J. D. 2003. Genome-wide mapping of protein-DNA interactions by chromatin immunoprecipitation and DNA microarray hybridization. *Methods Mol Biol*. 224: 99–109.

Lipshutz, R. J., S. Fodor, T. Gingeras and D. Lockhart. 1999. High-density synthetic oligonucleotide arrays. *Nat. Genetics* 21 (Suppl.): 20–24.

Lockhart, D. J. et al. 1996. Expression monitoring by hybridization to high-density oligonucleotide arrays. *Nat. Biotechnol.* 14: 1675–1680.

McGall, G., J. Labadie, P. Brock, G. Wallraff, T. Nguyen and W. Hinsberg. 1996. Light-directed synthesis of high-density oligonucleotide arrays using seminconductor photoresists. *Proc. Natl Acad. Sci. (USA)* 93: 13555–13560.

McGuire, A. M., J. Hughes and G. M. Church. 2000. Conservation of DNA regulatory motifs and discovery of new motifs in microbial genomes. *Nat. Genetics* 10: 744–757.

Oleksiak, M. F., G. A. Churchill and D. L. Crawford. 2002. Variation in gene expression within and among natural populations. *Nat Genet.* 32: 261–266.

Perou, C. M. et al. 2000. Molecular portraits of human breast tumors. *Nature* 406: 747–752.

Petricoin, E. F. III. et al. 2002. Medical applications of microarray technologies: a regulatory science perspective. *Nat. Genet* 32 (*Suppl.*): 474–479.

Quackenbush, J. 2002. Microarray data normalization and transformation. *Nat. Genet* 32 (*Suppl.*): 496–501.

Rajewsky, N., M. Vergassola, U. Gaul and E. D. Siggia. 2002. Computational detection of genomic *cis*-regulatory modules applied to body patterning in the early *Drosophila* embryo. *BMC Bioinformatics* 3: 30.

Rasmussen, R., T. Morrison, M. Herrmann and C. Wittwer. 1998. Quantitative PCR by continuous fluorescence monitoring of a double strand DNA specific binding dye. *Biochemica* 2: 8–11.

Reinke, V. and K. P. White. 2002. Developmental genomics approaches in model organisms. *Annu. Rev. Genom. Hum. Genet.* 3: 153–178.

Ross, D. T. et al. 2000. Systematic variation in gene expression patterns in human cancer cell lines. *Nat. Genetics* 24: 227–235.

Roth, F. P., J. Hughes, P. Estep and G. M. Church. 1998. Finding DNA regulatory motifs within unaligned noncoding sequences clustered by whole-genome mRNA quantitation. *Nat. Biotechnol.* 16: 939–945.

Rouillard, J.-M., C. J. Herbert and M. Zuker. 2002. OligoArray: genome scale oligonucleotide design for microarrays. *Bioinformatics* 18: 486–487.

Rustici, G. et al. 2004. Periodic gene expression program of the fission yeast cell cycle. *Nat. Genetics.* 36: 809–817.

Schadt, E. E. et al. 2003. Genetics of gene expression surveyed in maize, mouse, and man. *Nature* 422: 297–302.

Schena, M. et al. 1995. Quantitative monitoring of gene expression patterns with a cDNA microarray. *Science* 270: 467–470.

Spellman, P. et al. 1998. Comprehensive identification of cell cycle-regulated genes of the yeast *Saccharomyces cerevisiae* by microarray hybridization. *Mol. Biol. Cell* 9: 3273–3297.

Storey, J. D. and R. Tibshirani. 2003. Statistical significance for genome-wide studies. *Proc. Natl Acad. Sci. (USA)* 100: 9440–9445.

Tamayo, P. et al. 1999. Interpreting patterns of gene expression with self-organizing maps: Methods and application to hematopoietic differentiation. *Proc. Natl Acad. Sci. (USA)* 96: 2907–2912.

Tusher, V. G., R. Tibshirani and G. Chu. 2001. Significance analysis of microarray data applied to the ionizing radiation response. *Proc. Natl Acad. Sci. (USA)* 98: 5116–5121.

van't Veer, L. J. et al. 2002. Gene expression profiling predicts clinical outcome of breast cancer. *Nature* 415: 530–536.

Velculescu, V. E., L. Zhang, B. Vogelstein and K. Kinzler. 1995. Serial analysis of gene expression. *Science* 270: 484–487.

Velculescu, V. E. et al. 1997. Characterization of the yeast transcriptome. *Cell* 88: 243–251.

Velculescu, V. E. et al. 1999. Analysis of human transcriptomes. *Nat. Genetics* 23: 387–388.

Wada, R., C. Tifft and R. Proia. 2000. Microglial activation precedes acute neurodegeneration in Sandhoff disease and is suppressed by bone marrow transplantation. *Proc. Natl Acad. Sci. (USA)* 97: 10954–10959.

White, K. P., S. Rifkin, P. Hurban and D. S. Hogness. 1999. Microarray analysis of *Drosophila* development during metamorphosis. *Science* 286: 2179–2184.

Whitfield, C. W., A.-M. Cziko and G. E. Robinson. 2003. Gene expression profiles in the brain predict behavior in individual honey bees. *Science* 302: 296–299.

Wolfinger, R., G. Gibson, E. Wolfinger, L. Bennett, H. Hamadeh, P. Bushel, C. Afshari and R. S. Paules. 2001. Assessing gene significance from cDNA microarray gene expression data via mixed models. *J. Compu. Biol.* 8: 625–637.

Yang, L., D. K. Tran and X. Wang. 2001. BADGE, Beads Array for the Detection of Gene Expression, a high-throughput diagnostic bioassay. *Genome Res.* 11: 1888–1898.

Yvert, G. et al. 2003. Trans-acting regulatory variation in *Saccharomyces cerevisiae* and the role of transcription factors. *Nat. Genetics* 35: 57–64.

Zhang, L. et al. 1997. Gene expression profiles in normal and cancer cells. *Science* 276: 1268–1272.

5 *Proteomics and Functional Genomics*

Proteomics may be defined loosely as the study of the structure and expression of proteins, and of the interactions between proteins. Our discussion of proteomics begins with a description of how proteins are annotated computationally, then moves on to a survey of how data on the expression and identity of proteins in cells are obtained. We also describe several methods that are used to study the structure of proteins.

    Functional genomics, which here refers to documentation of the functions of large numbers of genes using mutational and recombinant molecular biological approaches, is included in this chapter to emphasize that it shares with proteomics the common goal of ascertaining biological function.

Functional Proteomics

Protein Annotation

Gene annotations now routinely link to a variety of protein databases that provide information on the structure and function of the encoded protein product. The standard Ensembl UniProt or NCBI Entrez protein reference sequence annotations are very similar to nucleotide sequence entries. They supply the predicted protein sequence, note the derivation of the sequence, and annotate the location of key features of the protein. The most relevant key features are the predicted protein domains, which in most cases are derived by comparing the sample sequence with databases of all known protein sequences.

    The major databases for curated protein sequences are Swiss-Prot (http://www.expasy.ch; Figure 5.1) and Protein Information Resource (PIR: http://pir.georgetown.edu). These databases are extensively crosslinked with one another and with other genomic resources such as transcript and SNP

Figure 5.1 Proteomics tools on ExPASy.
The layout of ExPASy, the Expert Protein
Analysis System Web site, is documented
at **www.expasy.org/sitemap.html**.

www.expasy.ch

Database	Primary structure	Structure prediction
▶ SWISS-PROT/TREMBL	▶ Physicochemical properties	▶ *de novo* helical/sheet prediction
▶ ProSite	▶ Prediction of coiled regions	▶ Secondary structure prediction
▶ SWISS-2DPAGE	▶ Identification of PEST sequences	▶ Tertiary structure prediction
▶ SWISS-3DIMAGE	▶ Representation of hydrophobicity	▶ Identification of TM domains
▶ Enzyme nomenclature	▶ Domain recognition by alignment	▶ Superposition of 3-D structures
▶ Links to other servers	▶ Prediction of modification sites	

data. Swiss-Prot is supplemented by TrEMBL, a computer-generated annotation based on nucleic acid sequences that have not yet been curated by a
human.

Protein domain predictions are derived from these databases by a variety of different algorithms. PROSITE is the standard database of protein
domains extracted from the Swiss-Prot database. PROSITE uses a pattern-
recognition algorithm, along the lines discussed in Box 5.1. In essence, a
set of residues commonly found in a cluster of related proteins is identified and used as the basis for comparison with other protein sequences. For
example, I-G-[GA]-G-M-[LF]-[SA]-x-P-x(3)-[SA]-G-x(2)-F is the consensus
signature for aromatic amino acid permeases, which means that all proteins
with this domain have a sequence resembling "isoleucine then glycine then
glycine or alanine then glycine then methionine then leucine or phenylala-
nine," and so on (with x representing any amino acid). Such consensus elements are not always obvious, so a more general approach is to generate a
weight matrix (also known as a **profile**) of position-specific amino acid
weights and gap penalties.

The options for exploring protein similarities are bewildering. Since pattern and profile search methods are not guaranteed to yield the same results,
it is generally a good idea to compare multiple domain search methods. All
new protein sequences deposited in a protein database are compared with
the complete list of profiles in the various databases by way of hidden

Markov model algorithms. For example, the PROSITE database represents a collection of protein sequences that are structurally related to one of over 1,700 known functional protein domains. Similar approaches are used to represent the Pfam, TIGRFAM, and Panther families of proteins and protein domains. Other domain-finding methods linked to the UniProt annotation of a protein include SMART, PRINTS, and ProDom, which may give alternate perspectives on the structure and function of the sequence. The TRANS-FAC database brings together information on transcription factors, while SENTRA serves a similar role for prokaryotic signal transduction proteins.

In addition to assigning each gene product to a family of similar proteins, protein annotations aim to provide information on the primary, secondary, and tertiary structures of proteins. As shown in Figure 5.2, primary structure refers to the sequence and modifications of the amino acids. Secondary structure is the local folding into alpha helix, beta sheet, or coiled coil. Tertiary structure is folding of these elements into domains, which can also fold into quaternary conformations and multi-protein complexes.

(A) Primary

```
MKVLLRLICFIALLISSLEADKCKEREEKIILVSSANEIDVRPCPLNPNEHKGTITWYKD
DSKTPVSTEQASRIHQHKEKLWFVPAKVEDSGHYYCVVRNSSYCLRIKISAKFVENEPNL
CYNAQAIFKQKLPVAGDGGLVCPYMEFFKNENNELPKLQWYKDCKPLLLDNIHFSGVKDR
LIVMNVAEKHRGNYTCHASYTYLGKQYPITRVIEFITLEENKPTRPVIVSPANETMEVDL
GSQIQLICNVTGQLSDIAYWKWNGSVIDEDDPVLGEDYYSVENPANKRRSTLITVLNISE
IESRFYKHPFTCFAKNTHGIDAAYIQLIYPVTNFQKHMIGICVTLTVIIVCSVFIYKIFK
IDIVLWYRDSCYDFLPIKASDGKTYDAYILYPKTVGEGSTSDCDIFVFKVLPEVLEKQCG
YKLFIYGRDDYVGEDIVEVINENVKKSRRLIIILVRETSGFSWLGGSSEEQIAMYNALVQ
DGIKVVLLELEKIQDYEKMPESIKFIKQKHGAIRWSGDFTQGPQSAKTRFWKNVRYHMPV
QRRSPSSKHQLLSPATKEKLQREAHVPLG
```

(B) Secondary

α-helix

β-sheet

(C) Tertiary

Figure 5.2 Primary, Secondary, and Tertiary Protein Structure. The primary structure is the sequence of amino acids, some of which can be identified as likely sites for modifications such as phosphorylation (shown in color for S, T, and Y residues). The secondary structure is the local folding of alpha helix and beta sheet, for example. These fold into tertiary structure, and very often into distinct domains highlighted in blue, green, and red.

EXERCISE 5.1 *Structural annotation of a protein*

Using the conceptual translation of the gene that you identified in Exercise 1.2, perform the following steps designed to annotate structural features of the protein.

Access the UniProt file for the protein through the Ensembl site, and find:

a. The Ensembl translation ID and UniProt accession number of the protein.

b. List the start and endpoints of the Prosite and Pfam domains.

c. What is the predicted isoelectric point, charge, and mass of the peptide?

d. What is the Prosite Consensus Pattern for this protein?

e. If you can, find an image of the 3D structure of the predicted protein.

ANSWER: From the Ensembl home page, **http://www.ensembl.org**, choose Human as the species, and then search for the name of your protein—for example, IL13. Link out to the appropriate gene (ENSG00000169194) and scroll down to the transcript/translation summary. Clicking on "Protein information" brings you to the Ensembl ProteinView. (a) The Protein ID is ENSP00000304915. (b) Toward the bottom of the page, the interleukin domain is listed as residues 43-70 for Prosite, and 15-57 for Pfam. (c) Next to the Peptide Sequence, various Stats are listed, including Isoelectric point (8.23), Charge (4), and Molecular weight (15,788). (d) Just above these data is a link to the UniProt viewer, which lists this protein as P35225. Find the Cross-Reference to the PROSITE NiceSite view, which documents a 25-28 amino acid consensus pattern for the protein as [LI]-x-E-[LIVM](2)-x(4,5)-[LIVM]-[TL]-x(5,7)-C-x(4)-[IVA]- x-[DNS]-[LIVMA]. (e) Now return to the Ensembl Protein View and click on the "Pfam" link to see a ribbon structure of the protein such as the one shown here. (You can also access the PDB to see a different view.)

It should be appreciated that much of the information associated with a given protein annotation is generated by computer prediction and/or comparison with similar proteins that have been studied in other organisms. Even the primary protein sequence is often a best guess in the sense that the start methionine is usually assumed to be encoded by the first AUG identified by the gene-finding algorithm or by mapping of the 5' end of a transcript. Furthermore, alternative splice variants are not accurately described by predictive algorithms or EST sequence surveys, and can be missed even after extensive molecular characterization of the gene. It is also generally assumed that post-transcriptional editing has not occurred, and that rare amino acids have not been inserted, even though examples of both these phenomena have been described in a wide variety of organisms. (One prominent example is RNA editing of the vertebrate glutamate receptors; Bass 2002.)

Annotation of the primary sequence generally only indicates the predicted sites of post-translational protein modification. This is a particularly acute problem for eukaryotic proteomes, since protein modification is ubiquitous and subcellular structure is so rich. Software is available online to suggest likely sites for phosphorylation by a variety of kinases; for glycosylation, acetylation, methylation, and myristoylation; for peptide cleavage (including ubiquitin-conjugation sites that target proteins for degradation); and for targeting of proteins to intracellular compartments such as the mitochondrion, lysosome, chloroplast, or nucleus. These tools have variable accuracy. Methods exist for confirming that each type of modification or intracellular targeting actually occurs, but they rely on prior generation of monoclonal antibodies against the protein and/or purification of native protein, both of which are time-consuming endeavors. Yet because the activity of a protein is so often regulated by chemical modifications, annotating these effects is just as important as annotating the distribution of protein expression. High-throughput methods for detecting post-translational modification are lagging well behind most other areas of genome science research.

Protein annotations often link to tools that enable graphical portrayal of the predicted or described protein secondary structure and other protein features, such as those shown in Figure 5.3 for a *Drosophila* octopamine receptor. Numerous moderately successful algorithms have been developed for prediction of the distribution of alpha helix, beta sheet, or coiled coil, based on properties such as the distribution of charge and bulk of amino acid side chains. Each of the programs listed on the ExPASy (Expert Protein Analysis System) proteomics tools site adopt distinct heuristics that incorporate alignments of similar sequences to improve predictions. If random assignment of protein secondary structure yields a score of 33 on an accuracy scale from 0 to 100, current predictive algorithms will tend to increase the score only to 75%. Hydrophobicity plots are at the core of good methods for predicting membrane-spanning domains; their performance is increased by including the presence of charge bursts on either side of the membrane. Gene annotations also provide direct links to representation of tertiary structure in the form of ribbon or cylinder diagrams, described later in this chapter.

BOX 5.1 Hidden Markov Models in Domain Profiling

One of the more active areas of current investigation is the identification of **sequence motifs** from genomic data. This task spans a wide variety of specific applications. Examples include finding tRNA genes, identifying α-helix regions in proteins, and detecting transcription factor binding sites in the upstream regions of genes. Many of these seemingly disparate tasks can be approached through the use of **profile hidden Markov models.** We introduced hidden Markov models in Box 2.3; here we build on those basic concepts to describe profile HMMs.

Profile HMM methods begin with the construction of a probabilistic description of the variation in a motif of interest—a protein secondary-structure element, for example, or a DNA binding site. Typically, the parameters and structure of the model are obtained by training, using an empirical data set of examples of the motif in question. For example, a profile HMM for a DNA binding site is constructed by aligning many examples of that binding site and extracting characteristic properties.

Profile HMMs are a probabilistic extension of the notion of a **consensus sequence,** as illustrated in this simple example:

```
    1 2 3 4 5 6 7

A   A C T - - T A

B   A C T C - T G

C   A T T - - T T

D   T C T C T T G

E   A T T C - G G
```

One possible consensus sequence for this multiple alignment is:

$$[AT][CT][T][CT]^*[GT][AGT]$$

This consensus sequence successfully captures much of the information in the alignment. It indicates the main columns of the motif, identifies the potential locations of insertion/deletion variation, and describes potential variation at sites in the motif.

Consider the two sequences ACTCTG and TTTTGA. Both match the consensus sequence, but it is clear after inspecting the entire alignment that the second sequence does not appear to be an example of the motif. The alignment contains more information than the consensus sequence, and it is that information that profile HMMs try to capture and exploit.

A simple profile HMM describing this alignment is shown in Figure A. Let us consider how the HMM in this figure generates sequences. The sequence TCTTA can be generated if the model emits a T from main state 1 (with probability 0.2), moves to main state 2 (with probability 1), emits a C (probability 0.6), moves to main state 3 (probability 1), emits a T (probability 1), moves to main state 4 (probability 0.4),

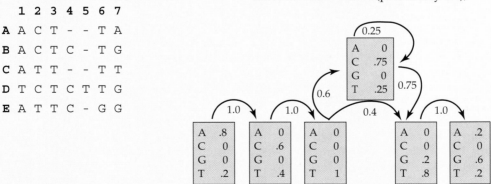

Figure A Simple profile HMM.   The model contains five *main states* (along the bottom row) that are represented in every example of the motif. Insertion/deletion variation is handled, in this case, by a single *insert state* (the top row). The *emission probabilities* (beside each base) match the empirical frequencies of bases in each column of the alignment used for training. The *transition probabilities* (curved arrows) reflect the empirical locations of indels, which are observed only between main states 3 and 4.

emits a T (probability 0.8), moves to main state 5 (probability 1), and emits the final A (probability 0.2). Thus, the probability of this HMM generating this sequence is:

$$Pr(TCTTA) = 0.2 \times 1 \times 0.6 \times 1 \times 1 \times$$
$$0.4 \times 0.8 \times 1 \times 0.2 = 0.00768$$

The presence of the insert state allows an arbitrary number of bases to be inserted between main states 3 and 4. Thus, the probability of the sequence TCTCCTA is:

$$Pr(TCTCCTA) = 0.2 \times 1 \times 0.6 \times 1 \times 1 \times$$
$$0.6 \times 0.75 \times 0.25 \times 0.75 \times 0.75 \times 0.8 \times$$
$$1 \times 0.2 = 0.001215$$

The movement from main state 3 to the insert has probability 0.6; once there it emits the C with probability 0.75; and it remains in the insert state with probability 0.25. We see that this is much greater than the probability of this sequence occurring in random sequence with equal frequencies of each base ($Pr = 4^7 = 0.00006$).

Finally, consider the sequence AGTTG. Although it appears to match the alignment fairly well, it has probability 0 under this HMM because G has never been observed at the second position. When training, it is generally undesirable to have 0 emission probabilities, to avoid such situations. The use of **pseudocounts** insures that all emission probabilities are nonzero.

The current model only allows for indel variation between main states 3 and 4. This assumption may or may not be biologically reasonable. In order to allow insertions at any location in the sequence, in more general HMMs every main state is linked to an insert state. Furthermore, to allow examples of the motif that are missing one or more main states, delete states are added to the HMM. The general form of a profile HMM is shown in Figure B.

Main states can be thought of as the core of the motif. Insert states, as their name implies, allow for the insertion of residues beyond the motif core, while delete states provide a mechanism for the (typically rare) circumstance where a main state is missing from a representative of the motif being modeled. Training and determination of the final form of the model are explained in more detail by Durbin et al. (1998).

The profile HMM has numerous uses. Given a specific sequence, one can ask whether or not it is an example of a motif by comparing its probability of being generated by the profile HMM with the probability that the sequence would be observed in a random stretch of DNA. Extending this notion, the PFAM database consists of profile HMMs constructed for many protein families. A query sequence can be compared to each of the HMMs in the database to see if it is representative of any of those families. Profile HMMs can also be used to probe genomic sequences and identify likely examples of the motif. Finally, the transition and emission probabilities within a profile HMM help us understand the molecular biology of the motif being modeled.

Figure B General form of a profile HMM.   Every main state is linked to an insert state and delete states are added.

(A) Alpha-helical content prediction

(B) Transmembrane domain prediction

(C) Phosphorylation site prediction

Figure 5.3 Protein structural profiles. Profiles of the *Drosophila* octopamine receptor OAMB protein, as determined using software linked to ExPASy. (A) Prediction of
α-helical content, using the Chou-Fassman algorithm with ProtScale. (B) Prediction of
location of the seven transmembrane domains typically found in G-protein linked receptors, using TMpred. (C) Prediction of probable serine, threonine, and tyrosine phosphorylation sites, using NetPhos.

Protein Separation and 2D-PAGE

The general strategy used to characterize the proteome of a cell type is to separate the proteins from one another and then determine the identity and, if possible, the relative quantity of each protein. There is also an analytical challenge to document differences in concentration and state of modification of tens of thousands of different proteins in hundreds of different cell types under a variety of environmental conditions. Further description of the state of activity of proteins is not yet within the realm of genomics, but it is possible to monitor the subcellular localization of proteins in vivo using green fluorescent protein tags that allow monitoring of the subcellular distribution of proteins (Kelso et al. 2004). Protein separation has conventionally been achieved using **two-dimensional polyacrylamide gel electrophoresis,** or **2D-PAGE** (Figure 5.4), in which separation according to charge is followed by separation according to mass. While this procedure does not guarantee separation of all proteins, it is nevertheless possible to identify several thousand different spots on a carefully prepared gel. The proteins are detected most simply by staining the gel with either silver stains or fluorescent dyes such as SYPRO Ruby; for some applications, a radioactive group can be incorporated to increase sensitivity.

Separation according to charge, or more technically, **isoelectric point (pI),** originally relied on carrier molecules known as ampholytes to establish a pH gradient in a column of acrylamide gel during electrophoresis. This procedure is not sufficiently reproducible for side-by-side comparison of gels prepared from two different tissue samples, so it has largely been replaced by immobilized pH gradients (IPGs) that are built into commercially supplied gel slices. These IPGs are now being produced to cover overlapping narrow pH ranges, so greater resolution can be achieved by running subsamples under a series of different conditions that are optimized for a particular comparison.

Separation according to molecular mass is performed by running samples out of the isoelectric focusing gel into an SDS-PAGE slab gel. The SDS detergent both masks any charge differences and denatures the proteins so that they migrate through the pores in the gel according to size of the protein, much like electrophoretic separation of DNA molecules.

2D-PAGE is not quantitative over the complete range of protein concentrations, which are known to cover several orders of magnitude. There are at least three sources of error: non-stoichiometric extraction of proteins from various cellular constituents, failure of proteins to absorb into or migrate out of the isoelectric focusing gel, and non-linear responses of the dye used for detection. Highly charged and low-abundance proteins tend to be underrepresented due to poor extraction, though new buffers with special nonionic detergents and other reagents improve performance. Similarly, proteins sequestered in organelles and membranes or bound up in nuclear or extracellular matrix are not necessarily represented in proportion to their

Figure 5.4 Two-dimensional polyacrylamide gel electrophoresis. (A) In 2D PAGE, protein extracts are applied to the center of an isoelectric focusing strip and allowed to diffuse along the ionic gradient to equilibrium. The strip is then applied to an SDS gel, where electrophoresis in the second dimension separates proteins into "spots" according to molecular mass. The size of each spot is proportional to the amount of protein. (B) A partially annotated human lymphoma 2D gel from the ExPASy database.

(A)

(B)

abundance, while differences in abundance between treatments may merely reflect altered cellular conditions that affect the efficiency of extraction. Very large proteins may not migrate into the SDS gel, and very small ones may migrate off it under conditions that separate the majority of cellular constituents. Fluorescent dyes reportedly have greater linear range than silver stains for monitoring concentration differences, but they may not accurately report concentration differences over more than two orders of magnitude.

Web sites are available that allow individual researchers to download images and documentation of characteristic 2D-gels from a variety of tissues in a variety of species under well-defined conditions. The ExPASy Web site, for example, offers the SWISS-2DPAGE database with a variety of options for searching for proteins on gels, or querying gels to identify proteins (Figure 5.4B). Side-by-side comparison of 2DPAGE gels is not straightforward due to variation in migration rates and local imperfections, but can be facilitated by software that distorts images so that they can be overlaid. For example, the Melanie package (**http://ca.expasy.org/melanie**), developed by the Swiss Institute of Bioinformatics, provides statistical data on the probability that a particular spot corresponds to one that has already been annotated on a reference gel.

Semiquantitative estimation of protein concentrations can be made by integration of pixel intensities over the surface area of a spot, allowing crude calculation of relative abundance between treatments after appropriate normalization. Comparisons between gels are only suitable for contrasts of very similar tissues, as variation in post-translational modification can cause subtle shifts in migration of the same protein prepared from different tissues. However, new two-color fluorescence detection methods have been introduced that allow direct comparison of protein abundance from two conditions that are separated on the same gel (von Eggeling et al. 2001). This approach is also known as 2D-Difference Gel-Electrophoresis (DIGE).

Several other methods available for protein separation offer distinct advantages, both for thorough characterization of a proteome, and for isolation of native protein complexes. Standard chromatographic methods similar to those classically used for protein purification can be employed to separate subfractions of proteins prior to 2D-PAGE. This step allows more concentrated samples to be loaded, and also increases the resolution of low-abundance proteins.

Affinity chromatography, which is illustrated in Figure 5.5, is a proven technique for purifying groups of proteins that form physical complexes within a cell. The idea is to reversibly link one of the components of the protein complex to a chromatographic matrix, then wash the cellular constituents over the column under gentle conditions that allow normal protein-protein interactions to occur. Once the bulk of the cellular proteins have washed through, the remaining proteins can be eluted with a different buffer, and then characterized by a variety of methods.

One method of transient crosslinking is to use a monoclonal antibody directed against the known component. Another is to use recombinant DNA

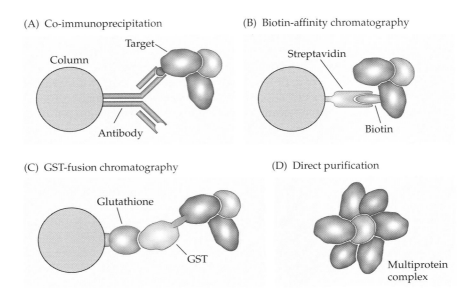

(A) Co-immunoprecipitation

Column

Target

Antibody

(B) Biotin-affinity chromatography

Streptavidin

Biotin

(C) GST-fusion chromatography

Glutathione

GST

(D) Direct purification

Multiprotein
complex

Figure 5.5 Affinity chromatography. A variety of methods exist for purifying proteins that interact with a target protein. (A) In co-immunoprecipitation, an antibody tethered to the chromatographic column matrix binds to the target protein, which in turn binds loosely to interacting partners. (B) Biotin can be chemically crosslinked to the target protein, which will then bind to streptavidin on the column. (C) GST fusion proteins synthesized using recombinant DNA methods consist of glutathione S-transferase, which binds to glutathione on the column, translated in frame with the target protein. (D) Large macromolecular complexes can be purified directly on sedimentation gradients or by other chromatographic methods.

technology to tag the "bait" protein with a peptide fragment such as polyhistidine, maltose binding protein (MBP), or glutathione S-transferase (GST) that will bind to commercially supplied columns. The sequences that encode these tags are built into cDNA cloning vectors so that when expression of the protein is induced in bacterial cells, the tag is incorporated at either the N-terminal or C-terminal end, where it is unlikely to interfere with protein folding. A library of cDNAs in such vectors is known as an **expression library.** Combinations of two or more tags together on the same protein can increase efficiency and purity of purifications, and is particularly useful where thousands of different proteins are being purified. For nucleic acid–binding complexes, affinity chromatography has been performed with biotinylated RNA or DNA probes that bind to streptavidin matrices. Examples of this technology include description of up to one hundred proteins that constitute molecular machines such as the spliceosome, nuclear pore complex, and spindle pole body.

Mass Spectrometry and Protein Identification

Protein identities are routinely determined using a combination of peptide sequencing and mass spectrometric (MS) methods that achieved high-throughput scale in the late 1990s (Chalmers and Gaskill 2000). The basis of this approach, shown in Figure 5.6, is that each protein can be identified by the overlap between a set of identified and predicted signature peptides. Rather than directly determining the amino acid sequence by chemical means, peptide fragments generated by trypsin digestion of whole proteins are separated according to their mass-to-charge ratio (m/z). Each m/z peak corresponds to a peptide 5–20 amino acids in length, whose precise mass is a function of the actual sequence of amino acids. Any given peak may correspond to dozens of possible peptides and so does not uniquely identify

Figure 5.6 Basis of mass spectrometry peptide sequencing.   (A) Proteins isolated from a 2D gel or affinity column are digested by trypsin into short peptide fragments. (B) The fragments are separated by time-of-flight (MALDI-TOF or ESI-TOF) mass spectrometry. (C) Matching the profiles of multiple peptides in the observed mass spectrum with all of the predicted peptide fragments for the proteome allows protein identification.

a protein; but several peaks derived from one or even several proteins provide a statistically supported identification of the protein (Fenyö 2000).

A **mass spectrophotometer** consists of three units. The first is an *ionization device* that moves individual peptide fragments from the solid phase into a gaseous ion phase after extraction from a 2D gel or elution from a chromatography column. Two common ionization units are the matrix-assisted laser desorption ionization (MALDI) and electrospray ionization (ESI) devices. The second unit is a *separation chamber*, in which the ions move through a vacuum according to their charge and mass and are separated on the basis of time-of-flight (TOF). The third device is a *detector* with the sensitivity and resolving power to separate peaks for over 10,000 separate molecular species, with a mass accuracy as low as 20 parts per million. Just a few picograms of fragments, corresponding to several million molecules, can be detected with resolution of differences of less than one-tenth of a Dalton in relative mass of molecules 2,000 Da in size.

Protein identities are computed automatically from peptide m/z spectra by comparison with a database of the predicted spectra in the proteome of the organism under study. These databases are assembled by *in silico* trypsin digestion of conceptually translated EST or cDNA sequences, and/or predicted protein coding regions in genomic DNA. Ambiguities due to an incomplete database, to overlap of the spectra of two or more possible peptides, and to differences in post-translational modification can be resolved by obtaining actual peptide sequence data using the **tandem mass spectrometry** (**MS/MS**) technique (Figure 5.7). In this procedure, before allowing a species of ion to strike the detector, it is shunted into a collision chamber in which nitrogen or argon gas molecules at low pressure collide with the ions. This causes the peptide backbone to break, liberating subfragments that are then separated by time-of-flight and detected, Because the subfragments are derived from a single peptide, the distances between the peaks define the mass differences of the subfragments, and hence identify the constituent amino acids sequentially from both ends of the peptide. Computation of just a fraction of the residues can be sufficient to resolve ambiguities.

In order to characterize the entire proteome of a cell population, **liquid chromatography** can be combined with tandem mass spectrometry (**LC/MS/MS**), obviating the need for prior electrophoretic separation of spots. The combination of partial sequence data with peptide masses and appropriate analytical software is sufficient to allow protein identification in complex mixtures of tens of proteins that emerge from each fraction of the chromatograph. Using a more complex two-dimensional cation exchange/reversed phase liquid chromatography approach known as Multidimensional Protein Identification Technology (MudPIT), Washburn et al. (2001) were able to detect and identify 1,484 yeast proteins, including many rare species and integral membrane proteins that are not detected with other methods. Furthermore, the identity of the individual peptides represented in the MS/MS spectra provides information on the likely folding pattern of membrane-spanning regions, which are not exposed to chemical cleavage. MudPIT provides enhanced resolution of peptide mixtures, and is being used

(A)

(B)

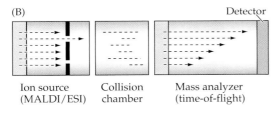

Ion source
(MALDI/ESI)

Collision
chamber

Mass analyzer
(time-of-flight)

(C)

Figure 5.7 Peptide sequencing by tandem mass spectrometry. (A) Any peptide mass corresponds to multiple fragments from different proteins in the proteome, with the number of possible matches increasing with size of the proteome. This plot shows the predicted number of fragments of each size for *E. coli*, yeast, and human. (B) The problem of aligning peptide fragments with proteins is simplified by partial sequencing of fragments using tandem mass spectrometry. (C) The peaks of MS/MS profiles are separated by characteristic widths that correspond to the amino acid that is knocked off either end of the peptide sequence in the collision chamber. (A from Fenyö 2000.)

to study protein interactions as well as the protein content of organelles, viruses, molecular machines, and affinity-purified protein complexes (Liu et al. 2002).

Neither 2D gels nor MS are quantitative methods, in the sense that the relative intensity of spots on a gel or peaks in a spectrum are at best qualitative representations of protein concentration. This is true of comparisons within and between samples, and relates to factors such as non-stoichiometric extraction of proteins from tissues, overlap of spots and peaks, and the non-linearity of protein-dye response. However, an emerging method for quantifying protein expression between samples or treatments is to label one of the samples with a heavy isotope such that the MS peaks lie immediately adjacent to one another. In this case, the ratios of the heights of the peak are indicative of relative expression level. Heavy isotopes include ^{15}N

EXERCISE 5.2 *Identification of a protein on the basis of a MS profile*

Given the protein sequence:

MALWMRGFFYTPKPGAGSLQPLRALEGSLQKGIVEQCCTKSICSR

compute the number of tryptic digestion products and their lengths in amino acids. Given the approximate amino acid masses listed below, also compute the expected approximate masses of the different peptides, and determine whether the protein is likely to be present in MS profile (A) or (B) shown in the figure.

Ala	A	89	Arg	R	174	Asp	D	133	Pro	P	115
Gly	G	75	Lys	K	146	Glu	E	147	Phe	F	165
Met	M	149	Gln	Q	146	Cys	C	121	Val	V	117
Leu	L	131	Asn	N	132	Ser	S	105	Trp	W	204
Ile	I	131	His	H	155	Thr	T	119	Tyr	Y	181

(A)

(B)

Mass:Charge ratio

Mass:Charge ratio

ANSWER: Trypsin cleaves at arginine and lysine residues, so the predicted fragments are:

MALWMR	(889)	ALEGSLQK	(970)
GFFYTPK	(966)	GIVEQCCTK	(1,123)
PGAGSLQPLR	(1,156)	SICSR	(636)

Profile B includes peaks corresponding to each of these fragment sizes, so is most likely to correspond to the protein.

and ^{2}H, which can be incorporated into the proteins either in vivo or after extraction. In vivo labeling involves growing cells in a medium supplemented with the heavy isotope, which is incorporated during normal biosynthesis. Incorporation following extraction can be performed by direct coupling to reactive amide groups on lysine residues in particular.

Isotope-coded affinity tag (ICAT) reagents crosslink to cysteine residues on proteins, and include a biotin group that allows purification of labeled proteins. The commercially supplied reagents carry eight light (hydrogen) or heavy (deuterium) atoms substituted on the carbon side chain, ensur-

(A)

Biotin Linker Thiol-specific
(heavy or light) reactive group

Heavy reagent; d8-ICAT (X = deuterium)
Light reagent; d0-ICAT (X = hydrogen)

(B) Mass difference from stable isotopes

(C) Identify peptide by sequence information (tandem MS)

(D)

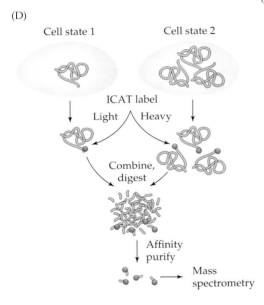

Figure 5.8 Quantitative proteomics using ICAT reagents. (A) Uniform labeling of two different protein samples is achieved using commercially available ICAT reagents that have eight light or heavy hydrogen atoms incorporated onto the carbon chain of a side-group that is crosslinked to reduced cysteine residues. (B) After tryptic digestion, ICAT-labeled fragments are purified by affinity chromatography against the biotin group on the label. (C) Mass spectrometry is used to identify differentially regulated peaks, and (D) tandem MS is subsequently performed to determine the identity of chosen differentially regulated fragments. (After Gygi et al. 2000.)

ing uniform separation of labeled fragments during MS (Figure 5.8). This procedure was first applied to the characterization of differences between yeast cells grown on two different carbon sources (Gygi et al. 1999), but has since been used to study such phenomena as trypanosome development, apoptosis, liver disease, and prostate and brain cancer.

Only a handful of direct comparisons of protein and mRNA levels in the same types of cell have been performed, but such studies highlight the relatively low correspondence between the contents of proteomes and transcriptomes (Celis et al. 2000). There are several reasons for this, including:

• Gene expression is regulated at several levels, including translation and degradation, which leads to uncoupling of transcript and protein levels (both RNA and protein turnover can be extremely rapid, taking only minutes in some cases).

- Alternative splicing and protein modification lead to misrepresentation of the total levels of both classes of gene product when single spots are measured.

- Proteins are present over a large range of concentrations inside cells, whereas detection methods are not sensitive over more than two or at most three orders of magnitude, while low level proteins are often undetected.

- Microarrays are more suited to comparing relative expression levels, whereas 2D gels tend to capture steady-state levels of protein expression (although pulse-chase experiments can also yield a representation of protein turnover).

These observations imply that some caution should be placed on the interpretation of the biological significance of differences in mRNA abundance detected by microarrays and chips. It should also be recognized that protein expression levels are not necessarily indicative of protein function. This is because post-translational modification, subcellular localization, and association with small molecules and other proteins all greatly affect protein function.

A convenient way to visualize specific proteins, either in extracts or in whole cells and tissues, is through the use of **antibodies.** Antibodies are secreted immunoglobulin proteins that are a part of the adaptive immune response mounted by vertebrates against foreign agents. When a purified protein is injected into a mouse or rabbit, for example, it becomes an antigen, and the animal responds by generating a series of antibodies that recognize a variety of epitopes on the protein. These polyclonal antibodies can be converted into **monoclonal antibodies** (**Mabs**) by fusion of the mouse B cells with a myeloma cell line to create an immortal hybridoma that expresses a single class of immunoglobulin. Mabs recognize a single **epitope,** or short peptide, and so are highly specific for individual proteins or even modified protein isoforms. Binding of a Mab to a protein is detected indirectly: a commercially supplied secondary antibody that is conjugated to some type of label is bound to the constant region of the Mab, as shown in Figure 5.9. The label can be a radioactive tag such as ^{125}I, a fluorescent dye, or an enzyme that catalyzes a pigment-forming reaction.

The use of Mabs to detect proteins that have been separated according to mass on a gel and transferred to a nylon membrane is known as **Western blotting** (by analogy with Southern and Northern blotting for DNA and RNA hybridizations). A particularly sensitive type of detection performed in the wells of a microtiter plate is known as **enzyme-linked immunosorbant assay,** or **ELISA,** and is useful for high-throughput and semi-quantitative analysis of protein expression in cell extracts.

The localization of proteins to tissues or whole organisms either in whole mount or thin sections is known as **immunohistochemistry.** Subcellular localization of protein distribution can be performed using **immunogold labeling,** in which gold particles of various sizes are bound to the antibodies and visualized by electron microscopy. Co-localization of two or three

(A)

Reporter

Epitope

2° antibody

▶ ^{125}I radioactivity

▶ Fluorescent dye

▶ Enzyme (HRP, AP)

▶ Gold particle

1° antibody

Protein

(B)   (C)   (D)

Western blot   Immunohistochemistry   Immunogold labeling
with fluorescent dyes

Figure 5.9 Antibodies and immunohistochemistry.   Proteins are detected using primary antibodies directed against an epitope in the target protein. The primary antibody is detected by a secondary antibody conjugated to one of a variety of labels, including radioactivity, fluorescence, gold particles, and enzymes. Labeled protein can then be detected in blots (B), tissue preparations (C), and thin sections (D).

different proteins in a tissue can be achieved by clever choice of primary and secondary antibodies from different animals, and is very useful for directly assessing whether the same proteins are expressed in the same cells.

Protein Microarrays

Protein microarrays have not yet seen wide usage. Proof of principle has, however, been achieved, including successful development of methods for high-throughput production and purification of micromolar amounts of protein. It has also been demonstrated that proteins can be displayed and visualized on a glass surface in such a way that they retain the capacity to bind to other molecules in a native manner.

Proteins have traditionally been produced by cloning random cDNA fragments into an inducible expression vector in *E. coli*. The bacteria are plated on a filter, grown, and then lyzed to expose the expressed protein. Since eukaryotes use different translation initiation signals, the translation start site is supplied by the cloning vector and, so long as the fusion preserves the open reading frame, a nearly full length protein can be expressed. The major drawbacks of the method are that *E. coli* does not modify proteins correctly, other proteins in the bacteria may interfere with function, and miniaturization is not feasible. Some of these problems have been overcome by adopting different host cells, including yeast, lepidopteran, and mammalian cell lines, or plant protoplasts.

Several new methods for high-throughput protein purification have emerged. One, called **phage display,** is to express the protein as a fusion to a phage capsid protein, so that it is expressed on the surface of readily purified phage particles. Another is to express full-length cDNAs in reticulocyte lysates, which are a cell-free protein synthesis system. As described above, perhaps the most common method is to incorporate polyhistidine or glutathione S-transferase fusions into expression libraries, as these protein fragments can be recognized by affinity chromatography.

Proteins can also be arrayed as desired either on aldehyde-coated glass slides (which form covalent crosslinks to amine residues at the N-terminus or with exposed lysine residues) or nickel-coated slides (that crosslink to polyhistidine tags). With appropriate care to maintain the hydration of samples during printing, and using the same arraying robots that are used to print cDNA microarrays, protein microarrays can be used to interrogate a variety of interactions (Figure 5.10; MacBeath and Schreiber 2000). Specific protein-protein interactions can most simply be detected by labeling a probe protein (for example, an antibody) with a fluorescent tag.

Similarly, small molecules will bind to microarrayed proteins (a fact that provides a potent assay for the recognition of candidate drugs). Substrates for different kinase classes have been detected by allowing the kinase to transfer a radioactively labeled phosphate specifically to the target protein, demon-

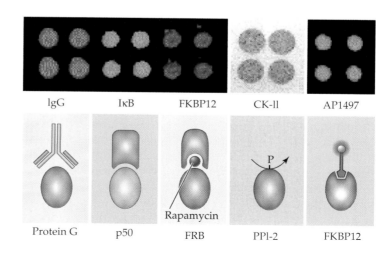

Figure 5.10 Protein microarrays. Proof-of-principle experiments have shown that immobilized proteins can be bound to antibodies (Protein G–IgG), protein cofactors (p50–IκB), and protein/small molecule complexes (FRB–FKB12). In addition, specific phosphorylation will occur (PPI-2 by CKII), and proteins will bind dye-labeled small molecules (AP1497–FKB12). (After MacBeath and Schreiber 2000.)

IgG   IκB   FKBP12   CK-ll   AP1497

Protein G   p50   FRB   PPI-2   FKBP12

Rapamycin

Figure 5.11 Self-assembling NAPPA protein microarrays. Each spot on a nucleic-acid programmable protein microarray crosslinks a plasmid vector and anti-GST antibody to the glass slide. Rabbit reticulocyte lysate is used to transcribe and translate the protein, which is trapped by the antibody. In this way, protein is exposed on the array without any need for high-throughput purification.

strating that enzymes can be induced to act on proteins on an array. Methods for determining the abundance of expressed proteins in complex extracts by binding to microarrays of printed specific antibodies, or of mixtures of antibodies to arrayed antigens, are under development (Haab et al. 2001).

The first demonstration that protein arrays could be used to identify novel protein interactions was provided by Zhu et al. (2001), who expressed and printed 5,800 yeast ORFs as GST/polyHis fusion proteins. This protein array was used to interrogate over 80% of the yeast proteome with calmodulin, and it revealed 39 calmodulin-binding proteins, 33 of which were previously unknown. Interrogation with six different types of lipid identified a total of 150 phosphatidylinositol (PI) binding proteins, some of which are specific for particular PI lipids and many of which are likely to have roles in membrane signaling.

A promising approach utilizing on-chip protein synthesis may circumvent some of the problems that have limited the adoption of protein microarray technology, namely inefficiencies of high-throughput protein purification and crosslinking of proteins to the glass surface. Nucleic-acid programmable protein arrays (NAPPA; Figure 5.11) are built by spotting a protein expression library onto the slide, and using cell-free reticulocyte lysate to transcribe and translate proteins in situ. These are captured locally by a monoclonal antibody against the GST epitope at the C-terminal end of each protein, most of which is left available in native conformation to interact with other proteins and reagents. Ramachandran et al. (2004) introduced the technology in a study of the pairwise protein–protein interactions among 29 components of the human DNA replication initiation complex.

Protein Interaction Maps

A dramatically different approach to determining which proteins interact with other proteins is to allow the interactions to occur in vivo and then to detect an artificial physiological consequence of the interaction. This is the strategy behind **two-hybrid screens.** The gene encoding a protein that normally requires two physically adjacent domains to function is split into two genes. Each of these is fused to a library of cDNA fragments that produce hybrid proteins. Function is only restored if the two hybrid proteins physically interact, bringing the two functional domains of the original protein back together (Figure 5.12).

The original yeast two-hybrid (Y2H) method (Fields and Song 1989) used a transcriptional activator, GAL4, which consists of a DNA-binding domain (BD) and an activation domain (AD). Neither domain is capable of activating transcription on its own, since the BD requires an activation sequence, while the AD must be brought to the promoter of a gene through the BD. Classically, the BD is fused to a "bait" protein for which interacting partners are sought, while the AD is fused to a library of random cDNA clones— the "prey." When the two pieces are brought together in the same yeast cell, any prey peptide that binds to the bait will bring the BD and AD together, restoring the capacity of the GAL4 transcription factor to induce expression of a reporter gene.

With modifications, the basic yeast two-hybrid method can be used with other systems, such as *E. coli* and mammalian cells, while alternative reporter assays are under development. The biggest limitation of Y2H technology is that it is prone to a high level of false positives (some baits activate transcription alone; some interactions occur by chance) and false negatives (the assay may not be sensitive enough; the fusion proteins lose their appropriate structure). Despite the fact that the interactions are induced to occur in the nucleus, Y2H has led to the successful documentation of numerous cytoplasmic and membrane-bound interactions, and is a powerful tool when appropriate controls are performed.

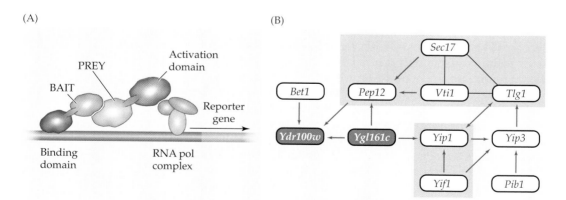

Figure 5.12 Yeast two-hybrid (Y2H) screens.   (A) Y2H works on the principle that interaction between two fusion proteins can reconstitute some biochemical function, such as transcriptional activation. (B) A protein interaction map for vesicular transport in yeast. Red arrows link interactions detected by whole genome Y2H; the black lines indicate interactions characterized by traditional methods. Darker shaded boxes show previously known interaction clusters. The two blue genes have unknown functions. (After Ito et al. 2000.)

For genome-scale screens, Y2H has been applied using a matrix approach to search for interactions between all possible combinations of proteins in an organelle, or even an organism (Ito et al. 2000). In the matrix approach, two distinct libraries of bait and prey proteins are constructed, and these are mated together in an ordered manner. Budding yeast is ideal for this purpose, since one library can be maintained in **a** mating-type haploid cells, the other in α cells, mating of which leads to co-expression of the fusion genes in **a**/α diploid cells. The matrices can be screened in a systematic manner, in which each individual bait clone is screened in replicate against the matrix of prey clones, and every interaction is assayed individually. This process results in increased resolution of false negatives and positives, but is laborious and time-consuming relative to the mass-mating approach, in which whole plates (or rows and columns of plates) are crossed en masse (Uetz et al. 2000). For this purpose, a selectable marker such as drug resistance is used as the target for transcriptional activation rather than a visible reporter gene. Positive clones can be characterized either by sequencing the cDNA inserts, or by tracing the interaction in a more systematic set of crosses based on which clones were present in the initial pool.

Interactome maps involving tens of thousands of interactions between thousand of genes have been constructed for flies and nematodes and have been used to support a conceptual prediction of the human protein interactome (Lehner and Fraser 2004). Even though only as few as 10 percent of the interactions detected by high throughput two-hybrid screens may be physiologically relevant, these approaches have been shown to reproduce many known complexes of interactions, and to suggest interactions that have been shown to occur upon experimental verification. By examination of similarities in the GO functional terms associated with multiple predicted interaction patterns, it is also possible to assign putative functions to unknown proteins. Similarities in these predictions across organisms should further increase confidence in such predictions. Some of the theoretical issues surrounding visualization and comparison of interaction networks are discussed in Box 5.2.

EXERCISE 5.3 *Formulating a network of protein interactions*

A yeast two-hybrid experiment detects the following protein–protein interactions when the cell surface receptor genes CSR1 and CSR2 are used as bait:

> CSR1 interacts with STK1, STK2, TFP4, CSR2, and CSR3, while CSR2 interacts with CSR1, CSR3, and STK1.

By contrast, after purifying protein complexes associated with signal transduction kinases STK1 and STK2, the following proteins are identified by MS:

(continued on next page)

Exercise 5.3 (continued)

STK1 complexes with TFP3, TFP4, CSR2, and CSP7, while
STK2 complexes with TFP7, CSR1, CSR3, and CSP4

Draw a network showing all possible pairwise interactions. Comment on
which interactions you feel are most robustly supported. If gene expression profiling indicates that CSR3 is never expressed in the same tissue
as CSR1 or CSR2, what conclusion might you reach about the function of
the protein interaction network?

ANSWER: The following network displays all of the yeast two-hybrid
interactions in blue, and the affinity chromatography MS interactions in
red. The most strongly supported interactions are between CSR1 and
CSR2, STK1 and CSR2, and STK2 and CSR1, as each of these pairs of interactions are supported by two independent data points. The microarray
result would imply that the interactions between CSR3 and the other
two CSR genes are probably false-positives that occur because each of
these proteins has a similar structure. A reasonable hypothesis is that
CSR1 and CSR2 form heterodimers that signal through STK1, whereas
STK2 signals from either CSR1 or CSR3.

Several other bioinformatic methods for describing protein interaction
maps have been proposed; to date most of these have been for the characterization of bacterial proteomes. One, known as the **Rosetta stone approach,**
is to ask which pairs of genes in one species are found as a single gene in
another species (Marcotte et al. 1999). The assumption is that two genes are
unlikely to fuse to form a single protein unless they are involved in the same
physiological process. This strategy has resulted in the assignment of putative functions to hundreds of previously unknown bacterial genes.

A similar approach is to ask which genes are always found in the same
cistron, indicating that they are co-regulated. Identification of cistrons
(stretches of DNA in microbes that are transcribed as a single mRNA but
encode multiple proteins) is not trivial, but physical proximity is a good
starting point for comparisons among divergent taxa. A third approach
based on evolutionary comparison is to simply document the patterns of
presence and absence of families of genes across a wide range of taxa (Pellegrini et al. 1999). To the extent that loss of one gene signifies that the biochemical pathway it participates in is dispensable in that organism, other

genes in the pathway might also be expected to be lost. Consequently, co-segregation of genes that show variable phylogenetic distribution provides a hint that these genes may interact.

Structural Proteomics

Objectives of Structural Proteomics

Structural proteomics strives to be able to predict the three-dimensional structure of *every* protein. Physical solution of all protein structures is not feasible, but it is thought that if high-resolution structures are obtained for a sufficiently large number of proteins, then essentially all possible protein structures will be within modeling distance of at least one solved structure (Stevens et al. 2001).

In mid-2004, the central repository for structural data, the Protein Data Bank (PDB; Berman et al. 2000; http://www.rcsb.org/pdb), held just under 27,000 protein structures, and over 3,000 structures are added each year. Only around 10 percent of these represent novel domains. It is estimated that there are between 4,000 and 10,000 distinct protein folds in nature, and that between 10,000 and 20,000 more protein structures may be sufficient to cover the full range of domain space (Burley 2000).

The concept of a **protein domain** is a difficult one, with slightly different meanings for biochemists, structural biologists, and evolutionary biologists. For our purposes, a domain is *a clearly recognizable portion of a protein that folds into a defined structure*. Most proteins are thought to resemble globular beads on a string, where the "beads" are domains that generally range in length from 50–250 residues, each of which performs a specific biochemical function. Some protein activities, however, are performed at the interface between two or more protein domains—often on two different proteins. If the two molecules are the same protein, the structure is a homodimer; otherwise it is a heterodimer (or heteromultimeric if there are multiple molecules in the protein complex). Folding of protein domains into quaternary structures and intermolecular complexes is largely beyond the purview of current structural biology, but clearly will be an important aspect of future structural proteomics.

Comprehensive structure determination will require broad sampling of protein sequence space. Traditional approaches have focused on proteins of known biological interest in humans, microbes, and model organisms. Greater sampling depth of these proteins—notably kinases, proteases, phosphodiesterases, nuclear hormone receptors, phosphatases, G-protein coupled receptors, and ion channels—is being pursued by industry, as these classes of molecule are proven targets for drugs and other pharmaceuticals. Public efforts are expected to focus on sampling breadth—for example, extending to the complete proteomes of one or more small microbial genomes, and representative sampling of novel ORFs from higher eukaryotes (vertebrates, invertebrates, and plants). The latter class includes pro-

BOX 5.2 Biological Networks In Genome Science

The initial thrust in genome research was heavily focused on cataloging information: What are the sequences of genomes? What genes are present in genomes? What are the functions of genes? What are the structures of proteins? Which genes are expressed in which tissues? As the field moves into the "post-genomic" era, research is moving beyond these single dimensional questions to topics addressing higher-order interactions of biomolecules. How are metabolic pathways regulated through interactions of DNA, RNA, and protein sequences? Which proteins have physical interactions, and what localized domains enable those interactions? These types of questions require methods for describing, visualizing, and statistically analyzing interactions between molecules. Graph theory and *topological networks* provide a common mathematical framework for these endeavors.

Networks are convenient structures for displaying interactions of many types, and over the years, they have been applied successfully to problems in many fields. For example, Internet traffic flow has been studied using network methods to identify aspects of "closeness" among major hubs on the Internet (for example, are users of nytimes.com also likely to surf their way to ESPN.com?). Here we describe the basic principles of graphs and networks, briefly outline how they can be applied to answer some important problems in genome science, and present some representative results from actual data analyses.

From the computational viewpoint, networks consist of *nodes* and *edges*. A simple example is provided in Figure A. This simple network might describe a situation where transcription factor A positively induces expression of gene B, which in turn induces expression of gene C, which then acts as a negative regulator of A. The actual molecules that (potentially) interact are the nodes in the graph (A, B, and C), and the nature of their interactions are indicated by the edges. In this example, the edges are *directed* since they imply that one node affects the other asymmetrically; in other cases,

such as protein–protein interactions, the edges may be undirected and simply indicate a physical interaction. A major effort in the gene expression arena has been the application of both Boolean and Bayesian networks to extract and understand genetic pathways. In a Boolean network, interactions take the form of on/off switches. For instance, a link between A and B may take the form "If A is expressed, then B is also expressed." Bayesian networks extend this notion to allow some degree of randomness. For instance, "If A is expressed, then B is expressed 40% of the time." More quantitative variants also exist: "If A is expressed with level X, then the probability of B being expressed at level Y is 90%."

Figure A

The challenges for constructing networks are many. While it seems straightforward to document who interacts with whom, it is less obvious how to determine if the pattern of observed interactions fits some hypothesized network structure, or if it deviates from the structure observed in another tissue or organism. Assuming that an acceptable measure of fit between observed and hypothesized network topology has been identified, the search through all possible networks is computationally demanding, as the number of potential networks grows explosively with the number of nodes. Given a network and some data, there are both theoretical and computational challenges in carrying out formal statistical procedures. Models of random networks must be developed, starting with such properties as levels of *connectivity* (on average, how many nodes are connected to a randomly chosen node in the network?). The distribution of connectivity in biological networks often turns out to be nonrandom, in the sense that there may be a small number of nodes with an excessively large number of connections. As discussed further in Chapter 6, such observations may be very informative with regard to inferring the fate of duplicate genes, or predicting the consequences of genetic and environmental perturbations.

One of the immediate benefits of network analysis is helping molecular biologists to focus their attention on clusters of genes that have been shown to be related because of physical interactions between their protein products, shared transcriptional profiles, or genetic interactions in double mutant combinations. An example of the latter illustrates some of the practical aspects of network analysis, and is provided by Tong et al. (2004), who constructed a gene interaction network involving approximately 1,000 genes and 4,000 interactions. The phenotypes of the 132 core mutants that they studied were modified by interaction with between 1 and 146 other mutations, for an average of 34 genetic interactions, which is four times as many as the observed average number of direct protein–protein interactions. Figure B shows sample topologies for interaction networks centered on three specific genes from their study (located in the center of each network). The interactions are represented by undirected gene linkages in the networks.

Taking this information one step further, Tong et al. formed networks showing the linkages of gene *functions* assessed according to gene ontology attributes (see Box 2.5). First, for more than ten percent of 756 GO attribute classes, genes sharing similar GO attributes are more likely to interact genetically with one another than with randomly selected genes. Reciprocally, over one quarter of all interactions were observed between genes with similar GO attributes. Figure C shows, even more interestingly, that particular classes of GO have increased likelihood of connections with one another. Thus, genes involved in DNA synthesis and repair tend to interact with one another, and also with genes in mitosis and chromosome movement, which in turn

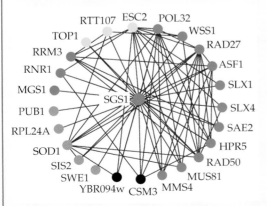

Figure B   (Figure from Tong et al. 2004.)

(*Continued on next page*)

BOX 5.2 *(continued)*

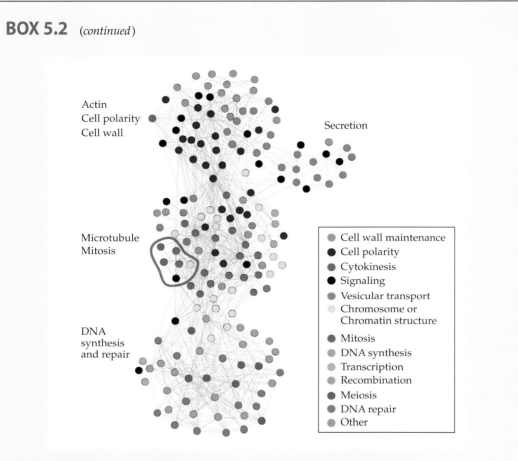

Actin
Cell polarity
Cell wall

Secretion

Microtubule
Mitosis

DNA
synthesis
and repair

- Cell wall maintenance
- Cell polarity
- Cytokinesis
- Signaling
- Vesicular transport
- Chromosome or Chromatin structure
- Mitosis
- DNA synthesis
- Transcription
- Recombination
- Meiosis
- DNA repair
- Other

Figure C   (Figure from Tong et al. 2004.)

interact with genes involved in cell polarity and cell wall maintenance.

Two of the major challenges in network theory are to develop the statistical tools for comparing network topologies, and to use these tools to compare networks assembled from genetic, proteomic, transcriptional, and other databases. The unification of networks is expected to provide a more holistic picture of physiological and developmental networks than can be provided by classical genetic analysis.

teins with no sequence similarity to other proteins, and constitutes up to one-third of the predicted proteome of every species whose genome is sequenced.

International efforts are expected to lead protein biochemists to the point where they can begin to model the structure and function of essentially any protein by the year 2010. Coordination of initiatives can be tracked through the "Structural Genomics" link of the PDB. Several large-scale high-through-put centers with automated pipelines are expected to generate 200 new structures a year, while a network of smaller research centers develops new technologies and tackles specific types of proteins expressed in unusual organisms or in diseased tissue.

The utility of a protein structure is a function of the novelty of the domain and the level of knowledge of its function (Figure 5.13). Highly refined structures of known classes of proteins with known functions are of most relevance in biomedical research, since they support rational drug design by increasing our understanding of the mechanism of catalysis, establishing the constraints on structure-function relationships between proteins and cofactors/ligands, and assisting in the interpretation of the effects of targeted mutations. Excellent examples include the design of HIV protease inhibitors and influenza neuraminidase inhibitors on the basis of structural data. The pharmaceutical sector is also particularly interested in better def-

Figure 5.13 Applications of structural proteomics. As structural biologists articulate and refine the structures of increasing numbers of proteins, the knowledge is put to practical use in other areas of biology and medicine.

inition and engineering of features that increase structural stability, the potential of proteins to form multicomplex interactions that may enhance or inhibit function, and optimization of protein performance. Hegyi and Gerstein (1999) estimated that the average protein fold in Swiss-Prot has 1.2 distinct functions (1.8 for enzymes), while the average function can be performed by 3.6 different folds (2.5 for enzymes). This suggests that structure will generally be quite useful for predicting function, and that novel structures might be expected to define novel functional families.

For proteins of unknown function, structures may suggest the location of active sites and hence promote rational site-directed mutagenesis to investigate function; or structures might suggest the use of particular enzyme assays based solely on the arrangement of residues at the putative active site. Ten common "superfolds," including TIM barrels and the αβ-hydrolase, Rossmann, P-loop NTP hydrolase, and ferredoxin folds, account for several functions each. Further structural data enhances our ability to draw inferences using comparative methods. And, as the universe of known protein domain structures increases, possibilities for *in silico* docking and modeling also increase.

Protein structures are deposited upon publication in the PDB, which since 1998 has been maintained by the Research Collaboratory for Structural Biology, consisting of groups from Rutgers University, the San Diego Supercomputing Center at UCSD, and the National Institute of Standards and Technology (NIST). Database management is facilitated by an ADIT (AutoDep Input Tool) Web-based interface that uses an internationally agreed upon macromolecular crystallographic information file (mmCIF) controlled vocabulary of 1,700 terms to avoid ambiguity. This tool also helps the data bank meet the bioinformatic challenges associated with archiving an exponentially growing set of data, complete with functional annotation and control and verification of structure quality. In addition to atomic coordinates, structures are deposited with journal references, functional information, and attributes of the experimental procedures used to determine the structure, all of which must be checked and formatted in such a way that the data is compatible with sharing over the internet.

Protein Structure Determination

Determining even a single protein structure is a labor-intensive effort that has traditionally required several years for each structure. Consequently, automation of each of the steps outlined in Figure 5.14 is essential to structural genomics. Recent advances include the utilization of genome sequence information to define targets and design primers for PCR-based cloning of ORFs into expression vectors; improved affinity chromatography methods for purifying fusion proteins in sufficient quantities for crystal growth; development of high-throughput robotic methods for crystal growth; cryogenic storage and robotic retrieval and orientation of crystals at synchrotron facilities to minimize delays and human error during the gathering of structural data (Abola et al. 2000); and greater automation of the computational methods used to solve structures.

The two experimental methods used to solve protein structures are **X-ray crystallography** and **nuclear magnetic resonance (NMR) spectroscopy.** Solving protein structures by crystallography proceeds in a series of steps, starting with data collection and moving through image processing, phasing, model building, model refinement, validation, and publication. Two types of data must be obtained in order to construct the electron density maps from which the structure is derived: the *amplitudes* and the *phases* of each diffracted X-ray. Amplitudes can be determined directly from the diffraction data, but

Figure 5.14 Flow diagram for solving a protein structure.   Current efforts are focused on streamlining the process by automation, but failures and bottlenecks require human input at each step.

phasing is a difficult problem. It can be solved computationally if extremely high-resolution data are available, or if data from very similar structures are used, but it generally requires additional diffraction data from crystals into which heavy atoms such as the transition metals mercury or gold have been incorporated. Recent advances in synchrotron beam tuning and the synthesis of seleno-methionine substituted proteins allow structural determination from a single crystal using a technique known as multiple anomalous dispersion (MAD). An introduction to the analytical prcedures can be found at http://www-structure.llnl.gov/Xray/101index.html, and a list of the wide range of software used at various phases of structural determination is given in Lamzin and Perrakis (2000). The PDB provides a range of online tools for interactive study of three-dimensional structures (Figure 5.15).

The two major advantages of NMR over X-ray crystallography are that NMR is performed in solution, so there is no requirement that the protein crystallize (which considerably expands the number of structures that can

Figure 5.15 The *Drosophila* Engrailed homeodomain dimer bound to DNA.
Four different three-dimensional views. (A) Swiss-3D image from http://ca.expasy.org/cgi-bin/sw3d-search-ac?P02836. (B) Close-up of binding site for the third helix in the major groove of DNA. (C) Still ribbon and (D) cylinder views of the same interaction from the PDB Structure Explorer, accessed by entering the accession name 3HDD at http://www.rcsb.org/pdb.

be studied), and the structure can be determined under conditions that resemble physiology and can readily be manipulated to mimic changes in pH or salt concentration. Typically, about 0.5 ml of 1 mM protein solution enriched in ^{13}C or ^{15}N is used to determine the structures of small proteins (10–30 kDa). This size range corresponds to only 25% of all yeast ORFs and a smaller percentage of higher eukaryotic proteins, but it will be possible to determine solution structures of isolated domains from larger proteins. Furthermore, recent refinements suggest that protein-lipid micelle analyses will allow structural determination of protein domains within membranes. A major constraint is that high-resolution structures often require that data be obtained over a period of several weeks, requiring that the proteins remain stable and soluble, and placing a limit on the number of structures that can be determined with a single machine. Computational advances are expected to address some of these difficulties, and technical advances are also expected to increase resolution so that NMR structures approach the 2.0–2.5 Å limit typical of X-ray crystal structures (Montelione et al. 2000).

Currently, human intervention is required for resolution of at least 50% of all structures and for the refinement that is needed for precise modeling of protein function. Given the tens of thousands of structures that need to be determined to cover the bulk of protein space, this poses a bottleneck, but the industrialization of the process is expected to accelerate the pace of structural solution. At the other end of the scale, resolution of macromolecular structures such as the 30S subunit of the ribosome complexed with antibiotics (Carter et al. 2000), RNA polymerase II (Cramer et al. 2001), and of the 12-protein subunit, 127-cofactor photosystem I of cyanobacteria (Jordan et al. 2001) demonstrates the power of crystallography to illuminate complex protein–protein and protein–drug interactions. High-resolution electron microscopy also plays an important role in elucidation of the organization of macromolecular complexes.

Protein Structure Prediction and Threading

Protein structure prediction plays an increasingly important role in attempts to infer function from sequence data alone (Baker and Sali 2001). The lower limit for detection of potential structural homologs by sequence comparison is 30% sequence identity. Above this level, comparative modeling can be used to estimate the likely structure of a protein by overlaying the unknown structure onto the known. Below this level, at least three classes of strategy are used to try to fit an initial model of the most likely protein structure: *ab initio* prediction, fold recognition, and threading. The success of these methods is judged at biannual CASP (Critical Assessment in Structure Prediction) conferences at which theoretical and experimental solutions of previously unsolved structures are compared (Jones 2000).

***Ab initio* protein prediction** starts with an attempt to derive secondary structure from the amino acid sequence, namely predicting the likelihood that a subsequence will fold into an α helix, β sheet, or coil, using physicochemical parameters or HMM and neural net algorithms trained on existing probabilities that similar short peptides adopt the particular structures. Such methods have been claimed to accurately predict three-quarters of all local structures. Subsequently, secondary structures are folded into tertiary structures, again using algorithms based on physical principles. Model quality is tested by fitting predictions against known structures and calculating the root mean square distance between the predicted and actual locations of α-carbon atoms on the peptide backbone. Attempts to fit the location and orientation of side chains are not yet within the realm of *ab initio* modeling, which is thus more concerned with generating hypotheses of the general shape of a polypeptide. Hypotheses can then be tested using site-directed mutagenesis and other biochemical approaches.

Fold recognition, or **structural profile,** methods attempt to find the best fit of a raw polypeptide sequence onto a library of known protein folds (such as the CATH or SCOP libraries), or onto a "periodic table" of possible forms that helices and sheets can fold into (Taylor 2002). A prediction of the secondary structure of the unknown is made and compared with the secondary structure of each member of the library of folds. The locally aligned sequences are also compared for sequence and/or profile similarity, and the two measures of structural and sequence similarity are condensed into a fold assignment confidence z-score that represents the probability of a match relative to random comparisons. A threshold chosen by application of the same algorithm to known structures is then used to identify likely matches in the manner in which an unknown domain folds relative to known domains (Figure 5.16).

Application of this approach to the whole predicted proteome of *Mycoplasma genitalium* resulted in putative functional assignment to an extra 6% of the ORFs that were not possible to assign based on sequence alignment alone (which had assigned function to just 16% of the 468 ORFs in the *M. genitalium* genome; Fischer and Eisenberg 1997). As the number of unique folds in PDB increases, the performance of this approach is expected to improve greatly.

Threading takes the fold recognition process a step further, in that empirical-energy functions for residue pair interactions are used to mount the unknown onto the putative backbone in the best possible manner. Gaps are accommodated and the best interactions are maximized in an effort to derive the most likely conformation of the unknown fold, and to discriminate among different possibilities. Threading of the complete predicted proteomes of 33 prokaryotes and 4 animals involving over 165,000 genes with the THREADER algorithm assigned folds to 70% of the bacterial proteins and 60% of the animal ones (Cherkasov and Jones 2004). Similarly, Zhang and Skolnick (2004) added an extra side-chain fitting optimization step to their PROSPECTOR algorithm and accurately predicted the structure of two-thirds of 1,360 medium-sized proteins in *E. coli* relative to PDB structures. An example of a practical application of threading is the demonstration that a gene that has a quantitative effect on fruit size in tomato, *fw2.2*, is likely to encode a member of the heterotrimeric guanosine triphosphate-

(A)

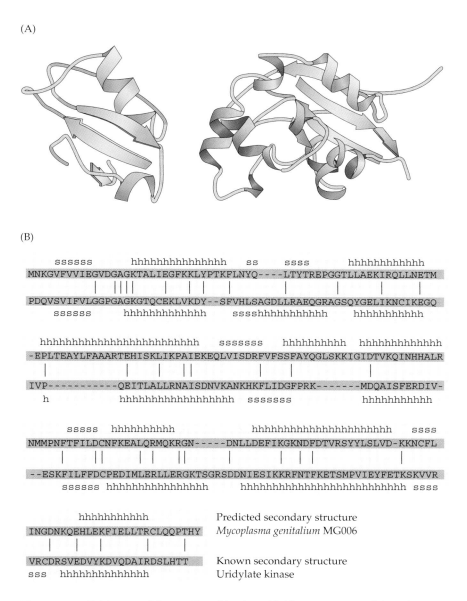

(B)

```
        sssss        hhhhhhhhhhhhhh   ss    ssss      hhhhhhhhhhh
 MNKGVFVVIEGVDGAGKTALIEGFKKLYPTKFLNYQ----LTYTREPGGTLLAEKIRQLLNETM
      |    ||||    |     |    |       |           |        |     |
 PDQVSVIFVLGGPGAGKGTQCEKLVKDY--SFVHLSAGDLLRAEQGRAGSQYGELIKNCIKEGQ
        sssss       hhhhhhhhhhhh     sssshhhhhhhhhh    hhhhhhhhhh


    hhhhhhhhhhhhhhhhhhhhhhhhhhh    sssssss    hhhhhhhhh   hhhhhhhhhhhh
 -EPLTEAYLFAAARTEHISKLIKPAIEKEQLVISDRFVFSSFAYQGLSKKIGIDTVKQINHHALR
         |       |     |       |                  |       |
 IVP-----------QEITLALLRNAISDNVKANKHKFLIDGFPRK-------MDQAISFERDIV-
     h          hhhhhhhhhhhhhhhhhh  sssssss          hhhhhhhhhh


       sssss  hhhhhhhhh            hhhhhhhhhhhhhhhhhhhhhhh   ssss
 NMMPNFTFILDCNFKEALQRMQKRGN-----DNLLDEFIKGKNDFDTVRSYYLSLVD-KKNCFL
       |    ||    |  |    ||      |     |  |  |                 |
 --ESKFILFFDCPEDIMLERLLERGKTSGRSDDNIESIKKRFNTFKETSMPVIEYFETKSKVVR
      sssss  hhhhhhhhhhhhhhhh      hhhhhhhhhhhhhhhhhhhhhhhhhhh  ssss
```

```
        hhhhhhhhhh            Predicted secondary structure
 INGDNKQEHLEKFIELLTRCLQQPTHY   Mycoplasma genitalium MG006
    |    |    |     |    |
 VRCDRSVEDVYKDVQDAIRDSLHTT     Known secondary structure
  sss  hhhhhhhhhhhhhh          Uridylate kinase
```

Figure 5.16 Fold recognition.   The objective of fold recognition and threading is to determine whether a domain is similar to a fold found in a known protein. Despite sequence identity of only 20% over the full length of the protein (A), this unidentified *M. genitalium* ORF MG006 (sequence shown in purple) was assigned as a probable member of the uridylate kinase family (sequence in green) on the basis of similarity of predicted secondary structure (B), which predicts a homologous folding pattern. (A after Fischer and Eisenberg 1997. B based on R. Altman, online lecture http://cmagm.stanford.edu/biochem218/16threading.html, no longer available.)

binding RAX family of proteins, and hence to play a role in controlling cell division during fruit growth (Frary et al. 2000).

The eventual aims of structural biology include not just the determination of protein structures, but also modeling of protein function at the atomic level. This will entail advances in modeling a number of aspects of protein function, including: protein–small molecule interactions; the effects of site-directed mutagenesis on structural conformation; the movements in active sites during the milliseconds over which a catalytic event takes place; the effect of side-chain bulk and orientation on the specificity of molecular interactions; the formation of active sites at the interface between protein domains; and the docking of two or more proteins to form multisubunit complexes. Undoubtedly such endeavors will require integration not just of theory and experimental structural proteomics, but also constant feedback with functional genomics and molecular biology.

Functional Genomics

Functional genomics encompasses all research aimed at defining the function of each and every gene in a genome. "Function" can be defined at several ontological levels, from biochemistry to cell biology on up to organismal phenotype. Two homologous genes that retain the same molecular structure and biochemical function may nevertheless have very different physiological roles in different organisms. They will often interact with a similar suite of proteins in different organisms, and at different phases of development, though the precise nature of the interactions can be highly labile. Nevertheless, the characterization of the function of a protein domain in one organism will generally provide a hint as to its function in another organism. Consequently, one of the first goals of functional genomics is to identify mutations that affect the activity of as many genes as possible in the major model organisms.

In this section we will consider approaches to functional genetics based on the generation and analysis of mutations. There are three basic approaches, each with its own goals:

1. The goal of **forward genetics** is to identify a set of genes that affect a trait of interest. This is pursued initially by random mutagenesis of the whole genome, screening for new strains with an aberrant phenotype that is transmitted stably to subsequent generations in Mendelian proportions. Traits of interest include morphology, physiology, and behavior.

2. The goal of **reverse genetics** is to identify phenotypes that might be caused by disruption of a particular gene or set of genes. In this case, the starting point is the DNA sequence of interest—perhaps previously uncharacterized open reading frames, perhaps a cluster of co-expressed genes, but ultimately each and every predicted gene. A variety of strategies for the systematic mutagenesis of specific genes is now available.

3. The goal of **fine-structure genetics** is to manipulate the structure and regulation of specific genes in such a way that novel functions and interac-

tions can be characterized, or so that hypotheses arising from in vitro analysis or structural comparison can be tested in vivo. This is a rapidly evolving field of research at the interface of molecular genetics and genomics.

Saturation Forward Genetics

Forward genetic strategies start with a phenotype and work in toward the gene or genes that are responsible for it. **Saturation mutagenesis** refers to genetic screens of such large scale that a point is reached where most new mutations represent second or multiple hits of previously identified loci (Figure 5.17). Classical genetic screens were based on identifying a phenotype, so to a broad approximation, the **saturation point** defines the subset of genes required individually for the development of a trait. For example, saturation mutagenesis screens for embryonic recessive lethal mutations in *Drosophila* defined a core set of fewer than 50 genes that are necessary for segmentation (Nüsslein-Volhard and Wieschaus 1980). It was later realized that such screens missed genes whose products are expressed maternally, as well as genes whose function is redundantly specified by other loci. Yet the subsequent molecular characterization of these loci provided the raw material for research that rapidly led to understanding of the process and the identification of other interacting loci by different strategies.

Many phenotypes, such as those related to behavior, physiology, or the morphogenesis of internal organs, cannot be studied by screening directly for visible aberrations. Consequently, screens in the genomic era increasingly rely on the isolation of random mutant chromosomes, followed by characterization of a battery of phenotypes. Because of the expense involved in mutagenesis of vertebrates, it is cost-effective to pool resources by screening hundreds of phenotypes simultaneously, in different laboratories that have expertise in different areas.

Mutant chromosomes can be isolated either by **insertional mutagenesis**—screening for the expression of a marker gene carried by a transpos-

Figure 5.17 Frequency of mutations recovered in "saturation" screens.   Even large screens (several thousand mutated genomes) typically fail to approach saturation. This figure shows the number of genes with 1, 2, 3, 4, 5, or more than 5 alleles recovered in two large zebrafish embryonic phenotypic screens (Driever et al. 1996; Haffter et al. 1996), as well as a series of P-element insertional mutagenesis screens in *Drosophila* that are summarized in Spradling et al. (1999).

TABLE 5.1 *Mouse Mutants from F₃ Recessive Screens*

Defect (phenotype)	Frequency of mutation in F₃ (% of offspring)	
	Germany[a]	England[b]
Craniofacial and skeletal defects	40	15
Coat color abnormalities	18	19
Eye defects, including cataracts	9	10
Behavioral abnormalities (circling, head tossing, etc.)	21	34
Skin or hair abnormalities	13	8
Growth, weight, size defects	4	17
Deafness	9	16
Clinical chemistry abnormalities	23	4
Immunological and allergic deficits	39	NA

[a]Data from Hrabe de Angelis et al. (2000).
[b]Data from Nolan et al. (2000).

able element—or by administering a high enough dose of mutagen to guarantee the generation of at least one visible mutation in each gamete. Table 5.1 summarizes the range of phenotypes scored and frequencies of mutations recovered in a pair of mouse mutagenesis screens (Hrabe de Angelis et al. 2000; Nolan et al. 2000).

Several types of mutagen are used to generate different classes of mutation, as shown in Figure 5.18. Point mutations (those affecting a single nucleotide) are best generated by chemicals such as ethylnitrosourea (ENU) or ethylmethanesulfonate (EMS). Such screens are easily performed, relatively unbiased, and result in multiple classes of effect, from non-sense and mis-sense coding mutations (that is, those introducing premature stop codons, frameshifts, or amino acid replacements) to splicing defects and disruption of regulatory sequences. Different point mutations affecting the same gene can result in very different phenotypes, potentially helping to define pleiotropic functions. The drawback of point mutations is that they are laborious to map, so cloning of the gene responsible for a mutant phenotype can take several years with current technology.

The tilling technique introduced in Chapter 3 (see Figure 3.17) may accelerate this process, as it provides a rapid means of scanning for newly induced mutations using thousands of PCR reactions to screen much of the genome (Henikoff et al. 2004). The tilling technique compares the parental chromosome before exposure to mutagen with the mutant chromosomes. An enzyme is used to cleave heteroduplex mismatches that are formed by annealing of PCR products that differ between normal and mutant DNA. The cleaved product migrates to a new position on a gel, and once identified can be sequenced to pinpoint the lesion.

Larger insertions and chromosome rearrangements are commonly produced by irradiation with X-rays or gamma rays. These abnormalities either

Figure 5.18 The three major types of mutagen.   Gamma irradiation induces chromosome inversions, translocations, and large deletions. Chemicals such as ENU induce point mutations and deletions of a few bases. Transposable element insertions can occur anywhere in a gene, but are often found in upstream regulatory regions. Advantages (red arrowheads) and disadvantages (green arrowheads) of each mutagen are indicated.

remove a gene (or genes) entirely, resulting in a null phenotype, or juxtapose novel regulatory sequences adjacent to a gene, resulting in loss of expression in the normal tissues and/or gain of expression in ectopic tissues. Cloning of the breakpoints can be performed by direct comparison of mutant and parental chromosomes, so does not necessarily require a large number of meioses for genetic mapping, but can nevertheless be laborious.

Transposable elements are an increasingly popular class of mutagen; some features of these elements are discussed in more detail in the section below on fine structure genetics. **Transposon mutagenesis** offers the major advantages that (1) the phenotype can often be reverted to the wild-type state by inducing the inserted element to "jump" back out of the genome (providing formal proof that the insertion caused the phenotype); (2) cloning of the site of insertion can be completed in a few days by "plasmid rescue" (see Figure 5.24A), and (3) isolation of new insertions does not require a phenotypic screen. In this sense, transposon mutagenesis is unbiased, but on the other hand, it is often biased by the biological fact that some transposons have insertion site preferences, or "hotspots," and "coldspots." These preferential sites exist at two levels: some genes are relatively refractory to certain transposons, while others are unusually frequent targets; and some transposons preferentially insert close to the promoter of target genes, partially disrupting expression rather than knocking out the gene entirely.

A further useful feature of transposons such as the *Drosophila* P-element and maize Ds element is that their transposition can be controlled by introducing a source of the transposase enzyme through genetic crosses. This allows for collection of a large number of mutants in a single generation. Often, the movement of the transposon is local—to a site within a few hundred kilobases—providing a means for "walking" along a chromosome and generating insertions in all the open reading frames in a small region of the genome.

Dominant mutations can be isolated in the progeny of the mutagenized parents (the F_1 generation), but recessive mutations require more laborious and complex designs in which the mutants are not recovered until at least the F_3 generation (Figure 5.19). Dominant mutations can be due to gain of function (**hypermorphs**), generation of a novel function (**neomorphs**), production of a dominant negative function (**antimorphs**: i.e., the dominant protein interferes with wild-type function), or to haplo-insufficient loss of function (**hypomorphs**).

Recessive mutations are most often hypomorphs that either reduce protein activity level or down-regulate transcription. There are, however, no simple rules relating dominance and recessivity to loss or gain of function. For example, two heritable recessive versions of human long-QT syndrome result from specific sodium and potassium channel functions, both of which affect membrane polarization in a similar manner despite the fact that the channels pump ions in opposite directions. Studies from model organisms indicate that multiple mutations are usually required to define the function or functions of a gene product at even a superficial level. Consequently, identification of a single mutation is often followed up by generation of a panel of new mutations that fail to complement the original allele.

The first step in mutant classification is the assignment of each mutation to a **complementation group:** a set of alleles that fail to complement (provide the function of) one another. Two independent mutations affecting the same locus will often (but not always) combine to produce a recessive phenotype that resembles that of at least one of the alleles when homozygous. Thus, complementation testing usually reduces a large number of mutations to a two- or threefold smaller set of loci. Intergenic non-complementation can also arise when two mutations affect genes involved in the same biochemical pathway. Thus, physical or genetic map data is required to verify the inference that a complementation group defines independent mutations in the same gene.

Steps must also be taken to preserve the new mutations. Recessive mutations are often not viable as homozygotes, and even viable mutations can be lost in a few generations as a result of genetic drift in the stock. This problem is solved differently for different organisms: for plants, mutant seed can be stored; for mice, bacteria, and nematodes, germ cells (or whole animals) can be frozen; fly geneticists use "balancer" chromosomes, which are described below. Colony maintenance and stock tracking require great care and attention to detail and are not taken lightly. Most genome projects have stock centers with budgets in the tens of millions of dollars, yet even so must make difficult decisions about which stocks to keep in the face of the incredible volume of genetic resources currently being generated.

Once a set of loci affecting a trait have been defined and mapped to chromosomal regions on the basis of recombination frequencies, molecular genetic procedures must be employed to identify the gene that is disrupted.

These can include searching for new single nucleotide mutations in a chromosomal region that maps to the genetic lesion; looking for altered transcript expression profiles; and attempts to complement the mutation by making transgenic organisms with extra wild-type copies of any candidate genes. Subsequently, identifying the molecular nature of each of the alleles that define a locus can be very revealing as to the function of the gene product. For example, alleles that cause similar phenotypes often map to the same regulatory region or protein domain, and hence serve to define those elements. These alleles may also provide clues as to the biochemical mechanisms by which distinct protein functions and interactions are mediated.

Somatic chromosomal rearrangements associated with many different cancers can be identified relatively quickly using comparative genome hybridization (CGH). Whereas large deletions and amplifications can be

(A)

(B)

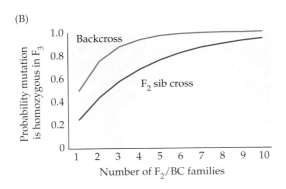

Figure 5.19 F$_1$ and F$_3$ genetic screens. After sperm mutagenesis, dominant mutations can be recovered directly in the F$_1$ progeny (A). For recessive mutations, crossing a heterozygous F$_1$ individual (orange) to a wild-type individual (white) results in transmission of each mutation to half of the F$_2$ grandchildren. Homozygotes (light red) are only recovered in one-quarter of the offspring produced by crossing two F$_2$ siblings, and only one-quarter of the individuals are mutant in such families. The probability of making any mutation homozygous is increased by setting up multiple families from each F$_2$. If a backcross between the heterozygous F$_1$ and an F$_2$ can be set up (B), the probability of homozygosity—and hence the efficiency of the screen—increases further.

mapped to a resolution of around 5 Mb by hybridization to metaphase chromosome spreads, probing of DNA microarrays with labeled genomic DNA facilitates mapping of breakpoints to just tens of kilobases (Mantripragada et al. 2003). The DNA spots on the microarray typically consist of fragmented and amplified BAC or P1 clones, but can also include cDNA clones or selected low-redundancy fragments. A difference in the ratio of fluorescence signal of normal to tumor DNA pinpoints a region of the genome that has been reduced or duplicated in the tumor; comparison of hundreds of samples highlights loci that are associated repeatedly with cancer progression.

High-Throughput Reverse Genetics

There are many situations in which random forward mutagenesis is either ineffective or inefficient at identifying genes involved in a trait. These include traits that are difficult to screen in live organisms, such as development of internal organs or complex aspects of physiology; quantitative continuous traits that are not disrupted in a Mendelian fashion, including life history and many behavioral traits; late-acting traits that cannot be screened directly because the genes that affect them disrupt an earlier aspect of development or kill the organism before the phenotype appears; and gene functions that are redundant either because there are multiple copies of the relevant genes or multiple pathways for achieving the observed phenotype. Some of these problems can be overcome with reverse genetic strategies that start with a gene of interest and work out toward a phenotype.

Systematic mutagenesis.   Systematic mutagenesis refers to the process of deliberately knocking out (mutating) a series of genes, one by one. It is typically employed in situations where the group of candidate genes to be targeted for mutagenesis is small enough that despite the extra work involved per gene, mutations can be recovered more efficiently than by screening large volumes of random mutants. These include cases where a group of transcripts expressed in conjunction with the development or physiology of a trait have already been defined but for which no functional data are available; or where a region of the genome has been shown to be important for the ontology of a disease and disruption of each putative gene in the region is warranted. For some species, the resources now exist for making a mutation in each and every gene in the genome, in which case the guarantee that each gene is disrupted provides a large advantage over random mutagenesis.

There are several ways to isolate a mutation in a known open reading frame. The classical genetic method is to screen for failure to complement a deficiency that removes a large portion of the genome that includes the gene of interest. A wild-type chromosome is mutagenized and crosses are performed to identify mutations that are sub-viable or that produce a mutant phenotype over the deficiency chromosome. The mutations are then placed in complementation groups and DNA sequencing is used to determine which set of mutations lie in the gene of interest.

Another method is to use PCR to screen a panel of mutant organisms for the insertion of a transposable element adjacent to the gene of interest. The forward primer is designed to hybridize to the terminus of the transposon, and a series of nested reverse primers are designed to hybridize to the gene. Genomic DNA from a number of different, potentially mutant stocks is pooled; if any of these carries an insertion near the gene, a unique amplification product will be observed. Sophisticated bar coding schemes have been designed for some model organisms to facilitate screening tens of thousands of insertion mutants such that just three or four reactions are required to precisely identify the relevant mutant stock. For *Drosophila* and *Caenorhabditis*, saturation transposon mutagenesis is performed in conjunction with direct sequencing of the insertion site by inverse PCR, in which case knowledge of the complete genome sequence leads directly to the assembly of a catalog of insertional mutants. Investigators need only query the database to obtain their mutation.

The most direct method for targeted mutagenesis is to specifically disrupt the ORF by homologous recombination (Thomas and Capecchi 1987). DNA that is carried into a cell by electroporation or chemical transformation tends to integrate at random locations, but if there is sufficient sequence homology with a piece of chromosomal DNA, it can align with the endogenous copy of the gene and replace it by two adjacent crossovers. This process is known as homologous recombination, and has been manipulated to allow precise targeting of mutations. The disruptive copy of the gene is engineered in bacteria and typically carries a selectable marker that allows isolation of insertion-bearing chromosomes (Figure 5.20A). New insertional mutagenesis cassettes have been designed to direct *lacZ* expression in place of the endogenous gene, allowing visualization of expression of the gene at the same time as the knockout is performed (Chen et al. 1997). The technique works efficiently in yeast and in mouse embryonic stem cells because of a high ratio of homologous recombination to unequal recombination (that is, random insertion anywhere in the genome). Mouse embryonic stem (ES) cells can be grown in culture, screened for the correct insertion, and then injected into a mouse blastocyst in such a way that they will populate the germ line and give rise to transgenic mice carrying the targeted mutation. Knockout mutations are first generated in heterozygous condition, and mutant defects are examined in homozygous or haploid progeny. The techniques used for constructing transgenic animals and plants are summarized in Box 5.3 (Gong and Rong 2003).

Knock-ins. Homologous recombination can also be used to perform "knock-ins," in which the wild-type copy is replaced with a specific modification (as opposed to a generic insertion) (Figure 5.20B). One application of knock-in technology is to place the expression of one gene under the control of the regulatory elements of another gene. For example, replacing the upstream regulatory region of the mouse *HoxD11* gene by the homologous region of the zebrafish *hoxd11* gene resulted in premature activation of expression of the gene and an anterior shift in the location of

(A)

(B)

Figure 5.20 Targeted mutagenesis.   (A) Gene knockout mutations in yeast or mammalian cells are usually constructed using a targeting vector that carries both positive and negative selectable markers. Recombination between identical sequences in the vector and chromosomal DNA (somewhere in the vicinity of exons 1 and 3 in this example) results in the replacement of exon 2 with the *G418* gene that encodes resistance to the drug neomycin. Enrichment for homologous recombination events is carried out by selecting for absence of thymidine kinase (tk) activity that arises when the whole construct inserts elsewhere in the genome. (B) Knock-in mutagenesis is performed similarly, except that there are two selection steps. There is no enrichment for homologous recombination in the first step, so sufficient clones must be screened to ensure that the targeted insertion is present. Subsequent expression of Cre recombinase results in removal of the two selectable genes between the two *loxP* sites (yellow). This event can be selected for on gancyclovir, which is toxic in the presence of the TK enzyme. In this example, exon 2 has been replaced by a modified form of the exon.

the sacrum (Gærard et al. 1997). Results from targeted replacement can be interpreted without concerns over whether sequences adjacent to the location of nontargeted insertions (that is, position effects) cause the premature activation or otherwise affect expression of the transgene.

Another application of knock-ins is to test the function of a particular point mutation in vivo, such as targeted mutagenesis of promoter sequences or testing of the requirements of specific amino acid residues in biological functions that cannot be measured in cell lines. **Gene therapy,** in which a mutant gene is replaced by a wild-type copy, also uses the knock-in approach to avoid potential deleterious effects of random mutagenesis or inappropriate regulation of transgenes that lie outside their normal chromosomal context.

RNAi screens. A simple alternative to homologous recombination is to induce transient loss of gene function using inhibitory RNA (RNAi) expression (Fire et al. 1998). The presence of just a few molecules of double-stranded RNA in a cell dramatically reduces the level of transcript of the associated gene, resulting in a loss-of-function phenotype. RNAi is a stretch of several hundred bases of both the sense and anti-sense strands of RNA corresponding to a gene, which hybridize together in a cell. Constructs that lead to expression of RNAi can be introduced into the embryo by injection, electroporation, or even (in the case of *Caenorhabditis*) by feeding the nematode bacteria that transcribe both strands of the RNAi. Alternatively, transgenic animals can be generated that produce RNAi by transcription of a construct that contains the two strands separated by a short loop, so that the molecule will self-hybridize once synthesized (Figure 5.21).

The mechanism of RNAi function is related to that of small regulatory microRNAs that have normal functions in development and physiology (Hannon 2002). Long dsRNA is first cleaved into 21–25-nt long small interfering siRNA molecules by the enzyme Dicer, and these are denatured and fed into an RNA-induced silencing complex (RISC) that aligns the antisense strand with the complementary mRNA. This results in post-transcriptional down-regulation through a variety of mechanisms, including mRNA degradation and inhibition of translation, that vary among plants and animals.

RNAi screens have been applied to whole chromosome mutagenesis in *C. elegans* by designing 16,757 constructs to specifically knock out expression of 86% of the identified open reading frames (Kamath et al. 2003). Mutant phenotypes were identified for 1,722 genes, two-thirds of which were previously undescribed. Annotation of cellular as well as whole-organism functions can be performed, depending on the choice of phenotypic assay. *Drosophila* and mammalian cell cultures have also been screened for various phenotypes with RNAi, and Kumar et al. (2003) reported a novel microarray platform for effective RNAi selection, in which spotted siRNAs are transfected into cells that report loss of translation of a green fluorescent protein reporter gene. RNAi is not guaranteed to identify all mutant phe-

(A)

(B)

Figure 5.21 Inhibitory RNA expression (RNAi). (A) For transient infection assays, double-stranded RNA (dsRNA) is synthesized either in *E. coli* or in vitro using T7 promoters oriented in opposite directions on either side of the cloned gene. The two strands self-anneal in the bacteria, and can be fed directly to nematodes or isolated and injected. (B) For most other eukaryotes, RNAi is pursued by cloning two copies of the gene head-to-tail under the control of a tissue-specific or inducible promoter. Transcription in transgenic organisms results in assembly of a dsRNA that is often inhibitory.

notypes (as evidenced by a small but significant failure to detect known phenotypes in the whole chromosome screens), but these screens are at least as efficient as classical mutagenesis for identification of novel functions.

Gain-of-function mutagenesis. Systematic gain-of-function genetics can be employed in numerous situations for which disruption of gene expression or protein structure is not the best way to study gene function (Figure 5.22). These include cases in which redundancy masks the loss-of-function phenotype, or where the earliest effects of gene disruption preclude ascertainment of a later function—such as when an embryonic lethal prevents study of the gene's function during adult organ development.

Generic methods for inducing ectopic gene expression employ inducible promoters, such as a heat-shock promoter or drug-inducible promoters. Some of these systems disrupt expression of a large fraction of the genome, which is a drawback, but appropriate controls can be performed to demonstrate that a phenotype is dependent on gain of expression of the transgene. A common criticism is that there is no way of knowing whether a gain-of-function phenotype is representative of the wild-type function of the protein, but from the perspective that mutagenesis screens are just the first step in the characterization of biological function, the criticism can be short-sighted. For instance, follow-up screens for genes that suppress the gain-of-function phenotype will typically identify proteins with similar biochemical, developmental, or physiological functions. Gain-of-function can also be an efficient method for characterizing the effect of modification of particular parts of a protein—for example, in defining the residues that are responsible for functional specificity of transcription factors or receptors.

Dominant loss-of-function is difficult to engineer, but will be an important tool for dissection of pleiotropic functions that are masked by early defects of mutations. One approach to these mutations is to design proteins that are expected to perform some but not all of the functions of the wild-type protein. For example, receptors that lack the intracellular domain will

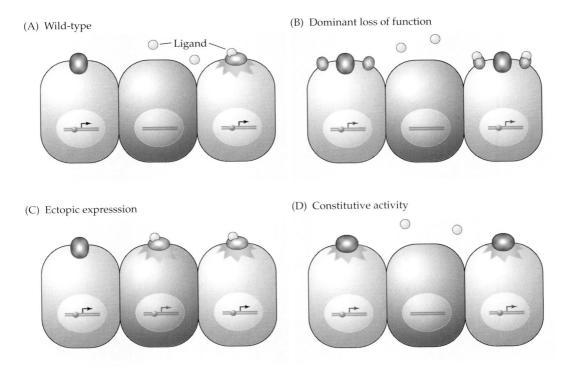

Figure 5.22 Targeted loss- and gain-of-function genetics. Wild-type gene activity can be modified through the construction of transgenic organisms in a number of ways. (A) Suppose a protein is expressed only in the light blue cells, and that it requires the yellow ligand in order to become active. (B) Dominant negative proteins (red) can be expressed in these cells to inhibit the activity of the endogenous protein—for example, by soaking up the ligand. (C) Gain-of-function can be engineered by inducing ectopic expression of the gene in other cell types or (D) by introducing a modified protein that is constitutively active even in the absence of ligand.

BOX 5.3 Transgenic Animals and Plants

Transgenic animals and plants are organisms that have been transformed by a foreign piece of DNA that is stably integrated into the genome and transmitted from generation to generation. The foreign DNA may be from the same or a different species; it may be simply a reporter construct, or it may encode a novel product or a gene that will be expressed in a novel pattern; and, it may be present in multiple copies, as a single copy, or may even replace the endogenous copy of the gene. In all of these cases, four major steps must be taken in order to construct a transgenic organism:

1. The foreign DNA construct must be synthesized in a form that can be shuttled into the host.

2. The transgene must be delivered into the host animal or plant cells.

3. The transgene must integrate stably into a germ line chromosome of the host organism.

4. Appropriate lines that are viable and express the transgene at desired levels must be identified and established as a permanent stock.

Bacteria and yeast can be transformed with vectors that replicate autonomously. Transformation of higher eukaryotes, however, generally requires that a host chromosome take up the DNA. Any piece of DNA that gets past the host defense systems can integrate at random anywhere in a genome, through the process of nonhomologous (illegitimate) recombination. In some eukaryote transformation procedures, there is little control over the number of copies that insert and, particularly in plants, it is not uncommon for tens of copies to integrate as a tandem array. This problem can be circumvented by carrying the transgene into the genome using a disabled transposable element. Genetic engineering is used to replace the TE's transposase gene (that normally catalyzes transposition) with the foreign DNA, and the transposase enzyme is supplied transiently on a helper plasmid only during the transformation step. Consequently, once inserted, there is no way for the transgene to hop back out. For example, *Drosophila* transgenesis is performed using modified P-elements (Rubin and Spradling 1982) or mariner elements; this system is now being applied to other insects of agricultural importance and even to some vertebrates.

Transposable elements usually insert at a single location, but there is no control over where this location is. Many insertion events knock out the function of an essential gene (and thus are deleterious), and most are affected by the transcriptional activity of the surrounding chromatin. The insertion site can be controlled by using artificial recombination systems such as the Cre-Lox system adapted from bacteriophage P1 for use in higher eukaryotes. As described in the text, homologous recombination can also be used to actually replace the endogenous gene.

A variety of different delivery systems are available for any given species. **Chemical transformation** or **electroporation** are convenient for delivering DNA into *cell cultures* such as protoplasts and stem cell lines, where there is no cell wall or chorion creating a barrier. *Large cells* such as oocytes can be directly **microinjected** with a DNA solution in fine glass needles, and gold particle guns have been developed to carry DNA into *plant cells and organelles* in a process known as **biolistics.**

Plant transformation is most simply performed using *Agrobacterium tumefaciens* bacteria as a vector (Birch 1997). The bacterium induces gall cells on the plant's tissue and injects part of Ti plasmid that has been engineered to carry the transgene. This T-DNA then incorporates into the plant's somatic cells, from which whole plants can be regenerated by manipulating the growth medium with appropriate hormones. Originally, this technology was only applicable to dicots, but increased understanding of the process now allows it to be used to

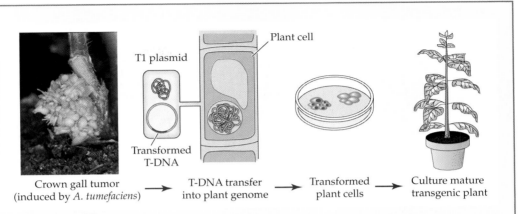

Crown gall tumor
(induced by *A. tumefaciens*) → T-DNA transfer into plant genome → Transformed plant cells → Culture mature transgenic plant

Figure A ***Agrobacterium*-mediated transformation.** *A. tumefaciens* induces gall cells on a plant, and injects T-DNA into the host cells. Transformed cells are then grown into whole plants by manipulating the hormones in the growth medium.

transform monocots, including agronomically essential grasses (Shen et al.1999).

Mammals can be transformed by injection of transforming DNA into their oocytes, as was exemplified dramatically by the production of the first transgenic rhesus monkey using a replication-defective retrovirus (Chan et al. 2001). Mice are more commonly transformed by first producing embryonic stem (ES) cells in culture (Capecchi 1989), then injecting the chosen cell line into a blastocyst (early embryo) from a host with a different visible genetic constitution (for example, coat color). Mature mice are chimeric, since the ES cells populate different tissue primordia. If they contribute to the germ line, transgenic progeny will be recovered and subsequent crosses are performed to establish a homozygous strain.

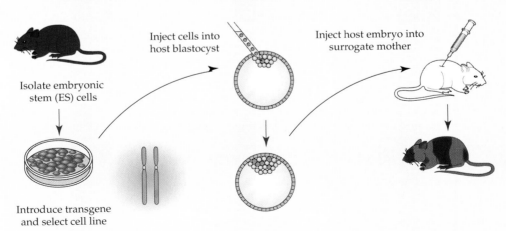

Figure B Transgenic mice via ES cells. A transgene is introduced into donor ES cells, which are then injected into a host blastocyst. The blastocyst is implanted in a surrogate mother and produces chimeric offspring that can in turn be propagated, producing homozygous transgenic offspring.

often absorb the ligand that normally activates the receptor, thereby interfering with the transmission of the signal by the endogenous protein as well. Inducible expression of the dominant negative form allows the function of the gene to be studied at any phase of development. Another approach is to screen a panel of constructs in yeast or tissue culture cells to identify dominant negative proteins prior to introduction of the transgene into the germ line of the model organism.

Viral-mediated transfection.   For many purposes, viral-mediated transfection can be more efficient than germ line transgenesis for introducing modified genes into a developing organism. Several broad host-range viruses have been developed for this purpose. Some viruses can be produced in both replication-competent and replication-incompetent forms. The former will infect a large field of cells in developing organisms, and thus can be used to study the effects of ectopic gene expression during embryogenesis of organisms for which traditional genetic analysis is not available. Notably, transfection with modified retroviruses has been used extensively to study appendage development and neurogenesis in chick embryos. Replication-incompetent virus is much safer to work with, and is more useful for studying the effects of ectopic expression on the differentiation of a small clone of cells. Viral transfection is better suited to specific hypothesis testing than general screening. Another application lies in comparative genomics, particularly for testing the evolution of gene function across a wide range of invertebrates.

Phenocopies.   It is worth emphasizing that gene function can also be "phenocopied," which means that mutant phenotypes can be mimicked by environmental disturbance. In fact, in pharmacology, phenocopies may be a much more direct indicator of gene function than mutagenesis or overexpression studies, owing to the plasticity of neuronal systems. Monoamine receptor knockout mice show fascinating and dramatic behavioral defects, but these do not always agree with the phenotypes observed after administration of receptor antagonists (e.g., Holmes 2001). In some cases, this may be because the drug recognizes and interferes with more than one protein. It may also be a function of sweeping changes in regulation of other receptor family members that occur in the absence of a particular gene function in the synapse. So, while gene disruption is the most obvious and direct way to study gene function, there are situations where it can be misleading, and functional genomic analysis generally calls for a variety of strategies for biological annotation.

Fine-Structure Genetics

Whole-genome mutagenesis is a very efficient approach to identification of a core set of genes that function in a particular process, but it is just a first step. Characterizing the function of every gene in a genome requires utilization of a suite of fine-structure genetic analysis techniques. Regional and pathway-specific screens have been devised to increase the efficiency of forward genetics. A battery of new procedures allow collections of genes to be expressed in situations where they are not normally active; facilitate expression of in vitro modified versions of genes; and are used to specifically disrupt gene function in particular tissues and at a certain times. Additionally, at least for microbes, the technique of genetic fingerprinting can be used to define subtle effects on growth that are not apparent in screens for heritable phenotypes.

Regional mutagenesis. Regional mutagenesis refers to genetic screens designed to saturate a small portion of the genome with insertions or point mutations. The aims are to generate lesions that disrupt each of the multiple functions of known genes, to identify genetic functions independent of evidence from expression data or prediction of coding potential, and to obtain a more thorough understanding of the genetic structure of chromosome regions than that offered by the preceding methods. A genome center with the capacity to process 1,000 mutant mouse strains in a year, for example, by focusing on a 10-cM region of a single chromosome could be expected to generate several mutations in every one of up to several hundred genes in the region. This strategy was used to saturate the albino region of chromosome 7 with 31 lethal mutations that fell into 10 complementation groups (Rinchik and Carpenter 1999).

Whereas point mutations and transposable element (TE) insertions are generated at random throughout the genome, they can be targeted to a particular region by only choosing to retain mutations that map to that region. This is most simply achieved through the use of **balancer chromosomes.** Balancers contain a recessive lethal that prevents the appearance of homozygous balancer individuals, one or more inversions that inhibit recombination with mutant chromosomes, and a dominant visible marker. *Drosophila* balancers contain multiple inversions that suppress recombination throughout the chromosome, but more restricted balancers can be constructed for other organisms in order to isolate a small region of the genome. Kile et al. (2003) used an inversion on mouse chromosome 11 to isolate 55 new mutations that are lethal at different stages of pre- and post-natal development, and 33 visible mutations with neurological, haematopoietic, morphological, and fertility defects.

When two heterozygotes for a mutant chromosome over the dominant marked inversion are crossed (mut/Bal × mut/Bal), failure to obtain progeny without the dominant phenotype (mut/mut) indicates that a recessive lethal is located opposite the inversion. Alternatively, complete co-segregation of the dominant marker and a dominant phenotype (for example, coat color) carried by a TE used to generate insertional mutants, indicates that the TE is inserted opposite the inversion and allows recovery of regional insertions in the absence of an overt phenotype. All other mutations or insertions can either be discarded, or sent to another group for analysis.

EXERCISE 5.4   *Designing a genetic screen*

Create a genetic screen designed to detect recessive mutations on chromosome 4 that are required in the female germline for viability of mice. Assume that you have available a strain of mice that carry a dominantly marked chromosome 4 with a large inversion (namely, a balancer chromosome).

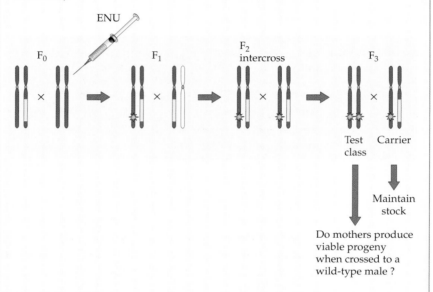

ANSWER: Progeny of mutagenized male mice are crossed to females that carry the chromosome 4 balancer, resulting in heterozygous mutant F_1 mice. These are crossed to another chromosome 4 balancer stock that carries another visible dominant marker on the other fourth chromosome shown in white. All progeny of this cross that do not have the white chromosome must be heterozygous for the mutagenized chromosome and the balancer. A cross between two of these F_2 mice results in homozygous mutant F_3 progeny. You would cross these female mice to a wild-type strain, and screen for the absence of viable progeny.

Mapping of the regional mutations proceeds by deficiency complementation (Figure 5.23). Model organism stock centers are accumulating overlapping deficiency collections, otherwise known as **segmental aneuploids.** The deficiencies can be induced by radiation, or can be generated in a site-directed manner using the Cre-Lox recombinase system to delete all of the DNA between two flanking TE insertions. Since up to a quarter of the genes in higher eukaryotic genomes are thought to be essential for viability, deficiencies removing tens of genes are generally homozygous-lethal and can even be sub-viable in heterozygotes as a result of the summation of haploinsufficient effects. With a set of overlapping deficiencies, it is possible to

Figure 5.23 Deficiency mapping. Recessive point mutations can be mapped by complementation testing. (A) Two classes of progeny—mutant and wild-type—are produced when a homozygous loss-of-function mutant is crossed to a line carrying a deficiency (Df) that removes the locus. If the deficiency does not remove the locus, all progeny will be wild-type. (B) By testing an ordered series of overlapping deficiencies, the extent of the region in which the mutation lies can be narrowed to tens of kilobases, at which point direct sequencing will often identify the mutation.

map point mutations and TE insertions with high precision, relative to both genetic and physical maps. This procedure also facilitates clustering of mutations into complementation groups.

Modifier screens. Where a phenotype is affected by a single known mutation, pathway-directed screens can be used to identify more genes that interact with the first mutation, in what are technically known as second-site modifier screens. Modifiers may either enhance or suppress a phenotype, and can do so in heterozygous or homozygous condition, depending on the sensitivity of the screen. **Synthetic lethal** screens identify pairs of loci that are singly viable but lethal in combination. Such screens have proven particularly useful in yeast: in one systematic series of crosses with 4,700 viable deletion strains, 291 interactions involving 204 genes were implicated in such processes as cytoskeletal organization and DNA synthesis and repair (Tong et al. 2001; see Box 5.2).

Synthetic visible defects are likely to be particularly important in uncovering redundant functions, as well as more generally for identifying genes that interact in the same pathway. It is important to recognize that a genetic interaction (that is, a phenotype produced by an interaction between two or more genes) can have a large number of mechanistic causes. Sometimes this is because two proteins encoded by the genes physically interact; sometimes it is because they lie in the same pathway—for example, one gene regulates another, perhaps with several intermediaries in between the two products; sometimes it is because they both utilize the same substrate. But the interactions can also be very indirect—for example, between a gene involved in neuropeptide secretion in the hypothalamus and one involved in liver or pancreatic homeostasis. Genetic backgrounds can also play a major role in modifying the observed interaction, and the more subtle a phenotype is, the more care must be taken to ensure that the genetic background is uniform.

An extremely attractive feature of pathway-based screening for functional genomics is that it can be performed in **heterologous systems.** That is to say, a gene identified in one organism that cannot be studied using classical genetics (for example, human, dog, or forest tree) can be screened for interactions in another organism (mouse, fly, or *Arabidopsis*). The finding that over 50% of known human disease genes have homologs in the invertebrate model organisms makes the fly and nematode popular systems for attempts to identify genes that interact with the primary disease locus (see Figure 1.15). The strategy is either to knock out the homologous gene in the model organism, or to express the gene ectopically in the model system, and then screen for modifiers of any aberrant phenotype. As an example, Presenilin is a causative agent in the onset of Alzheimer's disease, and when introduced into *Drosophila* produces a neurodegenerative phenotype that has been subject to modifier screens. These suggest an interaction with the well-characterized Notch signaling pathway (Anderton 1999).

For the identification of biochemical interactions, it is not even necessary that the phenotype in the heterologous organism show any relation to the disease phenotype. The great conservation of protein structures implies that interactions should be conserved at the level of gene families across the entire range of multicellular organisms, from plants to animals. By choosing different heterologous systems, it may be possible to screen a much wider array of potential phenotypes and interactions, and to adjust the sensitivity of the interactions to enhance the probability of detecting them.

Enhancer trapping and *GAL4*-mediated overexpression.   Access to genetic pathways that are not amenable to phenotypic characterization can be gained by searching for genes that are specifically expressed at the appropriate time and in the place of interest. While this can be done with microarrays, there are advantages to being able to observe expression in the whole organism. An important tool that allows such observation is **enhancer trapping.** This technique has the added feature that the gene is tagged with a TE insertion that either mutates the gene or can be used to induce a mutation. Enhancer trapping was initially developed in *Drosophila* (O'Kane and Gehring 1987), and derivative procedures are being adapted to most other model organisms.

An enhancer trap is a transposable element vector that carries a weak minimal promoter adjacent to the end of the element, which is hooked up to a reporter gene such as *lacZ* (β-galactosidase) or GFP (green fluorescent protein). When the TE inserts into a gene, the nearby enhancers now drive expression of the reporter gene through the minimal promoter that is inactive on its own, but faithfully reports aspects of the expression of one of the genes adjacent to the site of insertion (Figure 5.24B). This method can be used to identify genes that are expressed in just a small subset of cells or at a precise time in development, or after administration of a drug or behavioral regimen.

Various derivative methods, outlined in Figure 5.24C and D, have been developed for controlled gain-of-function genetics. Enhancers that are identified with enhancer traps can be harnessed to drive expression of other genes besides simply that of the innocuous reporter. Gain-of-function analysis is used to identify gene products that produce a novel phenotype when present at abnormal concentrations or in abnormal places. It can also be used to test the effects of in vitro modification of the gene; for example, modify-

Figure 5.24 Transposon mutagenesis and enhancer trap screens.
(A) Transposon mutagenesis occurs when a mobile element, such as the *Drosophila* P-element, inserts into the regulatory or coding regions of a gene. The former position will generally disrupt transcription in one or more tissues; the latter will usually produce a null allele. Cleavage of the inserted genomic DNA by a restriction enzyme that cuts in the P-element *and* somewhere in the flanking genomic DNA (vertical red lines) facilitates "plasmid rescue" of the DNA adjacent to the site of insertion. Modified P-elements carry a selectable marker (such as the mini-*white* eye color gene); plasmid sequences (labeled pUC), including an origin of replication and selectable drug resistance gene; and the transgenic sequences of interest. (B) Enhancer traps are transposons that insert immediately upstream of a gene in such a way that a reporter gene on the transposon (often *lacZ* or GFP) comes under the control of endogenous enhancers that normally drive expression of the genomic locus. (C) Instead of using a reporter, the enhancer trap can be used to drive expression of the transcriptional activator *GAL4*, which, when crossed to another line carrying a *GAL4*-responsive UAS promoter hooked up to a gene of interest, results in activation of expression of that gene where the enhancer is active. In this way, any gene can be turned on specifically in any tissue at a given time. (D) Systematic gain-of-function genetics is performed by establishing a library of transposon insertions that bring a UAS promoter adjacent to a genomic locus, expression of which is driven with *GAL4* under control of any desired promoter.

ing the active site of an enzyme or ligand-binding domain of a receptor. Systematic gain of function is achieved by substituting the reporter gene of an enhancer detector construct with the *GAL4* gene. *GAL4* encodes a potent yeast transcriptional activator that turns on any target gene with an upstream activator sequence (UAS) in its promoter. *GAL4* expressed in this way can be used to drive tissue-specific expression of any gene the investigator wishes to introduce under control of a UAS sequence.

For screening purposes, the first gene studied is a UAS-*lacZ* construct (since this shows where the enhancer is active), but once an enhancer that is active in some tissue of interest has been found, that line can be used for controlled gain-of-function analysis (Brand and Perrimon 1993). For example, a *GAL4* driver active in a particular nucleus of the hypothalamus might be used to turn on expression of a particular modified neurotransmitter receptor in that portion of the brain and hence to test the function of the modified receptor. The only requirement is that a different strain with the desired transgene under control of a UAS promoter be available for crossing to the driver. Crossing this line to a panel of different *GAL4* drivers can also be used to screen for effects of ectopic expression of the gene of interest in different tissues, and even to screen for novel phenotypes directly.

An alternative method of screening for the effects of ectopic expression is the **enhancer-promoter (EP) method.** This technique is used to find genes that result in a novel phenotype when expressed in a tissue of interest, as opposed to screening for effects of a particular gene in a range of tissues. The strategy is to replace the minimal promoter of the enhancer detector construct with UAS sequences, so that when the TE inserts upstream of a gene, the gene can be turned on under control of a *GAL4* driver of choice introduced on a different chromosome (Rorth et al. 1998). By mixing and matching different constructs maintained in different lines, it is possible to study phenotypes that are lethal to the organism but can be regenerated at will, since the individual UAS and *GAL4* constructs have no effect on their own.

Floxing.   Loss-of-function genetics can be supplemented by techniques that allow gene function to be disrupted specifically at a certain time of development, and/or in specific tissues. This is crucial for separating the different pleiotropic functions of a gene, as well as for examining the function of those genes required for embryonic viability. The powerful new technique used for this purpose in mouse genetics is known as **floxing** and is diagrammed in Figure 5.25.

Homologous recombination is used to introduce a *lox* site on either side of an essential exon of the gene. These sites are recognition-sequence targets for the P1 bacteriophage Cre recombinase enzyme, which is used to induce excision of a stretch of DNA between two adjacent *lox* sites. When mice homozygous for an exon flanked by *lox* sites are crossed to a line that expresses Cre under the control of a desired promoter, excision of the exon results in a temporally and spatially restricted mutation (Wagner et al. 1997). Viral infection has also been used to induce expression of the Cre recombi-

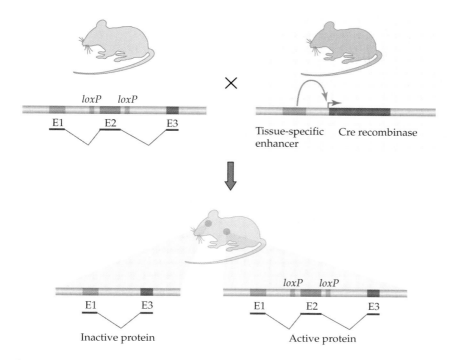

Figure 5.25 Floxing mice. One strain of mice is engineered by homologous recombination to carry *loxP* sites on either side of an essential exon of the gene that is being targeted for disruption. The other strain carries a transgene that expresses the Cre-recombinase gene specifically in the tissue of interest, and/or at a particular time. In F_1 progeny, excision of the exon between the *loxP* sites from genomic DNA in cells that express Cre recombinase results in a mutated gene that does not produce functional protein.

nase and consequent excision of sequences flanked by *lox* sites (Akagi et al. 1997).

Floxing has enormous potential for the dissection of the roles of genes in the development and pharmacology of the brain, as it will be possible to remove gene expression precisely from individual regions of the cortex. Temporal activity of the recombinase has also been modulated by fusing it to a mutated ligand-binding domain of the human estrogen receptor, with the result that it is responsive to tamoxifen that is supplied in the diet (Metzger and Chambon 2001).

Genetic Fingerprinting

One of the remarkable findings of genome sequencing has been that at least a quarter of all genes are species-specific, in the sense that homologous genes are not readily detected in other organisms. Many of these genes are fast-evolving, and it has been hypothesized that they usually will not be associated with critical functions and so may be refractory to both forward and reverse genetic dissection. One possibility is that they encode phenotypic modifiers that are required for subtle adjustment of traits, or only for responses to extreme environmental circumstances (where "extreme" means anything outside of the laboratory). No approaches have yet been devised for testing this proposition in higher eukaryotes. However, the **genetic fingerprinting** approach diagrammed in Figure 5.26 has already been used to demonstrate subtle and environment-specific functions of a large fraction of yeast and microbial genomes.

The idea behind genetic fingerprinting is to generate a population of thousands of independent mutations on an isogenic background, and then to perform an artificial selection experiment that will result in specific loss of mutations in genes that affect fitness (Smith et al. 1996). The mutations are introduced by mobilization or transfection of a transposable element, and are detected by PCR-mediated multiplex amplification of genomic fragments, using one primer in the TE and another in the genome. The microbes are grown in a **chemostat,** an apparatus that allows continuous feeding of exponentially growing cells with sterile medium for hundreds of generations. A selection differential as small as 1% will quickly amplify and will eliminate from the population a lineage carrying a mutated essential gene after just 20 generations of experimental evolution. The experiment is replicated multiple times to control for the fact that most mutations will actually be lost by genetic drift in one or a few of the replicates. Conducting the experiment under a variety of culture conditions allows detection of growth differentials on different nutrient sources, increasing the likelihood of identifying environment-specific functions. Mutations detected with the fingerprinting approach can subsequently be tested by more direct competition of wild-type and targeted mutant strains.

Summary

1. Several databases of protein domain families are available that use different profiling methods to predict similarity. Annotation of protein structure from sequence data often provides the first hint of gene function, but it is not definitive.

2. The simplest tool for characterizing the proteome of a cell is 2D electrophoresis, in which proteins are separated by charge in one dimension and mass in another. Methods for identifying each protein are available, but for comparisons of the same tissue type it is possible to compare an experimental gel with a canonical gel for the tissue that has been annotated previously.

3. High-throughput identification of proteins is performed by parallel peptide sequencing using time of flight mass spectrometry (MS). Each proteolytic fragment has a characteristic mass-to-charge (m/z) ratio, and comparison of a complex mixture of m/z peaks with the distribution predicted from the genome sequence allows inference of protein identities.

Figure 5.26 Genetic fingerprinting. (A) A complex mixture of tens of thousands of clones is constructed in which a single transposon is inserted in or adjacent to a different locus. (B) Each insertion is tagged, since the combination of a primer within the transposon and one in the genome will amplify a fragment of a diagnostic length. (C) The population is grown for a hundred generations. Samples are extracted every 10 generations for genotyping. Multiplex PCR is performed, followed by separation of fragments by electrophoresis. If the transposon is within a gene that affects viability under the growth conditions, the corresponding bands will disappear with time in replicate experiments (arrowheads). Some tags will also disappear by random sampling in a single replicate. (D) Averaging over experiments, a plot of fragment intensity against time is indicative of the selection pressure against the insertion. In this hypothetical example, selection against the red gene is more intense than against the blue gene.

4. Tandem mass spectrometry (MS/MS) is a method for partial sequencing of complex mixtures of peptides, based on random degradation of individual peptides deflected into an ionization chamber. Combined with liquid chromatography, it has emerged as the preferred method of proteome characterization, without the need to run 2D gels.

5. Quantification of protein levels can be performed by heavy isotope-labeling of peptide mixtures followed by MS, so that each peak appears side-by-side with a peak derived from an unlabeled sample, in proportion to the amount of protein. ICAT reagents facilitate reproducible labeling.

6. Protein microarrays consist of in vitro purified proteins or antibodies spotted onto a glass slide. Specific protein–protein, protein–drug, protein–enzyme, and protein–antibody interactions can be detected by fluorescent labeling.

7. Protein interaction maps are networks of protein–protein links indicating which proteins interact physically and/or functionally. Yeast two-hybrid (Y2H) screens are a direct way to determine which pairs of proteins assemble complexes under physiological conditions.

8. Protein structures are deposited in the Protein Data Bank (PDB). Structural proteomics aims to determine a representative structure of each of the hypothesized 5,000 distinct protein folds.

9. X-ray crystallography and NMR spectroscopy are the two major methods used to determine the three-dimensional structures of proteins and protein complexes.

10. Threading is a method for predicting protein tertiary structure by comparing predicted secondary structure and folding energies with those of solved structures. It allows structures to be predicted where there is less than 30% sequence identity, the lower cutoff for simple alignment.

11. Functional genomics is the process of assigning a biochemical and physiological/cell biological/developmental function to each predicted gene.

12. Forward genetics refers to the random generation of mutations followed by mapping and localization of the mutated gene. Saturation genetic screens can be based either on detection of visible phenotypes or on generation of large collections of transposable element insertions that may or may not disrupt a gene.

13. Reverse genetics refers to the process of generating a mutation in a specific open reading frame and then searching for a phenotype. Homologous recombination is a general method for knocking out gene function, but is difficult to perform on a high-volume scale and only works efficiently for mice and yeast.

14. Whereas a mutation that disrupts the function of a gene is called a "knockout," transgenic organisms can also be constructed in which the wild-type gene is replaced either by a different gene or by a modified form of the gene. These are called "knock-in" genes.

15. RNAi is a new method for abolishing gene function based on the ability of double-stranded RNA that is diced into short interfering siRNAs in the cell, to interfere with the synthesis of the gene product. Both transient and transgenic approaches can be taken.

16. An ever-increasing arsenal of methods for systematic manipulation of gene expression in vivo allow genes to be turned on or off in specific

tissues, at specific times, and in particular genetic backgrounds. This manipulative ability will allow dissection of the pleiotropic functions of genes in development, physiology, and behavior.

Discussion Questions

1. Why aren't transcript and protein expression profiles always in agreement?

2. What is a protein domain? To what extent is it possible to infer the function of a protein by sequence and phylogenetic comparison with domains present in other proteins?

3. Protein interaction networks can be assembled from yeast two-hybrid experiments, protein microarrays, and using bioinformatic methods. Discuss the limitations and biases inherent in each of these approaches.

4. What types of trait are amenable to dissection using random mutagenesis to identify genes that affect the phenotype? Under what circumstances is reverse genetic analysis more likely to lead investigators to "the genes that matter"?

5. How can functional genomics be applied to characterize the roles of human genes that do not have a known mutation?

Web Site Exercises

The Web site linked to this book at http://www.sinauer.com/genomics **provides exercises in various techniques described in this chapter.**

1. Build a zinc-finger protein profile and execute a PSI-BLAST exercise.

2. Characterize the secondary structure and modification sites of several related proteins.

3. Perform a simple threading exercise.

4. Identify the complementation groups in a hypothetical genetic data set.

Literature Cited

Abola, E., P. Kuhn, T. Earnest and R. Stevens. 2000. Automation of X-ray crystallography. *Nat. Struct. Biol.* 7 (Suppl.): 973–977.

Akagi, K. et al. 1997. Cre-mediated somatic site-specific recombination in mice. *Nucl. Acids Res.* 25: 1766–1773.

Anderton, B. H. 1999. Alzheimer's disease: Clues from flies and worms. *Curr. Biol.* 9: R106–109.

Baker, D. and A. Sali. 2001. Protein structure prediction and structural genomics. *Science* 294: 93–96.

Bass, B. L. 2002. RNA editing by adenosine deaminases that act on RNA. *Ann. Rev. Biochem.* 71: 817–846.

Berman, H. M. et al. 2000. The Protein Data Bank. *Nucl. Acids Res.* 28: 235–242.

Birch, R. G. 1997. Plant transformation: Problems and strategies for practical application. *Annu. Rev. Plant Physiol. Plant Mol. Biol. 48:* 297–326.

Brand, A. and N. Perrimon. 1993. Targeted gene expression as a means of altering cell fates and generating dominant phenotypes. *Development* 118: 401–415.

Burley, S. K. 2000. An overview of structural genomics. *Nat. Struct. Biol.* 7 (Suppl.): 932–934.

Capecchi, M. R. 1989. Altering the genome by homologous recombination. *Science.* 244: 1288–1292.

Carter, A., W. Clemons, D. Brodersen, R. Morgan-Warren, B. Wimberly and V. Ramakrishnan. 2000. Functional insights from the structure of the 30S ribosomal subunit and its interactions with antibiotics. *Nature* 407: 340–348.

Celis, J. E. et al. 2000. Gene expression profiling: Monitoring transcription and translation products using DNA microarrays and proteomics. *FEBS Lett.* 480: 2–16.

Chalmers, M. and S. Gaskell. 2000. Advances in mass spectrometry for proteome analysis. *Curr. Op. Biotechnol.* 11: 384–390.

Chan, A. W., K. Chong, C. Martinovich, C. Simerly and G. Schatten. 2001. Transgenic monkeys produced by retroviral gene transfer into mature oocytes. *Science* 291: 309–312.

Chen, X. et al. 1997. Targeting deletion of angiotensin type 1B receptor gene in the mouse. *Am. J. Physiol.* 272: F299–304.

Cherkasov, A. and S. J. Jones. 2004. Structural characterization of genomes by large scale sequence-structure threading. *BMC Bioinformatics* 5: 37.

Cramer, P., D. Bushnell and R. D. Kornberg. 2001. Structural basis of transcription: RNA polymerase II at 2.8 Ångstrom resolution. *Science* 292: 1863–1876.

Driever, W. et al. 1996. A genetic screen for mutations affecting embryogenesis in zebrafish. *Development* 123: 37–46.

Fenyö, D. 2000. Identifying the proteome: Software tools. *Curr. Op. Biotechnol.* 11: 391–395.

Fields, S. and O. Song. 1989. A novel genetic system to detect protein-protein interactions. *Nature* 340: 245–246.

Fire, A., S. Xu, M. Montgomery, S. Kostas, S. Driver and C. Mello. 1998. Potent and specific genetic interference by double-stranded RNA in *Caenorhabditis elegans*. *Nature* 391: 806–811.

Fischer, D. and D. Eisenberg. 1997. Assigning folds to the proteins encoded by the genome of *Mycoplasma genitalium*. *Proc. Natl Acad. Sci. (USA)* 94: 11929–11934.

Frary, A. et al. 2000. *fw2.2*: A quantitative trait locus key to the evolution of tomato fruit size. *Science* 289: 85–88.

Gérard, M., J. Zákány and D. Duboule. 1997. Interspecies exchange of a *Hoxd* enhancer in vivo induces premature transcription and anterior shift of the sacrum. *Dev. Biol.* 190: 32–40.

Gong, M. and Y. S. Rong. 2003. Targeting multi-cellular organisms. *Curr. Op. Genet. Der.* 13: 215–220.

Gygi, S., B. Rist, S. Gerber, F. Turecek, M. Gelb and R. Aebersold. 1999. Quantitative analysis of complex protein mixtures using isotope-coded affinity tags. *Nat. Biotechnol.* 17: 994–999.

Gygi, S., B. Rist and R. Aebersold. 2000. Measuring gene expression by quantitative proteome analysis. *Curr. Op. Biotechnol.* 11: 396–401.

Haab, B., M. Dunham and P. O. Brown. 2001. Protein microarrays for highly parallel detection and quantitation of specific proteins and antibodies in complex solutions. *Genome Biology* 2(2): 4.1–4.13.

Haffter, P. et al. 1996. The identification of genes with unique and essential functions in the development of the zebrafish, *Danio rerio*. *Development* 123: 1–36.

Hannon, G. J. 2002. RNA interference. *Nature* 418: 244–251.

Hegyi, H. and M. Gerstein. 1999. The relationship between protein structure and function: a comprehensive survey with application to the yeast genome. *J. Mol. Biol.* 288: 147–164.

Henikoff, S., B. J. Till and L. Comai. 2004. TILLING. Traditional mutagenesis meets functional genomics. *Plant Physiol.* 135: 630–636.

Holmes, A. 2001. Targeted gene mutation approaches to the study of anxiety-like behavior in mice. *Neurosci. Biobehav. Rev.* 25: 261–273.

Hrabe de Angelis, M. H. et al. 2000. Genome-wide, large-scale production of mutant mice by ENU mutagenesis. *Nat. Genetics* 25: 444–447.

Ito, T. et al. 2000. Toward a protein-protein interaction map of the budding yeast: A comprehensive system to examine two-hybrid interactions in all possible combinations between the yeast proteins. *Proc. Natl. Acad. Sci. (USA)* 97: 1143–1147.

Jones, D. T. 2000. Protein structure prediction in the postgenomic era. *Curr. Op. Struct. Biol.* 10: 371–379.

Jordan, P., P. Fromme, H. Witt, O. Klukas, W. Saenger and N. Krauss 2001. Three-dimensional structure of cyanobacterial photosystem I at 2.5 Å resolution. *Nature* 411: 909–917

Kamath, R. S. et al. 2003. Systematic functional analysis of the *Caenorhabditis elegans* genome using RNAi. *Nature* 421: 231–237.

Kelso, R. J., M. Buszczak, A. T. Quinones, C. Castiblanco, S. Mazzalupo and L. Cooley. 2004. Flytrap, a database documenting a GFP protein-trap insertion screen in *Drosophila melanogaster*. *Nucleic Acids Res.* 32: D418–20.

Kile, N. T. et al. 2003. Functional genetic analysis of mouse chromosome 11. *Nature* 425: 81–86.

Kumar, R., D. S. Conklin and V. Mittal. 2003. High-throughput selection of effective RNAi probes for gene silencing. *Genome Res.* 13: 2333–2340.

Lamzin, V. S. and A. Perrakis. 2000. Current state of automated crystallographic data analysis. *Nat. Struct. Biol.* 7 (Suppl.): 978–981.

Lehner, B. and A. G. Fraser. 2004. A first-draft human protein-interaction map. *Genome Biology* 5: R63.

Liu , H., D. Lin and J. R. Yates III. 2002. Multidimensional separations for protein/peptide analysis in the post-genomic era. *Biotechniques* 32: 898–911.

MacBeath, G. and S. Schreiber. 2000. Printing proteins as microarrays for high-throughput function determination. *Science* 289: 1760–1763.

Mantripragada, K. K., P. G. Buckley, T. Diaz de Ståhl and J. P. Dumanski. 2003. Genomic microarrays in the spotlight. *Trends in Genetics* 20: 87–94.

Marcotte, E., M. Pellegrini, H. Ng, D. Rice, T. Yeates and D. Eisenberg. 1999. Detecting protein function and protein-protein interactions from genome sequences. *Science* 285: 751–753.

Metzger, D. and P. Chambon. 2001. Site- and time-specific gene targeting in the mouse. *Methods* 24: 71–80.

Montelione, G., D. Zheng, Y. Huang, K. Gunsalus and T. Szyperski. 2000. Protein NMR spectroscopy in structural genomics. *Nat. Struct. Biol.* 7 Suppl: 982–985.

Nolan, P. M. et al. 2000. A systematic, genome-wide, phenotype-driven mutagenesis programme for gene function studies in the mouse. *Nat. Genetics* 25: 440–443.

Nüsslein-Volhard, C. and E. Wieschaus. 1980. Mutations affecting segment number and polarity in *Drosophila. Nature* 287: 795–801.

O'Kane, C. and W. J. Gehring. 1987. Detection *in situ* of genomic regulatory elements in *Drosophila. Proc. Natl Acad. Sci. (USA)* 84: 9123–9127.

Pellegrini, M., E. Marcotte, M. Thompson, D. Eisenberg and T. Yeates. 1999. Assigning protein functions by comparative genome analysis: protein phylogenetic profiles. *Proc. Natl Acad. Sci. (USA)* 96: 4285–4288.

Ramachandran, N., E. Hainsworth, B. Bhullar, S. Eisenstein, B. Rosen, A. Lau, J. C. Walter and J. LaBaer. 2004. Self-assembling protein microarrays. *Science* 305: 86–90.

Rinchik, E. M. and D. A. Carpenter. 1999. N-ethyl-N-nitrosourea mutagenesis of a 6- to 11-cM subregion of the Fah-Hbb interval of mouse chromosome 7: Completed testing of 4557 gametes and deletion mapping and complementation analysis of 31 mutations. *Genetics* 152: 373–383.

Rorth, P. et al. 1998. Systematic gain-of-function genetics in *Drosophila. Development* 125: 1049–1057.

Rubin, G. M. and A. C. Spradling. 1982. Genetic transformation of *Drosophila* with transposable element vectors. *Science* 218: 348–353.

Shen, W., J. Escudero and B. Hohn. 1999. T-DNA transfer to maize plants. *Mol. Biotechnol.* 13: 165–170.

Smith, V., K. Chou, D. Lashkari, D. Botstein and P. O. Brown. 1996. Functional analysis of the genes of yeast chromosome V by genetic footprinting. *Science* 274: 2069–2074.

Spradling, A. C. et al. 1999. The Berkeley *Drosophila* Genome Project gene disruption project: Single P-element insertions mutating 25% of vital *Drosophila* genes. *Genetics* 153: 135–177.

Stevens, R. C., S. Yokoyama and I. A. Wilson. 2001. Global efforts in structural genomics. *Science* 294: 89–92.

Taylor, W. R. 2002. A "periodic table" for protein structures. *Nature* 416: 657–660.

Thomas, K. and M. Capecchi. 1987. Site-directed mutagenesis by gene targeting in mouse embryo-derived stem cells. *Cell* 51: 503–512.

Tong, A. H. et al. 2001. Systematic genetic analysis with ordered arrays of yeast deletion mutants. *Science* 294: 2364–2368.

Tong, A.H.Y. et al. 2004. Global mapping of the yeast genetic interaction network. *Science* 303: 808–813.

Vetz, P. et al. 2000. A comprehensive analysis of protein–protein interactions in *Saccharomyces cerevisiae. Nature* 403: 623–627.

Von Eggeling, F., A. Gawriljuk, W. Fiedler, G. Ernst, U. Claussen, J. Klose and I. Romer. 2001. Fluorescent dual colour 2D-protein gel electrophoresis for rapid detection of differences in protein pattern with standard image analysis software. *Int. J. Mol. Med.* 8: 373–377.

Wagner, K. et al. 1997. Cre-mediated gene deletion in the mammary gland. *Nucl. Acids Res*. 25: 4323–4330.

Washburn, M. P., D. Wolters and J. R. Yates III. 2001. Large-scale analysis of the yeast proteome by multidimensional protein identification technology. *Nat. Biotech*. 19: 242–247.

Zhang, H., J. Kacharmina, K. Miyashiro, M. Greene and J. Eberwine. 2001. Protein quantification from complex protein mixtures using a proteomics methodology with single-cell resolution. *Proc. Natl Acad. Sci. (USA)* 98: 5497–5502.

Zhang, Y. and J. Skolnick. 2004. Automated structure prediction of weakly homologous proteins on a genomic scale. *Proc. Natl Acad. Sci. (USA)*. 101: 7594–7599.

Zhu, H. et al. 2001. Global analysis of protein activities using proteome chips. *Sciencexpress* www.sciencexpress.org. 26 July, 2001.

6 *Integrative Genomics*

An uncharitable view of genome research is that it is little more than a collection of technologies that can be used to pursue classical molecular genetics at an accelerated pace. Throughout this book, we have adopted the contrary viewpoint that in fact the conjunction of the new technologies with statistical and bioinformatic reasoning actually fosters an entirely different way of looking at biological systems. By extension, genome science will emerge as an independent discipline when a novel framework for studying and understanding biological complexity matures. Specifically, as comprehensive catalogs of genomic and cellular constituents are combined with computational strategies for studying how the parts assemble into working wholes, genome science will begin to address problems such as:

- Predicting microbial and cellular phenotypic properties.
- Ordering of genes and gene products into biochemical and developmental networks.
- Formulating *in silico* hypotheses for gene function that can be tested *in vivo*.
- Manipulating whole genomes to perform novel applications of biotechnological and biomedical importance.
- Integrating genome science with ecology and evolutionary biology.

An appreciation for the magnitude of this task can be seen from a glance at Figure 6.1, which plots the growth of knowledge in the sphere of genomics since the early 1990s. While the whole-genome sequencing phase is approaching completion for many organisms, the annotation of proteomes and transcriptomes is just ramping up, and construction of protein–protein and protein–DNA interaction maps is in its infancy. There is a vast literature on

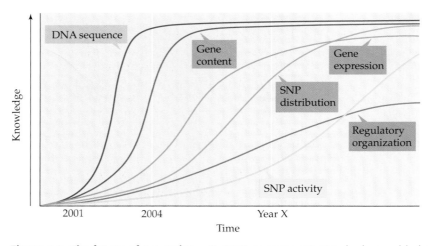

Figure 6.1 The future of genomics. By 2004, genome scientists had assembled a reasonably thorough picture of the sequence and gene content of representative genomes and were making inroads into description of gene expression profiles and protein function. Knowledge of SNP function and the organization of metabolic and other regulatory pathways is in its infancy. "Year X" is a debatable number of years into the future. (After Schilling et al. 1999.)

biochemical and metabolic pathways in bacteria and yeast, much of which is likely to apply to higher eukaryotes, but knowledge of the structure and design of developmental networks is arguably sporadic.

Yet some of the core questions in the biology of complexity, such as the molecular basis for homeostasis, the roles of constraint and pleiotropy in morphological evolution, and the structure of genetic variation in natural populations, all relate to the interactions among genes, their products, and the environment. To some observers, genome science offers the only real hope of tackling these deep biological problems.

This chapter begins by adding to the genomic triumvirate of genome, transcriptome, and proteome a description of methods for studying the metabolome. The **metabolome** may be defined as the sum total of substrates, metabolites, and other small molecules that are present in a population of cells. The second section of the chapter presents an overview of some of the *in silico* approaches being taken to turn parts lists into systems of dynamic equations and networks of interactions that allow us to predict and describe the properties of cells.

The title "integrative genomics" is misleading in the sense that very little work synthesizing the different aspects of genome science has yet been performed. This is largely because the discipline is so young; microarrays and tandem mass spectrometry are still emerging and expensive technologies, and there are only a few research groups in a position to integrate genomic platforms. It also reflects the predominantly reductionist framework of contemporary genetics, in which genomics is viewed as a means for identifying genes of interest which can then be studied using classical methods, more than as a discipline in its own right. Bioinformatics is emerging as an independent discipline, but more often uses data sets that have been generated by other researchers, and is just beginning to guide experimental design in a manner that leads naturally to integrative science. The methodologies discussed in the second half of this chapter are simply not part of standard genetics curricula, so are not yet sufficiently well known to support systems analysis.

A further reason for the disjoint nature of genomics is that a major part of the puzzle is still missing: namely, descriptions of the metabolites and regulatory interactions that are found in any organism. These topics are dealt with in the first part of this chapter, but as a separate set of research methods, because this research is also not yet integrated with the other areas of genome science. It is only possible to speculate on how the integration will occur, but one area will be in the prediction of gene function based on correlation between changes in expression profile and metabolite profiles, either in a panel of mutant lines or after environmental perturbation.

A major goal of evolutionary and ecological genomics lies in describing and interpreting the genomic responses to genetic variation, and to interactions between organisms. Host-parasite and host-symbiont interactions are generally mediated by small molecules, many of which have been characterized using painstaking biochemical approaches. But **metabolomics**—profiling the structure and distribution of metabolites (small molecules), particularly organic compounds—promises dramatic advances in this area. On the biochemical front, toxicology and dietary imbalance are just two areas where integration of transcript, protein, and small-molecule data will not just increase our understanding of metabolic mechanisms, but may quickly translate into clinical and nutritional recommendations. **Pharmacogenomics,** the study of interactions between drugs and genomes, is already represented by several new journals, and the subdiscipline of **psychogenomics,** dealing with the effects of mood-modulating chemicals, will see rapid advances once in vivo methods for parallel analysis of chemicals and gene products become established.

Metabolomics

Analysis of Cellular Constituents

Several spectrometric techniques (Table 6.1) are being applied to profile the complement of both organic and inorganic compounds found inside cells. The first three techniques (pyrolysis, infrared, and Raman spectrometry) are used to characterize differences in the total spectrum of cellular constituents, and can be used to screen for outliers in panels of mutants, natural isolates, or lines grown under different environmental conditions. When combined with mass spectrometry, the gas and liquid chromatographic techniques, as well as classical biochemical purification, provide a more focused view of which particular compounds vary among treatments. These techniques can

TABLE 6.1 *Metabolomic Methods*

Acronym	Method
PyMS	Pyrolysis mass spectrometry
FT-IR	Fourier transform-infrared spectrometry
DRS	Dispersive Raman spectrometry
NMR	Nuclear magnetic resonance spectroscopy
GC-MS	Gas chromatography mass spectrometry
LC-ESI-MS	Liquid chromatography electron spin ionization mass spectrometry
MDA	Multivariate data analysis: Chemometrics, artificial neural networks, genetic algorithms

be used to identify enzyme substrates and metabolites or to characterize quantitative variation among networks of compounds.

Pyrolysis refers to the thermal degradation of materials into volatile fragments (the pyrolysate) that can be separated according to mass-charge ratio (m/z) and analyzed by mass spectrometry, providing a chemical profile of the constituents in the complex sample (Irwin 1982). A few microliters of sample are dried onto a ferromagnetic foil, which is then rapidly heated either by a laser or by a current passing through a coil surrounding the sample. A temperature between 350° and 770°C is chosen to achieve a balance between lysis of carbohydrates and proteins. Products with $m/z < 50$ (such as methane, ammonia, water, and methanol) and large products with $m/z > 200$ are usually ignored, so that the biologically informative material lies between these ranges. Peak heights can be quantified and used to distinguish samples—for example, to identify microbial species in mixed isolates, to characterize the effects of growth conditions, to identify extreme metabolic outliers in pharmaceutical screens, and to detect contaminants in food products such as orange juice, Scotch whiskey, and virgin olive oil (Goodacre and Kell 1996).

Infrared spectrometry measures the absorbance of energy of specific wavelengths by highly polar bonds (such as C—N, N—H, C=O, etc.), which provides a fingerprint of the presence of the chemical groups present in particular molecules in a sample. Samples are typically dried onto the surface of a metal plate, or are mixed with KBr as a fine, dried powder that reflects infrared radiation passed through the sample. Automation of the procedure allows up to 100 samples to be analyzed in parallel, with hundreds of data points collected over the range of wavelengths from 4,000 to 600 cm^{-1} (Figure 6.2). **Chemometric** procedures are again used to extract information from the spectra and to distinguish outlier samples (Winson et al. 1997). Applications include monitoring effects of drugs on microbial and fungal

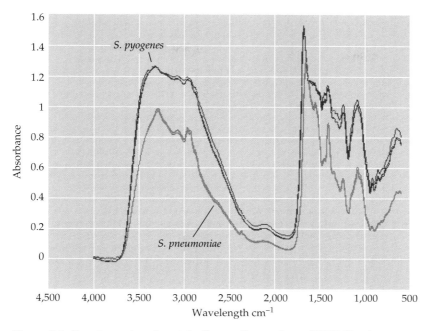

Figure 6.2 Spectrometry of metabolites. Comparison of FT-IR (Fourier transform-infrared spectrometry) spectra for two species of *Staphylococcus* reveals global differences in metabolites without directly identifying the actual molecules that are involved. (After Goodacre et al. 1996.)

metabolism, pharmaceutical screening, and quantitative typing of plant cell wall differences (often due to variation in lignins).

A promising approach to detecting a full range of metabolites is direct **nuclear magnetic resonance spectroscopy** of cellular extracts, or even of cellular suspensions. Different resonance spectra can be used to focus on carbon or phosphate metabolism, or to observe changes in ion gradients and energy metabolism. Proton NMR has been used to characterize overall changes in the metabolome without prior specification of which compounds are being observed, using a strategy dubbed FANCY, for Functional Analysis of Co-responses in Yeast (Figure 6.3; Raamsdonk et al. 2001). The idea is that many mutations may not produce a visible growth phenotype, since metabolic networks have the capacity to re-route metabolites to restore biomass production; yet such mutations might be detected through their effect on intracellular substrate concentrations. NMR spectra obtained from several replicates of each of six strains were processed by principal component analysis followed by discriminant function analysis (the former identified the major components of variation, the latter was used to remove within-strain variation), and were shown to result in clustering of two strains

Figure 6.3 Nuclear magnetic resonance spectroscopy with chemometrics. NMR spectra capture the intensity of free-induction decays due to specific atoms, typically protons. After Fourier transformation and normalization, principal component analysis is applied to identify the major components of variation in the overall spectra in a set of samples. Further statistical manipulation such as discriminant function analysis may separate the samples by removing within-group factors, in this case clearly showing that the orange and blue samples tend to cluster. (After Raamsdonk et al. 2001.)

known to disrupt 6-phosphofructo-2-kinase activity. While neither mutant had a visible phenotype, they apparently shared a distinct metabolic profile. In principle, this method might be used to sort through the entire genome, clustering mutants that disrupt similar metabolic processes.

Identification of individual molecules within a sample requires techniques in which the various compounds are first separated using standard methodologies such as gas, liquid, or HPLC chromatography, and then visualized either by stains or absorbance spectra, or more comprehensively using mass spectrometry (MS). Different systems are most appropriate for different classes of molecule. For example, amino acids do not require MS for characterization; lipids and lignins can be separated by gas chromatography (GC), complex carbohydrates by capillary electrophoresis (CE), and complex steroids and derivatives by HPLC. Where well-characterized standards exist—for example, in relation to lipid or monoamine profiles, which are crucial for research into cardiovascular disease and pharmacology, respectively—there is often no need for an MS step. In addition, certain metabolites and small hormones, such as albumin, cortisol, insulin, melatonin, and prolactin, can be detected immunologically, while a variety of enzymatic assays exist for detection of individual compounds such as glucose, ketones, lactate, and alcohols.

No single method currently allows each of the hundreds of different classes of metabolite within cells to be assayed together, but the combination of chromatographic and MS profiling under different conditions gives a remarkable and ever-improving display of the metabolome.

Another clever strategy for detecting mutants that disrupt metabolism is to look for a change in the capacity of cells to modify an "unnatural" substrate. Yarema et al. (2001) intercepted the sialic acid biosynthesis pathway in a human cell line by adding an analog of UDP-N-acetylglucosamine, called N-levulinoylmannosamine (ManLev), to the culture medium. After ManLev is passed along the biochemical pathway, it is presented at the cell surface, where the unique chemical reactivity of a ketone group allows the level of metabolism of the substrate to be assayed using a fluorometric assay performed in conjunction with cell sorting. After 10 cycles of sorting combined with expansion of the culture to introduce genetic diversity as mutations accumulate, several low- and high-expressing cell lines were characterized, with less than 0.3% and more than 600% of the original activity, respectively. These lines, which would not have been isolated using standard genetic screens, now provide models for the study of inborn errors of sialic acid metabolism, a condition that affects such physiological processes as development, immune function, synaptic plasticity, malignancy, and metabolite storage.

Metabolic and Biochemical Databases

Several Web-based compilations of metabolic and biochemical pathways have emerged in the past few years, as documented in Table 6.2. A very useful starting point is the ExPASy/Boehringer-Mannheim series of thumbnail sketches linked into a complete overview of metabolism on one page, and cellular biochemistry on another, along the lines shown in Figure 6.4. The BRENDA, EcoCyc, LIGAND, UMBBD, and WIT databases present aspects of metabolism in searchable formats that allow cross-genome comparison. PathDB is a feature at the National Center for Genome Research that is designed more as a set of tools for analyzing and visualizing metabolic pathways than as a database. The KEGG, PEDANT, and GOLD databases cover whole genome annotations more generally.

A second important class of bioinformatic databases in the context of integrative genomics are those that document protein–protein and protein–DNA interactions. Many of these sites are under development. The PathCalling site at Curagen Corporation presents tools and databases that allow browsing and assembly of protein–protein interactions established by a variety of methods, particularly physical interactions detected in a whole genome two-hybrid screen of yeast (Uetz et al. 2000). PathCalling also supports visualization of likely pathways in other species based on identification of orthologs. Protein–DNA interactions are supported by the TRANSFAC, MIRAGE, and TRRD sites, which in addition to providing searchable databases of known transcription factors and transcription-factor binding sites, link to an ever-expanding suite of tools for detection of novel DNA binding sites and conserved putative regulatory motifs in the regulatory regions of genes. Among the software programs available for this are MatInspector (Quandt et al. 1995), AlignACE (Hughes et al. 2000), CoBind (GuhaThakurta and Stormo

TABLE 6.2 *Databases of Biochemical Pathways and Networks*

Database	Description and URL
BRENDA	Comprehensive enzyme information system http://www.brenda.uni-koeln.de
EcoCyc	Encyclopedia of E. coli genes and metabolism http://ecocyc.org
EPD	Eukaryotic promoter database http://www.epd.isb-sib.ch
ExPASy	Thumbnail sketches of metabolism and cellular biochemistry http://www.expasy.ch/cgi-bin/search-biochem-index
KEGG	Kyoto encyclopedia of genes and genomes http://www.genome.ad.jp/kegg/kegg2.html
LIGAND	Database for enzymes, compounds, and reactions http://www.genome.ad.jp/dbget/ligand.html
MIRAGE	Molecular informatics resource for the analysis of gene expression http://www.ifti.org
PathCalling	Curagen's database of yeast protein-protein interactions http://portal.curagen.com/pathcalling_portal/index.htm
PathDB	Online tool for constructing and visualizing biological networks http://www.ncgr.org/pathdb
PEDANT	Computational analysis and annotation of complete genomes http://pedant.gsf.de
SCPD	Saccharomyces cerevisiae promoter database http://cgsigma.cshl.org/jian
STKE	Science's signal transduction knowledge environment http://stke.sciencemag.org/index.dH
TRANSFAC	Transcription factor database http://transfac.gbf.de/TRANSFAC
TRRD	Transcriptional regulatory regions database http://www.bionet.nsc.ru/trrd
UMBBD	University of Minnesota Biocatalysis/Biodegradation Database http://umbbd.ahc.umn.edu/index.html

2001), and various others that incorporate information from expression profiling, comparative genomics, and molecular biology assays.

As explained in Box 4.4, the major difficulty in compiling databases of regulatory sites is that DNA binding sites are often short (less then 10 base pairs) and binding specificity can be heavily influenced by cofactors and flanking DNA sequences. Gibbs sampling procedures are now being supplemented by neural networks that are less sensitive to mismatches. Sorting through the relationship between the significance of the match of a sequence to a motif, and the biological function of the putative site, is not within the realm of current bioinformatics approaches.

Figure 6.4 Visualization of metabolic pathways.   Two percent of the ExPASy representation of microbial metabolism, assembled from the adjacent tiles E7 and F7, showing aspects of fatty acid and arginine biosynthesis. Compounds are in black, cofactors in red, and enzymes in blue. (From **http://biochem.boehringer-mannheim.com/ prodinfo_fst.htm?/techserv/metmap.htm**.)

In Silico Genomics

There are two major approaches to modeling metabolic and genomic networks. The first is to attempt to derive the properties of enzymatic pathways from a knowledge of kinetic parameters associated with the activity of each component and the order in which the components act. The second is to search for global properties of networks based on static properties of the system such as stoichiometry and Boolean logic. The closing section of this book presents a brief introduction to some of the first applications of both classes of modeling approach to genome data.

Metabolic Control Analysis

Metabolic control analysis (MCA) refers to the study of the sensitivity of global properties of metabolic pathways, most simply flux and metabolite concentrations, to perturbation of individual enzyme activity. "Flux" refers to the rate at which metabolites pass through a pathway and is a function of the velocity of the reaction at each enzymatic step, which in turn can be computed using Michaelis-Menten kinetics. Kacser and Burns (1973) first developed the concept of a control coefficient, which is a relative measure of the effect of perturbation of one component on the system. At steady state, the summation of all the control coefficients in a system is unity, since an increase in activity of one enzyme must be compensated for by a decrease in another. As the number of enzymes increases, the contribution that each makes to regulation of the total flux diminishes, as does the effect of a mutational perturbation in the activity of an enzyme. For an online discussion as well as a mathematical derivation of MCA, see http://dbk.ch.umist.ac.uk/mca_home.htm.

Two of the first applications of MCA were in providing explanations for the dominance and near-neutrality of gene activity. The predicted relationships between enzyme activity and flux or metabolite concentrations are nonlinear, as shown in Figure 6.5. If a microbe has evolved to optimize its growth rate, then metabolic flux should be as high as possible, and the activity of each enzyme(which is a function of its level of expression as well as catalytic rate constant set by the protein structure) will take on a value that places it close to the point at which the maximal flux reaches a plateau. Under these conditions, halving the dosage of the enzyme (as in a heterozygote) will have little effect on the total flux, and hence the gene tends to be dominant (Kacser and Burns 1981).

In addition, slight changes in activity due to new mutations and polymorphisms will not greatly affect flux, so the genetic variants should have little effect on fitness, and their dynamics will be in accordance with the expectations of neutral theory. A considerable body of experimental data from *E. coli* in particular corroborates this conclusion (Dykhuizen et al. 1987).

Theoretical modeling becomes more complicated when one realizes that most biochemical pathways are branching, and that activity of permeases and external factors such as drugs and hormones can have a major impact on metabolic rates. It is also clear that certain loci, notably phosphoglucose isomerase in diverse organisms, exert unexpectedly high levels of control over metabolism, perhaps because they sit at branch points where selection is more concerned with the regulation of power and efficiency than raw flux (Watt 1986). A particularly interesting application of MCA is in investigating the reasons for covariation in enzyme activities and steady-state substrate concentrations, both in natural populations and as a result of mutation accumulation. Study of these correlations may suggest whether selection is more likely to act on flux or on metabolite pools, and may indicate the role of historical contingency in metabolic divergence (Clark 1991; Clark and Wang 1994). A number of tools for modeling are available online, notably Gepasi (Mendes 1993), SCAMP (Sauro 1993), and MetaModel (Cornish-Bowden and Hofmeyr 1991), as well as DBsolve (Goryanin et al. 1999), a more recent tool that is integrated with a database of enzyme EC codes and stoichiometric data. All of these can be found readily by searching the Web with Google or other search engines.

The extension of metabolic control analysis to other regulatory networks such as receptor-ligand interactions, signal transduction, transcription, cell cycle regulation, and synaptic function is much less advanced. Two major

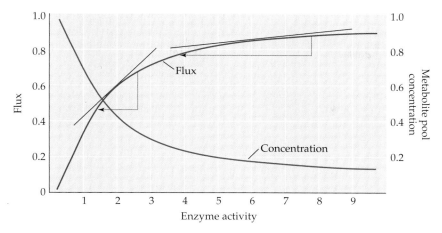

Figure 6.5 Metabolic control analysis of linear pathways. Parameters such as metabolic flux (blue) and metabolite pool concentrations (red) are not expected to be linear functions of enzyme activity. Metabolic systems are expected to evolve to a point where a large change in activity of the enzyme has only a very small effect on flux. The two black tangents to the flux curve show that the effect of halving enzyme activity has a much smaller effect on flux at higher enzyme activities. See Clark (1991) for a discussion of how fitness may affect the distribution of enzyme activities in natural populations.

difficulties encountered are that measurement of kinetic parameters for protein–protein interactions in vivo is extremely difficult, and mathematical models of reactions that do not involve metabolite conversion (for example, phosphorylation or initiation of transcription) are lacking. Nevertheless, some notable successes have been obtained. Barkai and Leibler (1997) identified a mechanism for the robustness of the bacterial chemotaxis network to perturbation of the concentrations and activities of components of the chemosensory apparatus. Ferrell (1997) demonstrated how a graded concentration of ligand can be converted into a switchlike intracellular response, and has begun to model the effects of variation on the transmission of the response down an intracellular kinase cascade. Threshold-dependent transcriptional regulation has been modeled as a consequence of the cooperativity of protein–protein–DNA interactions and applied to a thorough description of the lambda genetic switch (Ptashne 1992). Similarly, conversion of a morphogen gradient into a robust, stable, and accurate transcriptional response has been modeled using detailed knowledge of the molecular biology of the retinoic acid receptor (Kerszberg 1996). More comprehensively, Bhalla and Iyengar (1999) have initiated attempts to combine models of multiple biochemical pathways in series (Figure 6.6). They argue that networks of interaction will often give rise to emergent properties, such as the ability to integrate signals at different timescales and to generate distinct outputs dependent on the strength and duration of the input signal.

Numerous authors, starting with Jacob and Monod as early as 1962 and reviewed by McAdams and Arkin (1998), have remarked on similarities

(A)

$$A + B \underset{k_b}{\overset{k_f}{\rightleftarrows}} AB$$

$$A + B \underset{k_b}{\overset{k_f}{\rightleftarrows}} C + D$$

$$d[A]/dt = k_b [C][D] - k_f [A][B]$$

Figure 6.6 Regulatory network analysis.
(A) Like metabolism, regulatory networks can be modeled as systems of differential equations representing simple two-component interactions and reactions. (B) Two examples from Bhalla and Iyengar (1999) are shown, involving the phospholipase C and Ras signaling pathways, both of which are integrated through the epidermal growth factor receptor (EGFR).

(B)

between biological regulatory pathways and electronic circuit construction. Principles such as serial and parallel processing, feedback,* cooperative combinatorial control, and fan-out control of multiple subsystems are thought to be prevalent in the design of biological networks. Genome science promises to contribute to our understanding of biological circuits by cataloging temporal programs of gene activation, identifying the design principles of transcriptional enhancers and repressors, and establishing networks of protein–protein and protein–DNA interactions.

A practical extension of this work is in biological engineering, for example: the introduction of bistable genetic switches and oscillatory networks in microbes; designing efficient multistep degradative pathways into which checkpoints and feedback loops are incorporated; and the introduction of advanced biosensors. In most cases, the control of noise, sensitivity, and robustness associated with circuits designed *de novo* is unlikely to meet industrial specifications, but may be enhanced and optimized by artificial selection (Nielsen 1998). Subsequent research into the genetic changes accompanying selection promise to yield new insight into biological circuit construction.

---

*Both positive and negative feedback are included in this category, as are autoregulatory (where a protein affects its own expression) and indirect feedback (where the feedback is mediated by another protein downstream in the pathway).

An even more ambitious task for modeling based on the structure of known gene products is to simulate the cellular interactions that are required for pattern formation during development. A step in this direction has been taken by von Dassow et al. (2000) in their model of the *Drosophila* segment polarity network. They established a system of almost 50 nonlinear ordinary differential equations that describe parameters of the function of five genes expressed in a repetitive series of stripes in the cells of each embryonic segment (Figure 6.7A). These parameters are designed to allow for variation in factors such as the half-lives of the mRNA and proteins, their binding rates, and cooperativity coefficients over at least three orders of magnitude. Since the true values of the parameters are unknown, the objective was to determine whether random permutations of the parameters could support the emergence of the observed stripe patterns as the system runs to equilibrium, starting from a reasonable set of initial conditions. Astonishingly, 1,192 solutions were found among a set of 240,000 random parameter assignments, corresponding to 1 in 200 trials, and over 90% of individual possible parameter values were represented in the set of solutions. Just as remarkably, perturbation of individual parameters once a solution had been found showed that in the majority of cases a single variable could be perturbed over 10% of its allowed range—an indication of great robustness.

Such analyses suggest that the notion of a single optimal configuration of kinetic parameters for genetic components of complex regulatory networks is biologically unnecessary, and imply that developmental modules must be quite free to absorb genetic variation. The Ingeneue software developed by von Dassow et al. (2000) can be adapted to describe and model all manner of gene networks, and is available online at http://www.ingeneue.org.

Systems-Level Modeling of Gene Networks

The second broad class of genomic modeling eschews any explicit reference to the kinetic properties of defined pathways, and instead aims to find generic systems properties that emerge from the logic and connectivity of interacting networks. The most general of these approaches are not even concerned with the structure of actual networks, dealing rather with universal properties such as the propensity to evolve. The results are consequently less obviously amenable to experimental verification than are models that take data in the form of the stoichiometry and order of reactions as their starting point. Truly integrative genomics aims to merge observation with modeling, along the lines of classical hypothetico-deductive science.

The generic approach is perhaps most starkly illustrated by Kauffman's innovative ideas concerning the centrality of spontaneous order in biological homeostasis and evolution. Reducing genetic interactions to a network of Boolean switches, in which each gene interacts with k other genes, Kauffman (1993) argues first that stable attractor states are an expected property of complex systems and hence that there is a pervasive tendency toward order independent of natural selection. Second, he shows that as k increases, a phase transition occurs from a tendency toward the generation of order

(A)

(B)

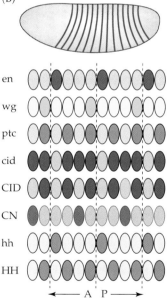

Figure 6.7 Modeling of pattern formation using Ingeneue. (A) Von Dassow and colleagues (2000) established a model for patterning of segment polarity in *Drosophila* embryos that used five genes (engrailed, *en*; wingless, *wg*; patched, *ptc*; cubitus interruptus, *ci*; and hedgehog, *hh*; ovals) and seven protein products (including the repressor fragment of CI, and the PTC-HH complex; boxes). Black arrows indicate activating interactions, and red lines indicate repressing interactions, each of which were modeled as a differential equation. Only when the two interactions shown as dashed lines (involving the CI repression of engrailed, and WG autoactivation) were included did Ingeneue simulations reproduce the target pattern of expression (dark shading indicates expression) of each of the genes and proteins shown for a real embryo (B).

to a tendency for chaos, such that there must be a limit to the amount of interaction that genomes can support. Third, between order and chaos is a phase in which biological systems may be poised on the edge of chaos, yet are most capable of adaptation and coevolution; and the average number of 3–5 interactions per gene that support this state is at least consistent with empirical data. Critics have argued that it is not clear that biological switches obey Boolean logic, and that the theory does not help explain quantitative phenomena, but the metaphors of rugged fitness landscapes and order out

(A)

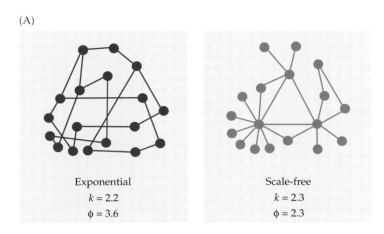

Exponential
$k = 2.2$
$\phi = 3.6$

Scale-free
$k = 2.3$
$\phi = 2.3$

(B)

Mean pathway length

Number of substrates

(C)

Mean number links per substrate

Number of substrates

Figure 6.8 Large-scale organization of metabolic networks. (A) Whereas random linkage of nodes into a network with just over 2 links per node results in most nodes having $k = 1, 2,$ or 3 connections (left), metabolism seems to be organized as a scale-free system, in which a few nodes with many more connections are found (right). In this hypothetical example of 18 substrates, the average pathway diameter ϕ (number of links between any two substrates/nodes) is significantly reduced in the scale-free network relative to the random/exponential one. (B) Forty-three organisms with varying number of known metabolic substrates were nevertheless found to have a constant mean pathway diameter, which comes about because there are more reactions predicted per substrate in more complex metabolomes (C). (After Jeong et al. 2000.)

of chaos are persuasive and may attract more attention as mathematical and biological sciences merge in the coming years.

Another example of the exploration of the large-scale organization of genomic systems based largely on their degree of connectivity lies in the demonstration that metabolic systems are scale-free rather than exponential (Jeong et al. 2000). As shown in Figure 6.8, this means that rather than each node (substrate) in a metabolic network being connected randomly with k other nodes with a Poisson-distributed probability P(k) ~ e^{-k}, the fraction of nodes with many links is actually much higher and follows a power law, P(k) ~ $ke^{-\gamma}$. This was shown to be true for 43 organisms that included archaea, bacteria, and a eukaryote, despite the fact that the number of known metabolic substrates ranges from about 200 to 800. The average path length (number of catalytic steps required to connect any two metabolites) was also seen to be constant at a "small world" linkage of just over three—which means that increases in substrate complexity are bought by increasing the mean number of connections per node. A consequence is that metabolic networks are quite resistant to metabolic disturbance, since removal of one enzyme can be compensated for by shunting the reaction through a different pathway. Further research is addressing the question of whether and how other cellular networks, such as signal transduction and control of cell division, are structured along these lines.

It is proving possible to model the optimization of metabolic pathways and rerouting of substrate utilization in response to mutation, or alteration of nutrient sources, by assembly of *in silico* genotypes (Figure 6.9; Edwards and Palsson 2000). The metabolic databases allow the assembly of a universal stoichiometric matrix of m metabolites by n reactions, as well as a vector of all known fluxes through the network, from which an organism-specific matrix can be extracted given the sequence of the genome. For example, the minimal *E. coli* matrix consists of 436 metabolites and 720 reactions. Various constraints can be imposed based on established features of metabolism and desired inputs. A desired metabolic objective such as maximal growth rate (flux) is also imposed, and linear programming is then used to find a feasible and/or optimal set of parameters describing the flux through each reaction. Setting chosen fluxes to zero allows prediction of how the organism will respond to a mutation by redistributing fluxes, in the majority of cases with no major impact on the metabolic objective. Sixty-eight of 79 predictions made by this metabolic reconstruction approach were confirmed in vivo, with just seven enzymes involved in central metabolism being indispensable, and only nine others reducing biomass yield by more than 10%. The same approach can be used to predict the growth requirements and metabolic properties of any microbial genome in a variety of environments, with obvious applications in bioengineering.

Finally, a natural progression in genome science is the integration of genome sequencing, gene expression profiling, proteomics, functional genomics, and theory. Ideker et al. (2001) define a four-step procedure that they used to explore galactose utilization in yeast.

1. Characterize the genome, proteome, and metabolome of the organism and assemble a preliminary model of a cellular process of interest.

2. Perturb the system through mutation or by changing the growth conditions, and monitor the changes in mRNA and protein expression responses.

3. Integrate the expression profiles with the model, including data on genetic and physical interactions.

(A)

(B)

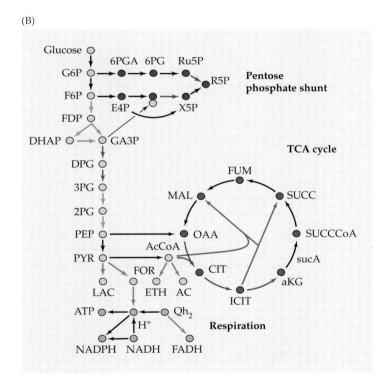

Figure 6.9 Metabolic reconstruction *in silico*. (A) *In silico* metabolomes are constructed by extracting the enzymes predicted to occur in a complete genome, synthesizing a stoichiometry matrix (S) and vector of possible metabolic fluxes (*v*). Solving for steady state defines the possible and optimal set of parameters for flux through each enzyme. (B) The diagram of *E. coli* central metabolism shows which enzymes are predicted to be essential for life (red arrows), which retard growth when deleted from the genome (orange), which are dispensable (black), and which are not utilized under aerobic conditions (gray). The green arrows indicate how flux is shunted across the TCA cycle when the *sucA* gene is deleted. (After Schilling et al. 1999.)

4. Modify the original hypothesis to accommodate differences between prediction and observation, and use standard molecular biological procedures to test the new hypothesis.

This approach, sometimes referred to as **systems biology,** goes well beyond simple estimation of gene function as a result of association between profiles of gene expression and physical interactions, to probing mechanisms of regulation, and in time promises to shed light on the deep structure of biological integration.

Summary

1. Genomics is not yet an integrated discipline with its own set of research questions, but the combination of technological advances and computational biology/bioinformatics is beginning to change the way that geneticists approach a wide range of research.

2. Metabolomics is the study of the structure and distribution of small molecules, particularly organic compounds, in tissues and organisms. A variety of spectrometric techniques are being developed for analysis of the metabolome.

3. Databases of protein–protein and protein–DNA interactions are an increasingly important online tool for assembling and studying networks of interacting gene products.

4. *In silico* genomics refers to computational research aimed at predicting the metabolic and physiological performance of cells, using genomic data to define the initial capabilities of the biological system that is being modeled.

5. Computational approaches can either explicitly attempt to model individual reactions and known networks of interactions (for example, metabolic control analysis), or be concerned with global properties of generic systems, such as the organization and robustness of networks.

6. Integrative genomic approaches will coordinate all areas of genome research with classical molecular biology, in a framework of hypothesis testing and formulation.

Discussion Questions

1. Describe how you would go about profiling the complete set of novel organic compounds found in a unique tropical plant that is found to have anticancer activity.

2. Experimental and theoretical physics have always been tightly linked disciplines, but that has not been the case in cell and developmental biology. How do you think the relationship between theoretical modeling and empirical research will change in the postgenomic era?

3. What is the relationship between robustness, redundancy, and homeostasis in genetic systems? How can it be studied?

4. How far into the future do you think Year X in Figure 6.1 will be? Is this a reasonable representation of the growth of knowledge in genome science?

Literature Cited

Barkai, N. and S. Leibler. 1997. Robustness in simple biochemical networks. *Nature* 387: 913–917.

Bhalla, U. S. and R. Iyengar. 1999. Emergent properties of networks of biological signaling pathways. *Science* 283: 381–387.

Clark, A. G. 1991. Mutation-selection balance and metabolic control theory. *Genetics* 129: 909–923.

Clark, A. G. and L. Wang. 1994. Comparative evolutionary analysis of metabolism in nine species of *Drosophila*. *Evolution* 48: 1230–1243.

Cornish-Bowden, A. and J. Hofmeyr. 1991. MetaModel: A program for modelling and control analysis of metabolic pathways on the IBM PC and compatibles. *Comput. Appl. Biosci.* 7: 83–93.

Dykhuizen, D., A. Dean and D. Hartl. 1987. Metabolic flux and fitness. *Genetics* 115: 25–31.

Edwards, J. and B. Palsson. 2001. The *Escherichia coli* MG1655 *in silico* metabolic genotype: Its definition, characteristics, and capabilities. *Proc. Natl Acad. Sci (USA)* 97: 5528–5533.

Ferrell, J. E. 1997. How responses get more switch-like as you move down a protein kinase cascade. *Trends Biochem. Sci.* 22:288–289.

Goodacre, R. and D. Kell. 1996. Pyrolysis mass spectrometry and its applications in biotechnology. *Curr. Op. Biotechnol.* 7: 20–28.

Goodacre, R., É. Timmins, P. Rooney, J. Rowland and D. Kell. 1996. Rapid identification of *Streptococcus* and *Enterococcus* species using diffuse reflectance-absorbance Fourier transform infrared spectroscopy and artificial neural networks. *FEMS Microbio. Lett.* 140: 233–239.

Goryanin, I., T. Hodgman, E. Selkov. 1999. Mathematical simulation and analysis of cellular metabolism and regulation. *Bioinformatics* 15: 749–758.

GuhaThakurta, D. and G. Stormo. 2001. Identifying target sites for cooperatively binding factors. *Bioinformatics* 17: 608–621.

Hughes, J., P. Estep, S. Tavazoie, and G. M. Church. 2000. Computational identification of *cis*-regulatory elements associated with groups of functionally related genes in *Saccharomyces cerevisiae*. *J. Mol. Biol.* 296: 1205–1214.

Ideker, T. et al. 2001. Integrated genomic and proteomic analyses of a systematically perturbed metabolic network. *Science* 292: 929–933.

Irwin, W. J. 1982. *Analytical Pyrolysis: A Comprehensive Guide.* Marcel Dekker, New York.

Jacob, F. and J. Monod. 1962. On the regulation of gene activity. *Cold Spring Harbor Symp. Cell Reg. Mech.* 193–209.

Jeong, H., B. Tombor, R. Albert, Z. Oltvai and A. Barabási. 2000. The large-scale organization of metabolic networks. *Nature* 407: 651–654.

Kacser, H. and J. Burns. 1973. The control of flux. *Symp. Soc. Exp. Biol.* 27: 65–104.

Kacser, H. and J. Burns. 1981. The molecular basis of dominance. *Genetics* 97: 639–666.

Kauffman, S. A. 1993. *The Origins of Order.* Oxford University Press, Oxford.

Kerszberg, M. 1996. Accurate reading of morphogen concentrations by nuclear receptors: A formal model of complex transduction pathways. *J. Theor. Biol.* 183: 95–104.

McAdams, H. H. and A. Arkin. 1998. Simulation of prokaryotic genetic circuits. *Annu. Rev. Biophys. Biomol. Struct.* 27: 199–224.

Mendes, P. 1993. GEPASI: A software package for modelling the dynamics, steady states and control of biochemical and other systems. *Comput. Appl. Biosci.* 9: 563–571.

Nielsen, J. 1998. Metabolic engineering: Techniques for analysis of targets for genetic manipulations. *Biotechnol. Bioeng.* 58: 125–132.

Ptashne, M. 1992. *A Genetic Switch,* 2nd Ed. Cell Press, Cambridge, MA.

Quandt, K., K. Frech, H. Karas, E. Wingender and T. Werner. 1995. MatInd and MatInspector: New fast and versatile tools for detection of consensus matches in nucleotide sequence data. *Nucl. Acids Res.* 23: 4878–4884.

Raamsdonk, L. M. et al. 2001. A functional genomics strategy that uses metabolome data to reveal the phenotype of silent mutations. *Nat. Biotechnology* 19: 45–50.

Sauro, H. M. 1993. SCAMP: A general-purpose simulator and metabolic control analysis program. *Comput. Appl. Biosci.* 9: 441–450.

Schilling, C., J. Edwards and B. Palsson. 1999. Toward metabolic phenomics: Analysis of genomic data using flux balances. *Biotechnol. Prog.* 15: 288–295.

Winson, M. K. et al. 1997. Diffuse reflectance absorbance spectroscopy taking in chemometrics (DRASTIC): A hyperspectral FT-IR-based approach to rapid screening for metabolite overproduction. *Analytica Chimica Acta* 348: 273–282.

Uetz, P. et al. 2000. A comprehensive analysis of protein-protein interactions in *Saccharomyces cerevisiae*. *Nature* 403: 623–627.

von Dassow, G., E. Meir, E. Munrow and G. Odell. 2000. The segment polarity network is a robust developmental module. *Nature* 406: 188–192.

Watt, W. B. 1986. Power and efficiency as indexes of fitness in metabolic organization. *Amer. Natur.* 127: 629–653.

Yarema, K., S. Goon and C. Bertozzi. 2001. Metabolic selection of glycosylation defects in human cells. *Nat. Biotechnology* 19: 553–558.

Glossary

***ab initio* gene discovery** A method for identifying likely genes in a stretch of genomic DNA sequence that does not depend on prior information such as similarity to a gene in another species or identity to a transcript. Most *ab initio* approaches use a **hidden Markov model** to search for sequence motifs that are commonly found in genes, such as long open reading frames, intron-exon boundary signatures, and conserved upstream regulatory motifs.

***ab initio* protein structure prediction** Prediction of protein tertiary structure based on computational modeling, given just the primary sequence of the protein.

acrylamide A polymeric compound used to make gels for electrophoretic separation of proteins or nucleic acids; hence PAGE (polyacrylamide gel electrophoresis).

admixture The mixture of two genetically distinct populations.

affinity chromatography A process whereby proteins can be separated or purified based on their affinity to the matrix in a chromatography column.

alignment The process of lining up two or more DNA or protein sequences so as to maximize the number of identical nucleotides or residues while minimizing the number of mismatches and gaps.

alternative splicing The combination of different sets of exons to produce two or more primary mature messenger RNAs from the same primary transcript. Commonly observed in higher eukaryotes, with the result that a single gene can generate multiple protein isoforms.

annotation In the most general sense, linking information from the literature to database entries for genes or proteins. In the context of genome sequencing, annotation refers to the identification of putative genes using a combination of *ab initio* methods, homology searches, and physical evidence.

antibody An immunoglobulin molecule that recognizes a specific protein, and is generated by exposing the immune system of a small mammal (e.g., mouse or rabbit) to the protein.

association mapping The search for genes that affect disease susceptibility by testing whether each of the alleles at a series of DNA polymorphisms tend to be present in affected individuals more or less commonly than expected by chance. Linkage disequilibrium between the sites both assists and complicates the mapping process.

balancer chromosome Chromosomes that have been deliberately engineered to carry multiple inversions that suppress crossing over, and so can be used to maintain recessive mutations in a genetic stock. Balancers usually also carry a recessive lethal and a dominant visible genetic marker.

base-calling The process of calling the series of nucleotides from a sequence trace, now usually automated, but with the facility for manual curation to resolve ambiguities.

bioconductor A project that brings together open-source software written in the R programming language, for the purposes of statistical analysis of genomic data.

BLOSUM A matrix of weights representing the probability with which each amino acid is substituted by another amino acid, based on the observed frequencies of such substitutions in a database. Used in scoring schemes for sequence alignment algorithms.

case-control association mapping An approach to screening for associations between genetic markers and disease status, based on a comparison of allele frequencies in a group of affected individuals, and in a similar control group of unaffecteds.

cDNA A DNA molecule that is complementary to a mRNA (messenger RNA) molecule. The first strand of cDNA is synthesized by **reverse transcriptase**, but the term is also used to describe a double stranded DNA clone that is derived from a transcript.

cDNA clones A complementary DNA copy of a full-length transcript, synthesized by reverse transcription of mRNA.

cDNA library A collection of cDNA clones, typically isolated from a single tissue.

cDNA microarrays An array of cDNAs on a glass microscope slide of nitrocellulose filter, that is hybridized to labeled mRNA for the purposes of profiling gene expression.

centiMorgan (cM) The standard unit of genetic map distance, corresponding to a 1% probability of a crossover occurring between two sites in any meiosis.

chain termination sequencing The most commonly used method for sequencing clones of DNA up to one kilobase in length, based on a method first devised by Fred Sanger, in which molecules of all possible lengths are produced by random termination of DNA polymerization when a dideoxynucleotide (ddNTP) is incorporated.

chemometrics A series of analytical methods for quantifying chemical profiles, including Principal Component Analysis and Artificial Neural Networks.

chemostat A device used to maintain a continuous culture of a microbe, commonly employed for studies of experimental evolution.

chromatin immunoprecipitation microarrays (ChIP chips) Microarrrays consisting of DNA corresponding to potential regulatory regions of genes that are used to detect sequences that bind to transcription factors.

chromosome painting A cytological procedure for aligning the chromosomes of two different eukaryotic species, based on fluorescence in situ hybridization (FISH). A set of chromosome-specific probes from one species are prepared using unique combinations of fluorescent dyes, and these "paint" the chromosomes in a mitotic chromosome spread from cells of the second species.

chromosome walking A procedure for cloning a large contiguous portion of a chromosome, in which a probe prepared from the end of one clone is used to identify overlapping genomic clones in a library. The process is iterated for as many steps as it takes to cover the region of interest.

clusters of orthologous genes (COGs) Sets of genes from a collection of species that are proposed to encode the same gene product, based on pairwise best-match sequence similarity. For eukaryotes, the acronym KOGS is sometimes used.

complementation group A set of mutations that fail to complement one another (that is, the mutant phenotype appears when one copy of each mutation is present in the same individual), and hence are likely to affect the same gene.

consensus sequence A hypothetical sequence consisting of the most common amino acid at each position in a multiple alignment of DNA or protein sequences.

contig A contiguous stretch of cloned DNA. May refer simply to a scaffold of overlapping clones that have been mapped physically, or to a long stretch of DNA sequence assembled by merging two or more sequences.

cosmid Large insert plasmids that generally exist as a single copy within host bacterial cells, but which contain cos sites that allow in vitro packaging of inserts as phage molecules if desired.

CpG islands Stretches of vertebrate DNA typically between 1 and 2 kb long that contain a ten-fold higher frequency of the doublet nucleotide CG than occurs elsewhere in the genome. They are commonly found at the 5′ end of genes.

Cre-Lox recombination system A combination of site specific recombinase (Cre) and its recognition site (*lox*) from the bacteriophage P1 that has been engineered into yeast, mouse and other eukaryotic genomes to facilitate targeted recombination.

C-value paradox The absence of any correlation between the number of genes and the amount of DNA in a genome. A corollary is that there is high variation in the DNA content of even closely related organisms, and no obvious relationship between complexity and DNA content.

cytological map A map of the location of genes or other DNA features relative to the banding patterns of the chromosomes of a species.

data normalization The process of removing systematic biases from microarray data that otherwise contributes to misinterpretation of apparent differences in transcript abundance.

dideoxynucleotide A nucleotide that is missing the hydroxyl groups on both the 2′ and 3′ carbon atoms of the sugar backbone, and which is incapable of covalently linking to the next nucleotide in a growing molecule of DNA. Used in chain termination sequencing.

DNA-binding motif A short stretch of DNA, usually between 8 and 12 nucleotides, that is recognized by a DNA-binding protein. Such motifs can be represented by a profile of the frequency of occurrence of each of the four nucleotides at each position in the sequence. They typically identify sequences important for the regulation of gene expression.

DNA library A collection of hundreds of thousands of clones, each of which contains a different piece of genomic or cDNA. If the clone is oriented in such a way that the clone can be transcribed and translated, the library is called an expression library.

EBI's Ensembl genome browser The major repository and resource for genomic data in Europe, run jointly by the Sanger Center in Cambridge, England and the European Bioinformatics

Institute within the European Molecular Biology Laboratory.

ectopic expression Activation of expression of a transgene in cells outside the normal domain of expression, that is, in an abnormal cellular context.

embryonic stem (ES) cell A cell line that can be transformed and manipulated in culture, and then injected into the blastula (early embryo) where it integrates with and contributes to the development of the adult animal. Injected embryos are chimeric, but if the ES cells populate the germ line, a transgenic organism is produced in the next generation.

enhancer An orientation and distance-independent regulatory sequence that increases levels of transcription in a spatial and temporal manner. Enhancers may occur anywhere in a gene, can act over hundreds of kilobases, and can sometimes affect the transcription of several genes.

enhancer trap A method for detecting genes expressed in a particular tissue at a particular time, based on random insertion of a modified transposable element that encodes a reported enzyme adjacent to the enhancers of a gene.

epistasis Genetic interaction between two or more loci; commonly used to imply that the phenotype associated with one allele is modified by the genotype at another locus.

epitope The short stretch of molecule, typically less than 10 amino acids, that is recognized by an antibody.

E-value The expected number of sequences in a database that would by chance produce an equivalent or better alignment score than the one under consideration.

expressed sequence tag (EST) A sequenced piece of cDNA. Whereas a full-length cDNA sequence defines the structure of a transcript, an EST is merely a tag that indicates that the particular sequence is a part of a transcribed gene.

expression library A library of cDNA clones in a vector that allows the gene products to be expressed (transcribed and translated) in a controlled manner.

expression vector A cloning vector that facilitates transcription and translation of a cDNA fragment that is inserted into the multiple cloning site.

expressivity The severity of the trait in individuals who have a susceptibility allele and show the trait.

F$_3$ designs Genetic screens designed to isolate recessive mutations which require that the phenotype be measured in third generation progeny of the mutagenized individual.

floxing A method for inducing a mutation at a precise time and place in an organism. When a mouse containing *loxP* binding sites on either side of an exon of the gene to be mutated (placed there by homologous recombination) is crossed to a strain expressing the Cre recombinase in the tissue of interest, the exon is excised solely in that tissue.

forward genetics Genetic analysis that moves from the phenotype to the gene.

fusion protein A hybrid protein produced by fusing parts of two genes together in an expression vector. The N-terminal part is often a tag such as polyhistidine or the small glutathione S-transferase (GST) protein domain, while the C-terminal portion is usually the protein of interest.

GAL4 A potent transcription factor from yeast that enhances gene expression only through a UAS sequence adjacent to the promoter. *GAL4* in the absence of UAS sequences has no effect on transcription in heterologous genomes, so it can be used specifically to drive expression of transgenes introduced into that genome.

gene knock-in Replacement of the endogenous gene with a different functional piece of DNA such that the inserted gene is expressed in place of the original gene. Germline gene therapy uses knock-in technology to replace a defective gene with an active copy. The replacements are performed using a positive-negative double selection strategy in embryonic stem cells.

gene knock-out A mutation that targets a specific gene, produced by using homologous recombination to replace an exon of the target gene with a piece of foreign DNA (which is sometimes the reporter gene *lacZ*). Insertional mutations can also cause gene knockouts.

genetic fingerprinting A strategy for testing the subtle effects of mutations on the fitness of microbial strains in competition with other strains during long-term culture.

genetic heterogeneity The phenomenon whereby the same disease or phenotype can be attributed to different genetic causes, either to different loci or different alleles

genetic map A map of the order of and distance between genes in a genome based on the frequency of recombination between markers. The markers may be physical (molecular variants) or visible (Mendelian loci). Mapping populations may be pedigrees, crosses between lines, or radiation hybrid cell panels. Genetic distances are given in Morgans (or centiMorgans).

germ line The population of cells in a multicellular eukaryote that are destined to undergo meiosis and become oocytes or sperm. The germ line is set aside very early in animal development, but can be specified at the time of flowering in plants.

haplotype A multi-site genotype consisting of two or more polymorphisms on the same chromosome. For example, individuals which are homozygous at one site for a G allele and heterozygous at a nearby site for A and T would have GA and GT haplotypes.

Hardy-Weinberg equilibrium Hardy-Weinberg equilibrium refers to the expectation that genotype frequencies will tend to be stable and predictable as a simple function of individual allelic frequencies, unless there is some evolutionary force such as migration, inbreeding, mutation, or selection tending to increase or decrease the number of heterozygotes.

heavy isotope labeling A method for quantification of protein expression differences between two samples. One protein sample is labeled with a heavy isotope such as deuterium, so that peptide fragments move slightly more slowly through a TOF spectrometer than do corresponding fragments from the unlabeled sample. ICAT reagents are used for uniform labeling of protein mixtures after extraction from cells.

heteroduplex DNA A double helix of DNA from two different alleles, so that it includes at least one mismatch.

heuristic search Algorithms that use time-saving methods to search for the most likely solution, usually reducing the search space by excluding unlikely solutions from the analysis. Heuristic methods are not guaranteed to find the optimal solution, but are often the only feasible way to perform a phylogenetic analysis or sequence alignment involving a large number of sequences.

hierarchical sequencing An approach to whole genome sequencing based on the principle that the genome is first divided into an ordered set of clones. As initially envisaged, the process consisted of cloning a genome into artificial chromosomes, then cosmid or BAC clones several hundred kilobases long, then plasmids up to 10 kb long. In practice, BAC clones are now more commonly directly sequenced using the shotgun strategy.

hidden Markov model A class of bioinformatics procedures for identifying sequence features. HMMs exploit the fact that many features have local amino acid or nucleotide usage patterns that are distinct from random sequence.

HKA test A test of neutrality named after its developers (Hudson, Kreitman, and Aguadé) based on the expectation that divergence between species and polymorphism within species are highly correlated in the absence of selection.

homogenous assay A SNP genotyping assay in which all of the steps are performed in a single sample tube or well of a microtitre plate. Since there is no need to transfer products from one reaction to another, there are savings in labor and materials.

homolog Biological features (ranging from molecules to traits) that show a similar structure due to the fact that they derive from a common ancestor. Homology thus refers to identity by descent. Similarity of structure may or may not reflect homology. Unfortunately, the term is often used merely to imply that two DNA or amino acid sequences are similar.

horizontal gene transfer The transfer of a gene from the genome of one individual to that of another, other than by sexual (i.e., vertical) transmission, usually used in the context of microbial genes that can be shown to have been picked up from a different species.

hybridization The annealing of one strand of DNA to its complementary strand to form a double helix, most often used in context where one strand is labeled by the investigator as part of a procedure for detecting arrayed transcripts or clones.

immunohistochemistry The detection of proteins in a tissue sample based on chemical detection of antibodies that recognize the protein.

inbreeding The process of mating siblings or close relatives repeatedly, leading to loss of genetic variability in the line. Near-isogenic lines (NIL) are almost homozygous throughout the genome as a result of 10 or more generations of inbreeding, and are commonly used in quantitative genetic analysis.

indel An insertion-deletion polymorphism. Indels range in size from one or a few bases, to several kilobases. Large indels often involve transposable elements.

in situ hybridization Detection of a mRNA in a tissue sample by hybridization of a section of whole mount of a tissue to a DNA or RNA probe that is complementary to the mRNA, and which is labeled with a fluorescent or radioactive group, or with a small compound such as biotin or digoxygenin that can be recognized by an antibody.

insertional mutagenesis The process of creating mutations by controlled insertion of a transposable element in the vicinity of a gene.

interference The observation that recombination is suppressed by the occurrence of nearby recombination events, with the result that the genetic map distances between more than two markers do not necessarily equal the sum of the distances between each pair of adjacent markers.

interval mapping A method for QTL mapping that uses the genotypes of two adjacent genetic markers to estimate the likely genotype at each point in the interval between the markers.

introgression Introduction of a small portion of one genome into the genetic background of another genome, by repeated backcrossing with selection for the region of interest.

isogenic Homozygous for the entire portion of the genome under consideration.

laboratory information management systems (LIMS) An automated tracking system implemented with bar codes, robotics, and software checkpoints to ensure accurate tracking of samples and data in high throughput genomics laboratories.

linkage disequilibrium (LD) The nonrandom segregation of genetic markers in a population. LD decays over time as a result of recombination, so tends to reduce as physical distance between the markers increases, but it can also be caused by a variety of other forces, including founder effects, admixture, and epistatic selection.

linkage disequilibrium mapping An approach to identifying the genes that correspond to a quantitative trait locus that is based on detection of linkage disequilibrium between a marker, the trait (or disease), and the causal polymorphism in an outbred population.

LOD score Logarithm of the odds score. A measure of statistical significance in association studies and linkage mapping, essentially the logarithm of the ratio of the probability of observing the data given an association to the probability under the null hypothesis.

MALDI-TOF Matrix-Assisted Laser-Desorption Ionization Time-of-Flight spectrometry. A method for ionizing and then separating small fragments of DNA or protein for the purpose of identifying the corresponding sequence.

mapping function A mathematical function that converts recombination frequencies into genetic map distances by accounting for the incidence of double crossovers between markers. Two commonly used functions are due to Haldane and Kosambi; the latter adjusts the data for interference.

marker-assisted selection An approach to improved animal and plant breeding based on selection of DNA markers in each generation that are associated with some desired trait, as opposed to selecting on the trait itself.

mass spectrometry (MS) A technique for determining the identity of a molecule based on comparison of the spectrum of molecules separated by mass/charge ratio, with a theoretical standard. In genomics and proteomics, separation is based on time-of-flight (TOF) of an ionized fragment through a vacuum, and is sensitive to mass differences as small as a few parts per million.

mate-pair sequences A pair of sequences derived from the two ends of a single clone. An essential component of shotgun sequencing as the distance between the pairs assists in resolving repetitive DNA sequences and in verifying the sequence assembly.

McDonald-Kreitman statistic A test for selection acting on protein sequences, based on comparison of levels of synonymous and replacement polymorphism and divergence.

metabolome The set of metabolites present in a cell or tissue, including such molecules as lipids, carbohydrates, steroids, and amino acids that mediate many of the physiological properties of organisms.

metabolomics High throughput methods for characterization of the metabolome.

metabolic control analysis (MCA) An approach to modeling of metabolism, based on biophysical and biochemical principles, that seeks to understand and predict the effects of genetic or environmental perturbation.

microsatellite A stretch of repetitive DNA made up of a variable number of several to one hundred or more tandem repeats of a small number of nucleotides, most commonly di- or trinucleotides. For example (AG)n or (CAG)n. Microsatellites tend to be highly polymorphic and heterozygous, and occur at high density (several per hundred kilobases) in the genomes of higher eukaryotes.

minimal genome   The smallest number of genes that are required to sustain life.

modifier   A mutation or polymorphism that slightly alters (modifies, either enhancing or suppressing) the phenotype associated with another mutation.

monoclonal antibodies (Mabs)   Antibodies derived from a single immunoglobulin gene that recognize a single epitope on a protein, generated by creating an immortal hybrid cell line that secretes the Mab.

motif   A short conserved sequence of nucleotides or amino acids, often suggesting conservation of function.

multiple cloning site   Also known as a **polylinker**; the site in a plasmid into which foreign DNA is inserted at one of a number of unique restriction enzyme sites.

multiplex PCR   Simultaneous amplification of multiple different fragments of DNA, by using several pairs of gene-specific primers in the PCR reaction.

National Center for Biotechnology Information (NCBI)   The major depository and resource for genomic data in the United States. NCBI is a part of the National Library of Medicine within the National Institutes of Health.

neutral theory   The null hypothesis explaining the distribution of molecular variation in natural populations in the absence of natural selection. Factors affecting rates of neutral evolution include mutation pressure, migration rate, population size, breeding structure, and recombination rate.

Northern blot   A method for characterizing gene expression based on transfer of mRNA to a nylon or nitrocellulose filter, which is then probed with chemically or radioactively labeled DNA corresponding to the gene of interest.

nucleotide diversity   The average proportion of nucleotide differences between all pairs of sequences in a sample. A measure of polymorphism that is a function of the number and frequency of variable alleles.

Online Mendelian Inheritance in Man (OMIM)   A website maintained by the NCBI and Johns Hopkins University in Baltimore that documents genetic and genomic information in relation to human disease.

orphan gene   A predicted gene that does not show sequence similarity to any other gene in the databases, and hence which cannot be assigned to a gene family.

orthologs   Two genes in separate species that derive from a common ancestor without duplication.

paralogs   Two genes that arose by duplication of an ancestral gene.

penetrance   The frequency of individuals with an allele who show the trait.

PERL   A simple programming language used to perform basic bioinformatic procedures such as extracting DNA sequences from a database.

pharmacogenomics   The study of the effect of the genomic basis of the response to drugs, toxins, and other pharmacological agents.

phenocopy   An environmentally induced phenotype that mimics a known mutation.

phylogenetic analysis   In the context of comparative genomics, a methodological approach to annotation of gene function based on the supposition that evolutionary history is a more reliable indicator of likely function than sequence similarity alone.

phylogenetic footprinting   The process of aligning the sequences of a stretch of DNA from multiple divergent species, typically for the purpose of detecting evolutionarily conserved elements that may encode genes or other important DNA sequences.

phylogenetic shadowing The process of aligning the sequences of a stretch of DNA from several closely related species, typically for the purpose of detecting unusually highly conserved DNA elements that may encode regulatory elements.

physical map A map of a genome consisting of an ordered series of large insert clones in which the distance between molecular features such as restriction enzyme sites and sequence-tagged sites (STS) is expressed in kilobases.

polony PCR colonies which are cell-free clones of a single molecule of DNA that accumulate as a concentrated spot of amplified DNA within an acrylamide matrix on a glass microscope slide.

population stratification Differences in allele or haplotype frequencies between populations. In many situations, the identity of the populations is not obvious as the population structure may be hidden, as for example historical populations that no longer correspond with geographic location or phenotypic attributes.

positional cloning Cloning of a gene that is responsible for a disease or trait on the basis of its position in the genome, generally using recombination mapping.

position effect A phenomenon often observed in transgenic animals and plant, in which the site of insertion has a large effect on the level of expression of the transgene.

profile A list of the frequencies of each amino acid in each position in a multiple alignment of protein sequences.

promoter The region immediately 5′ to the start site of transcription of a gene that serves as a binding site for the RNA polymerase initiation complex and gen-

erally also includes regulatory sequences.

protein interaction map A description of the network of interactions among a group of proteins, including physical associations detected using two hybrid screens and protein microarrays, as well as interactions inferred from biochemical and genetic analysis.

proteome The full complement of proteins that are found in a particular cell or tissue under a particular set of circumstances. May include information on their relative or absolute abundance.

pseudogene A DNA sequence that shares many of the structural features of true genes, but is not active. Many pseudogenes are produced by reverse transcription, so lack introns. They may never have been active, or may be in the process of decaying in the absence of any selection pressure to maintain function.

psychogenomics The study of behavior and psychosis using genomic approaches.

pyrolysis The thermal degradation of materials into volatile fragments, one of several spectrometric techniques used in profiling of the metabolome.

quantitative trait locus (QTL) A region of the genome that has a quantitative effect on a trait, meaning that it is only responsible for a portion of the genetic variance. QTL may affect continuous traits, or liability to discrete traits including diseases.

QTL mapping Determination of the location of quantitative trait loci (QTL) using statistically sophisticated procedures for detecting nonrandom associations between genetic markers and trait values. Conceptually similar to recombination mapping of several loci simultaneously.

QTN Quantitative trait nucleotides, the SNPs that actually contribute to the effect of a quantitative trait locus (QTL).

radiation hybrid mapping A method for assembly of genetic maps of vertebrates or plants in which fragments of the genome of one species are propagated in hybrid cell lines with another species. Co-segregation of sequences in multiple lines indicates that the two sequences are physically linked.

random mutagenesis The process of generating a large collection of new mutations, generally involving a screen for aberrant phenotype(s) or the insertion of transposable elements.

recombinant inbred line (RIL) A line derived from two genetically distinct parents that has been bred to be nearly homozygous throughout the genome (nearly isogenic) by several generations of inbreeding. Each member of a panel of recombinant inbred lines contains a different combination of fragments from the two parents, so RIL are very useful for mapping QTL.

recombination mapping Determination of the location of a gene that is responsible for a particular phenotype, disease, or quantitative trait, based on the co-segregation of linked genetic markers with the trait in a pedigree.

redundant gene A gene whose function can be supplied by another gene or genes if it is mutated. Redundancy can be due to multiple copies of the gene in a genome, or because the protein activity can be supplied by a different type of gene, or because the enzyme can perform the function with another genetic pathway.

reverse genetics Genetic analysis that moves from the gene to the phenotype.

restriction fragment length polymorphism (RFLP) A polymorphism that is detected as a difference in the length of restriction fragments formed when a piece of DNA is digested with a restriction enzyme.

reverse transcriptase The enzyme that converts RNA to single-stranded DNA; often encoded in the genome of RNA viruses.

saturation random mutagenesis program A forward genetic screen of a sufficiently large number of mutagenized chromosomes to guarantee that all genes that affect the trait of interest will be hit by at least one mutation.

sequenced-contig scaffold An alignment of sequenced contigs against a physical and eventually cytological map of the genome. The step in genome sequencing immediately preceding the finishing stage in which gaps between scaffolds are filled in.

sequence-tagged site (STS) Any sequenced fragment of DNA derived from a library of clones that is placed on the physical map of the genome.

serial analysis of gene expression (SAGE) A method for profiling gene expression based on the sequencing of very large numbers of unique tags corresponding to each gene in the genome.

shotgun sequencing Determination of the sequence of a long stretch of DNA by randomly breaking it into a redundant set of small clones that are sequenced en masse so that each fragment is represented between 5 and 10 times. The contig is then assembled by computer alignment of the overlapping sequences. Whole genome shotgun sequencing refers to the process of sequencing the entire genome in this way, without first dividing it into large clones as in hierarchical sequencing.

single-base extension (SBE) A SNP detection method based on minisequencing reactions that only detect the identity of the base adjacent to the sequencing primer.

single nucleotide polymorphism (SNP)
A site in the genome at which a single nucleotide is found to have two or more states in a collection of individuals of the same species. Most SNPs are substitutions involving just two nucleotides (for example A and G), but the term also applies to single nucleotide indels.

single sperm typing The determination of the genotype of multiple markers from a single spermatozoan, usually for the purpose of directly measuring recombination rates.

site-directed mutagenesis Deliberate modification using recombinant DNA technology of a cloned sequence so that the protein is altered when expressed in a transgenic organism, allowing testing of hypotheses as to the role of particular residues in protein function.

Southern blot A method for detecting DNA sequences in a restriction digest of genomic DNA that has been separated by electrophoresis and transferred to a nylon or nitrocellulose membrane. Used to detect differences in the genomic DNA encompassing a gene, and to see whether a gene is present in the genome of another species.

structural genomics Referred to as **structural proteomics** in this book, it refers to the study of the tertiary structures of the complete set of proteins encoded in genomes, generally using X-ray crystallography, NMR spectroscopy, and *ab initio* or homology-based prediction.

synteny The conservation of gene order between divergent lineages.

synthetic lethal Two mutations that are alone homozygous viable, but together are inviable. Synthetic lethality generally indicates the two genes function in a similar process.

systems biology An integrative approach to genome biology that involves theoretical modeling of large genomic, transcriptomic, proteomic, and metabolomic datasets to guide the generation and testing of hypotheses.

tagging SNPs Single nucleotide polymorphisms that are used to represent haplotypes, and which are chosen to capture the majority of DNA sequence variation in a population sample.

TaqMan assay A method for quantification of transcript levels, or for detection of SNPs, based on the digestion of a quenched fluorescent (FRET) primer during PCR amplification by the 5' exonuclease activity of the Taq polymerase.

threading An approach to prediction of protein structure based on a combination of secondary structure similarity and assessment of likely binding energies of potential folds.

TILLING (targeting induced local lesions in genomes) A method for rapidly identifying single nucleotide polymorphisms and mutations based on cleavage of heteroduplexes formed by two alleles of a target locus.

tiling path In an aligned set of large insert clones, the choice of a subset of clones that completely covers the contig with minimal redundancy.

transcriptome The complete set of transcripts expressed in a particular cell or tissue under defined conditions. A full description of a transcriptome includes data on the abundance of each class of transcript.

transient expression Activation of gene expression for a limited period of time, for example from plasmids injected into embryos, or after infection of a tissue with a replication-defective virus.

transmission disequilibrium testing A test of association based on an unequal ratio of allele frequencies in the affected children of heterozygous parents.

transposable elements Mobile pieces of DNA that can jump from one location in the genome to another, usually utilizing an enzyme encoded on the natural transposable element.

transposon mutagenesis Mobilization of modified transposable elements to create mutations that are tagged with the inserted piece of DNA.

UAS The binding site for the *GAL4* transcription factor, only found upstream of yeast genes. The *UAS-GAL4* combination can thus be used to drive transgene expression in plants and animals.

unigene set A set of unique cDNA clones derived from a cDNA library by filtering out duplicate copies of the same transcript.

unitig A unique contig, namely a set of overlapping DNA sequences that correspond to a single piece of genomic DNA that is represented multiple times in a shotgun sequence library.

Western blotting A molecular biology technique for detecting protein expression based on transferring proteins from an acrylamide gel to a membrane that is then probed with a labeled antibody.

yeast two-hybrid screen (Y2H) A method for detecting protein-protein interactions based on the reconstitution of transcription factor activity when one protein domain fused to a DNA-binding domain (the "bait") interacts with another protein domain fused to an activation domain (the "prey"). Initially developed in and most commonly performed in yeast cells.

List of Abbreviations

ADIT AutoDeposit Input Tool

ASO Allele-Specific Oligohybridization

BAC Bacterial Artificial Chromosome

BLAST Basic Local Alignment Search Tool

BLOSUM BLOcks Substitution Matrix

CASP Critical Assessment in Structure Prediction

CE Capillary Electrophoresis

DASH Dynamic Allele-Specific Hybridization

ddNTP dideoxyNucleotide Triphosphate

DGGE Denaturing Gradient Gel Electrophoresis

DHPLC Denaturing High Performance Liquid Chromotography

DOL Dye-labeled Oligonucleotide Ligation assay

EBI European Bioinformatics Institute

EC Enzyme Commission

ELISA Enzyme-Linked ImmunoSorbant Assay

ELSI Ethical, Legal, and Social Implications

EP Enhancer-Promoter

EST Expressed Sequence Tag

FISH Fluorescent In Situ Hybridization

FRET Fluorescence Resonance Energy Transfer

GAL4 Galactose-4 transcription activator protein

GC Gas Chromatography

GST Glutathione S-Transferase

HGP Human Genome Project

HKA test Hudson-Kreitman-Aguadé test of neutrality

HMM Hidden Markov Model

HPLC High-Performance Liquid Chromatography

ICAT Isotope Coded Affinity Tag

IDAT ImmunoDetection by Amplification with T7

IHGSC International Human Genome Sequencing Consortium

IPG Immobilized pH Gradient

LC-MS/MS Liquid Chromatography-Tandem Mass Spectrometry

LIMS Laboratory Information Management System

LINE Long Interspersed Nuclear Element

LOD Logarithm of the Odds

LTR Long Terminal Repeat

Mab Monoclonal Antibody

MAD Multiple Anomolous Dispersion

MAGEML MicroArray Gene Expression Markup Language

MALDI-TOF Matrix-Assisted Laser-Desorption Ionization Time-of-Flight spectrometry

MCA Metabolic Control Analysis

MIAME Minimal Information for the Annotation of Microarray Experiments

mmCIF macromolecular crystallographic information file

MPSS Massively Parallel Serial Sequencing

MS Mass Spectrometry

MudPIT Multidimensional Protein Identification Technology

NCBI National Center for Biotechnology Information

NMR Nuclear Magnetic Resonance

OMIA Online Mendelian Inheritance in Animals

OMIM Online Mendelian Inheritance in Man

ORF Open Reading Frame

OST ORFSequence Tag

P1 P1 large-insert bacteriophage clone

PAGE PolyAcrylamide Gel Electrophoresis

PCR Polymerase Chain Reaction

PDB Protein Data Bank

Q-PCR Quantitative Polymerase Chain Reaction

QTL Quantitative Trait Locus

QTN Quantitative Trait Nucleotide

RefSeq Reference sequence in GenBank

RFLP Restriction Fragment Length Polymorphism

RH Radiation Hybrid

RIL Recombinant Inbred Line

SAGE Serial Analysis of Gene Expression

SBE Single Base Extension

SDS Sodium Dodecyl Sulfate (detergent)

SINE Short Interspersed Nuclear Element

SNP Single Nucleotide Polymorphism

SSCP Single-Stranded Conformation Polymorphism

SSR Simple Sequence Repeat

STS Sequence Tagged Site

UAS Upstream Activator Sequence, used by Gal4

VDA Variant Detector Array

VNTR Variable Number Tandem Repeat

Y2H Yeast Two Hybrid

Index

About the Authors

Greg Gibson is Associate Professor in the Deparment of Genetics at North Carolina State University. He earned a Bachelor's Degree in Biology at the University of Sydney, and a Ph.D. in Cell Biology at the University of Basel (with Walter J. Gehring). Dr. Gibson has been awarded fellowships from the Helen Hay Whitney Foundation (1990–1993) and the David and Lucille Packard Foundation (1996–2001). His research is in developmental and pharmacological quantitative genetics and genomics, focusing on the fruitfly, *Drosophila melanogaster.*

Spencer Muse is Assistant Professor in the Department of Statistics, Bioinformatics Research Center at North Carolina State University. He earned a Bachelor's Degree in Statistics and a Ph.D. in Statistics and Genetics, both at NCSU (the latter with Bruce Weir and Trudy Mackay). Dr. Muse is the recipient of a Sloan Foundation Young Investigator award (1997–2001). His research is in molecular evolution and bioinformatics.

About the Book

Editor: Andrew D. Sinauer
Project Editor: Sydney Carroll
Production Manager: Christopher Small
Book and Cover Design: Jefferson Johnson
Book Layout and Production: Janice Holabird
Illustrations: S. Mark Williams, Pyramis Studios, Inc.
Index: Sharon Hughes
Book and Cover Manufacture: Courier Companies, Inc.